AN
INTRODUCTION
TO
ABSTRACT
MATHEMATICS

AN INTRODUCTION TO ABSTRACT MATHEMATICS

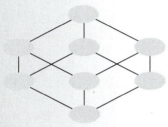

Robert J. Bond

William J. Keane

Brooks/Cole Publishing Company
I(T)P® An International Thomson Publishing Company

Pacific Grove • Albany • Belmont • Bonn • Boston • Cincinnati • Detroit • Johannesburg • London • Madrid
Melbourne • Mexico City • New York • Paris • Singapore • Tokyo • Toronto • Washington

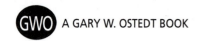 A GARY W. OSTEDT BOOK

Publisher: *Gary W. Ostedt*
Marketing Team: *Caroline Croley, Christina DeVeto*
Assistant Editor: *Carol Ann Benedict*
Advertising Communications: *Heidi Clovis*
Production Coordinator: *Kelsey McGee*
Production Service: *Tobi Giannone/Michael Bass & Associates*
Manuscript Editor: *Helen Walden*
Interior Design: *Lois Stanfield/LightSource Images*

Interior Illustration: *Asterisk Group, Inc.*
Design Coordinator: *Tobi Giannone*
Cover Design: *Roger Knox*
Cover Illustration: *Judith Larzalere*
Cover Photography: *David Caras Studio*
Typesetting: *Bi-Comp, Inc.*
Cover Printing: *Phoenix Color Corporation*
Printing and Binding: *R. R. Donnelley & Sons, Crawfordsville*

For more information, contact:

BROOKS/COLE PUBLISHING COMPANY
511 Forest Lodge Road
Pacific Grove, CA 93950
USA

International Thomson Publishing Europe
Berkshire House 168-173
High Holborn
London WC1V 7AA
England

Thomas Nelson Australia
102 Dodds Street
South Melbourne, 3205
Victoria, Australia

Nelson Canada
1120 Birchmount Road
Scarborough, Ontario
Canada M1K 5G4

International Thomson Editores
Seneca 53
Col. Polanco
11560 México, D. F., México

International Thomson Publishing GmbH
Königswinterer Strasse 418
53227 Bonn
Germany

International Thomson Publishing Asia
60 Albert Street
#15-01 Albert Complex
Singapore 189969

International Thomson Publishing Japan
Hirakawacho Kyowa Building, 3F
2-2-1 Hirakawacho
Chiyoda-ku, Tokyo 102
Japan

Printed in the United States of America

10 9 8 7 6 5 4 3 2 1

Library of Congress Cataloging-in-Publication Data

Bond, Robert J., [date].
    An introduction to abstract mathematics / Robert J. Bond and
William J. Keane.
        p. cm.
    Includes bibliographical references and index.
    ISBN 0-534-95950-7
    1. Mathematics.   I. Keane, William J., [date].   II. Title.
QA37.2.B63   1998
510–dc21   98-37192

To Ann and Sarah, for their love and constant support for what at times seemed an endless journey.

To Liz and Matt, for the first time, and to France, always.

CONTENTS

PREFACE FOR THE INSTRUCTOR

This book evolved from a course that has been taught at Boston College for many years in many different forms. The course, taken primarily by sophomore mathematics majors, has always been intended to prepare those students for the "abstract mathematics" of the title, for the rigor, careful argument, and logical precision that, although merely glimpsed in calculus, would be linchpins of all further study. Most students will have had two or three semesters of calculus and a semester of linear algebra so that they will have attained a reasonable degree of mathematical sophistication before starting this book.

Our choice of topics is motivated by the fact that students take introductory courses in abstract algebra and real analysis in their junior or senior years as mathematics majors. In order to be able to handle these courses effectively, they need to know the rules of logic and the rudiments of set theory as well as some basic properties of functions between sets: injective and surjective functions, image and inverse image, and invertible functions.

The book divides naturally into two (overlapping, to be sure) parts.

- Chapters 1–5 introduce the fundamentals of abstract mathematics. Logic, set theory, relations, functions, and operations lead to a careful study of one axiom system, the integers. These chapters form the core of the book. We always cover this material in our course, and we use it to introduce our students to the language of mathematics and to prepare them for their undergraduate careers in mathematics.
- Chapters 6–8 apply the ideas and techniques of the earlier material. These chapters for the most part are independent, and we choose which ones to cover as time permits and interests lead. The topics, though, are ones that we feel our students should know, will not find completely alien,

and will stand them in good stead later: infinite sets and cardinality; the properties of the real and complex numbers; and the important properties of polynomials over those fields, such as unique factorization.

Each section contains a wealth of examples and exercises. For the most part, especially in the early chapters, these concern familiar properties of the integers as well as some concepts from calculus, such as continuity and differentiability. (We do not define these latter topics rigorously, for they belong more properly in a course in real analysis.) A major goal of the text is for students to write mathematical proofs coherently and correctly, so we reinforce the logic developed in Chapter 1 throughout the book. In many exercises students must negate a statement or write a converse or a contrapositive. Some of the exercises build on the examples in the section, while others are a bit more challenging. We often leave some details of a proof as exercises, but only if, by then, students should find them routine. We think it is essential that instructors encourage (require?) students to fill in these gaps.

We have also included some "Discussion and Discovery Exercises" that are designed to have the students think a bit more deeply about what they have learned. In some of these exercises, they may propose a conjecture and then try to prove it or they may criticize a given proof for logical fallacies or lack of clarity. At the end of each section we have included "Historical Comments" or a "Mathematical Perspective," which attempt to put some of the subject matter in a wider historical and mathematical context. We do not mean them merely as supplemental biographies but intend them to give students some sense of the kinds of problems mathematicians have struggled with over the centuries, from the time of Euclid to the present day. The topics in these commentaries are the authors' personal choices, not meant to represent what every math major should know, but rather to be samples of some interesting material that can stimulate further study. Students can consult the list of references at the end of the book if they wish to learn more.

Chapter 1 introduces the rules of logic that students need to know in order to understand written proofs and to write their own. This means that we discuss nuts and bolts: statements, propositions, quantifiers, rules of inference, and the like. Since this book is not designed for a full-fledged course in mathematical logic, we strive not to be too technical, while maintaining an appropriate level of rigor. The goal from the beginning is for students to write mathematics with the proper style and notation and with the correct "grammar" and connecting phrases, so that the end product is lucid, logical, and mathematically correct.

Chapter 2 presents the rudiments of set theory: subsets, complements, union, and intersection, as well as infinite families of sets and partitions. An important goal of this chapter, in addition to a familiarity with the language of sets, is for students to learn how to prove (or disprove) that two sets are equal. Our experience tells us that students often have difficulty just *beginning* such a proof, so we provide many examples and exercises of this type.

(You will find such examples and exercises not only in this chapter, but also as they appear in context in other chapters.) If Chapter 6 is not covered, the instructor can omit Section 2.3 on infinite families of sets.

Chapter 3 begins the study of functions as mappings between sets. We use the more informal definition of function as a rule rather than as a set of ordered pairs because it is an approach students are familiar with, involves no significant loss of rigor, and is the way functions actually are used. We discuss the notions of injective, surjective, and invertible functions thoroughly since they are basic to any understanding of abstract algebra. We give examples involving image and inverse image since they are part of the language of analysis. We show how the tools of calculus (the Intermediate Value Theorem, for example) can be used to prove whether or not a function is invertible; we invite students to review some of the concepts from their calculus courses. We also want students to think in terms of *sets* of functions, so that binary operations on these sets will not seem so strange when we introduce them.

Chapter 4 presents binary operations and relations, two relatively independent topics that are prerequisites for any abstract algebra course and for Chapter 5. Some of the exercises in Chapter 4 are similar to problems students do in a beginning abstract or linear algebra course, problems that are fairly routine but can be difficult for some students when seen for the first time. For example, a standard problem in group theory is to show that the intersection of two subgroups of a group is itself a subgroup. This problem involves a special case of a problem in Section 4.1; namely, to prove that if two subsets of a set with a binary operation are closed with respect to that operation, then their intersection is closed. As an optional topic at the end of the chapter, we use equivalence relations to give a formal construction of the rational numbers from the integers.

In Chapter 5, we bring together many of the ideas of previous chapters in the context of the set of integers. We begin with an axiomatic approach so that students can appreciate the distinctions among definitions, axioms, and theorems. We introduce mathematical induction in Section 5.2 as a consequence of the Well-Ordering Principle which is taken as an axiom. We prove the Unique Factorization Theorem in Section 5.4 via the standard approach of introducing greatest common divisors and properties of primes. Section 5.5 discusses congruences in some detail, defines the set $\mathbf{Z}_n$ of congruence classes of integers mod n, and proves its algebraic properties. Section 5.6 gives the students an example of how mathematicians extend results: by trying special cases and then conjecturing suitable generalizations. The mathematics in this section is not difficult but it does give the students some exposure to the creative process.

Chapter 6 introduces the theory of infinite sets and discusses countable and uncountable sets and cardinality. The mathematics is admittedly more difficult in this chapter and students should have developed a good mastery of the topics in Chapters 2 and 3. Sections 5.5 and 5.6 are not prerequisite for this chapter. Induction (Section 5.2) needs to have been studied, of course; the Division Algorithm (Section 5.3) is mentioned in the proof of Theorem 6.1.8 and prime factorization (Section 5.4) in the proof of Theorem 6.1.11. (Some alternative proofs of the latter don't use prime factorization.)

Chapter 7 introduces ordered fields and compares and contrasts such fields with the integers as developed in Section 5.1. Section 7.2 on the real numbers gives the students some of the flavor of a beginning real analysis course, presenting topics such as the Least Upper Bound Axiom, the Archimedean Principle, and the fact that between any two real numbers are infinitely many rational numbers. Preparation for Section 7.2 does not require all of Section 7.1. Section 7.3 is a standard introduction to complex numbers including DeMoivre's Theorem and roots of unity. Except for the definition of field, this section is independent of Sections 7.1 and 7.2.

The goal of Chapter 8 is for students to learn some techniques for solving polynomial equations, a topic often neglected. The chapter begins with a study of polynomials over an arbitrary field and the relation between zeros of polynomials and their linear factors. We state and prove unique factorization and note analogies with the integers. We then discuss (without proof) the Fundamental Theorem of Algebra and give some methods for finding rational roots of a polynomial with integer coefficients.

Unquestionably, we provide too much material for a single semester. The majority of the core, Chapters 1–5, is essential; instructors can choose topics from the later chapters as time permits or can include them in an honors version of the course. But the goal of the entire book is to show our students how mathematicians think as well as some of the fascinating things they think about.

Acknowledgments

This book has been in development for several years and has been used in our transition course for almost as long. In that time, we have had the enormous benefit of our colleagues at BC who have taught from previous versions, contributed comments, suggested problems, and engaged in spirited, frank, and occasionally hilarious conversations. The book would be much different and much less, as would we, without them: Dan Chambers, Rob Gross, Richard Jenson, Peg Kenney, Charlie Landraitis, Harvey Margolis, Rennie Mirollo, Nancy Rallis, Ned Rosen, and John Shanahan. Special thanks are due to Rennie Mirollo for carefully reading the text and pointing out numerous inaccuracies. Any that escaped his diligent analysis are the responsibility of the authors.

We thank our independent reviewers, Edward Azoff, The University of Georgia; Gerald Beer, California State University–Los Angeles; Ron Dotzel, University of Missouri–Saint Louis; David Feldman, University of New Hampshire; Sherry Gale, University of North Carolina at Asheville; John

Robertson, Georgia College; Michael Stecher, Texas A & M University–College Station; and Mark Watkins, Syracuse University.

We also thank Gary W. Ostedt, our publisher at Brooks/Cole, for wise guidance that shaped the final version of this text and for his faith in us.

Finally, thanks to our production coordinator, Kelsey McGee, and our production editor, Tobi Giannone, who skillfully and cheerfully saw us through to publication.

An instructor's manual, consisting of solutions of all of the exercises, is available.

Robert Bond
William Keane

INTRODUCTION FOR THE STUDENT

This text and others like it are often described as *transition* books, primers for higher-level mathematics. What do we mean by a transition book and why is such a book necessary?

By now, you have seen a significant amount of mathematics, including at least a year or two of calculus and possibly some linear algebra. The mathematics in these courses is *quite* sophisticated. Calculus, for example, as developed by Newton and Leibniz, is the greatest mathematical achievement of the seventeenth century. The tremendous scientific advances of the last 300 years would not have been possible without the formulas and algorithms that follow from the theory of the integral and the derivative. Soon, you will take additional courses in such fields as probability, combinatorics, dynamical systems, linear programming, or topology, to list just a few examples. Given that calculus involves such high level mathematics, why does a math major need a *transition* course? Why not just plunge right into these so-called *advanced* courses?

One reason stems from the history of calculus itself. In the seventeenth and eighteenth centuries, mathematicians would manipulate infinite series much like ordinary finite sums. The results were usually quite correct, but the methods often led to errors. Here is an example:

The MacLaurin series expansion of $\ln(1 + x)$ is given by:

$$\ln(1 + x) = x - \frac{x^2}{2} + \frac{x^3}{3} - \frac{x^4}{4} + \dots \qquad (*)$$

This series converges for $-1 < x \le 1$. Differentiating both sides of $(*)$ gives:

$$\frac{1}{1 + x} = 1 - x + x^2 - x^3 + \dots \qquad (**)$$

If we substitute $x = 1$ in (∗∗), we get

$$\frac{1}{2} = 1 - 1 + 1 - 1 + \ldots .$$

The right-hand side of the equation, however, is not a convergent series. What has gone wrong here is the indiscriminate differentiation of a power series, term by term, as if it were the same as a finite sum. Sometimes this can be done and sometimes it cannot. In fact, (∗∗) is a true equation for all x such that $-1 < x < 1$. What becomes important is to *prove* under what conditions a power series *can* be differentiated term by term.

Another example is provided by letting $x = 1$ in (∗) above, giving the true equation:

$$\ln(2) = 1 - \frac{1}{2} + \frac{1}{3} - \frac{1}{4} + \ldots . \tag{∗∗∗}$$

Now if we rearrange the terms of the infinite series on the right-hand side, we obtain the equation:

$$\ln(2) = \left(1 + \frac{1}{3} + \frac{1}{5} + \ldots\right) - \left(\frac{1}{2} + \frac{1}{4} + \frac{1}{6} \ldots\right)$$

$$= \left(1 + \frac{1}{3} + \frac{1}{5} + \ldots\right) + \left(\frac{1}{2} + \frac{1}{4} + \frac{1}{6} \ldots\right) - 2\left(\frac{1}{2} + \frac{1}{4} + \frac{1}{6} \ldots\right)$$

$$= \left(1 + \frac{1}{2} + \frac{1}{3} + \frac{1}{4} \ldots\right) - \left(1 + \frac{1}{2} + \frac{1}{3} + \frac{1}{4} \ldots\right)$$

$$= 0.$$

Since we know that $\ln(2) \neq 0$, we have an apparent contradiction. The contradiction is resolved by noting that the right-hand side of (∗∗∗) is a conditionally convergent series; that is, the series converges but if the terms of the series are replaced by their absolute values then the resulting series diverges. It can be proven that a rearrangement of a conditionally convergent series will not necessarily converge to the same sum as the original series. In fact, a conditionally convergent series can be rearranged to converge to any given number or even to diverge.

In both of these examples, mistakes are made by treating an infinite sum the same as a finite sum. In trying to determine which rules that apply to finite sums also apply to finite series, it is necessary first to *define* carefully what we mean by an infinite series and then *prove* properties of series using that definition. As each property is verified, it can be used to prove subsequent properties.

Mathematicians of earlier centuries commonly manipulated formulas and symbols indiscriminately without regard for whether or not those manipulations were justified. Nevertheless, often these "missteps" actually led to true formulas or provided insights into *why* something was true. The great mathematician **Leonhard Euler** (1707–1783) is famous for making discoveries in a totally nonrigorous way. Here is an example.

You may recall that the infinite series $\sum_{n=1}^{\infty} \frac{1}{n^2}$ is a convergent series by the so-called *p*-test. But knowing that the series converges does not tell you *to what number* the series converges. In fact, this series converges to $\frac{\pi^2}{6}$.

Euler's "proof" of this fact goes like this: the MacLaurin series expansion of sin x is

$$x - \frac{x^3}{3!} + \frac{x^5}{5!} - \frac{x^7}{7!} + \ldots.$$

Dividing by x gives the equation:

$$\frac{\sin x}{x} = 1 - \frac{x^2}{3!} + \frac{x^4}{5!} - \frac{x^6}{7!} + \ldots.$$

If we set $\frac{\sin x}{x} = 0$, the roots are: $\pm\pi, \pm 2\pi, \pm 3\pi, \ldots$.

If we treat the infinite series as if it were a polynomial, as Euler did, then we can factor it as

$$\left(1 - \frac{x}{\pi}\right)\left(1 + \frac{x}{\pi}\right)\left(1 - \frac{x}{2\pi}\right)\left(1 + \frac{x}{2\pi}\right)\left(1 - \frac{x}{3\pi}\right)\left(1 + \frac{x}{3\pi}\right)\cdots$$

since this infinite product has the same roots and the same constant term as the infinite series.

So we get:

$$\frac{\sin x}{x} = 1 - \frac{x^2}{3!} + \frac{x^4}{5!} - \frac{x^6}{7!} + \cdots$$

$$= \left(1 - \frac{x}{\pi}\right)\left(1 + \frac{x}{\pi}\right)\left(1 - \frac{x}{2\pi}\right)\left(1 + \frac{x}{2\pi}\right)\left(1 - \frac{x}{3\pi}\right)\left(1 + \frac{x}{3\pi}\right)\cdots$$

$$= \left(1 - \frac{x^2}{\pi^2}\right)\left(1 - \frac{x^2}{4\pi^2}\right)\left(1 - \frac{x^2}{9\pi^2}\right)\cdots.$$

If we multiply out this last infinite product, as we would a finite product, we see that the coefficient of the x^2 term is the infinite series

$$-\frac{1}{\pi^2} - \frac{1}{4\pi^2} - \frac{1}{9\pi^2} - \cdots$$

On the other hand, the x^2 term of the MacLaurin series is $-\frac{1}{3!} = -\frac{1}{6}$. Multiplying both expressions by $-\pi^2$ gives us Euler's result.

We emphasize that Euler was not indifferent to the idea of convergence of an infinite series and, of course, he knew that a power series was not the same as a polynomial. His insight and cleverness produced significant mathematics. He was later able to give a proof that $\sum_{n=1}^{\infty} \frac{1}{n^2} = \frac{\pi^2}{6}$ that is considered rigorous by today's standards.

One of the important byproducts of finding a rigorous proof of a mathematical theorem is that it can often lead to new results or even to generalizations of the theorem, generalizations that may be impossible to discover by informal methods such as the ones employed by Euler.

For example, if you change the exponent in the series $\sum_{n=1}^{\infty} \frac{1}{n^2}$ from a 2 to another integer such as 3, 4 ... and so forth, you may well ask to what numbers these different series converge. There is a long history of attempts to answer this question.

First, we define a function $\zeta(s) = \sum_{n=1}^{\infty} \frac{1}{n^s}$ for s a real number greater than 1, so that $\zeta(2)$ is the series we considered above and $\zeta(2) = \frac{\pi^2}{6}$. (Note: the symbol ζ is the Greek letter zeta and the function $\zeta(s)$ is known as the Riemann zeta function, named after the mathematician **Bernhard Riemann** (1826–1866). You may recall the name Riemann from the study of Riemann sums in calculus.) Euler derived a formula for $\zeta(s)$ when s is an even positive integer. The formula involves a power of π and the so-called Bernoulli numbers, which we won't define here. No formula for $\zeta(s)$, when s is an odd positive integer, is known. In 1978, the French mathematician **R. Apery** proved that $\zeta(3)$ is irrational, but not much else is known about these numbers. This and countless other examples show that mathematics is not a closed subject. Many unsolved problems and even new areas of mathematics await the budding mathematician.

A transition book such as this one, then, is an introduction to the logic and rigor of mathematical thinking and is designed to prepare you for more advanced mathematical subjects.

We designed our course and this book with three goals in mind. First and foremost of these is to show you the elements of logical, mathematical argument, to have you understand exactly what mathematical rigor means and to appreciate its importance. You will learn the rules of logical inference, be exposed to definitions of concepts, be asked to read and understand proofs of theorems, and write your own proofs. At the same time, we want you to become familiar with both the grammar of mathematics *and its style*. We want you to be able to read and construct correct proofs, but also to appreciate different methods of proof (contradiction, induction), the value of a proof, and the *beauty* of an elegant argument.

A second goal is for you to learn how to do mathematics in a context, by studying real, interesting mathematics and not just concentrating on form. We have chosen topics that do not overlap significantly with other courses (such as the properties of the integers, the nature of infinite sets, and the complex numbers), that are essentially self-contained, and that will be useful to you later when you are exposed to more specialized, advanced mathematics.

Finally, we want you to realize that mathematics is an ongoing enterprise, with a long, fascinating, and sometimes surprising history. The notes sprinkled throughout the text are deliberately eclectic. "Historical Comments" give you pictures of the tremendous successes (and equally spectacular failures) of brilliant mathematicians of the past. "Mathematical Perspectives" may spotlight questions that are still unanswered and are the subject of current research, or that simply show an interesting further aspect of the material you have just studied.

AN
INTRODUCTION
TO
ABSTRACT
MATHEMATICS

MATHEMATICAL REASONING

Introduction: Early Mathematics

Mathematics in one form or another has probably existed in every civilization. Many cultures flourished in the area of the Tigris and Euphrates rivers known as Mesopotamia in what is now Iraq. The Babylonian civilization of that area has left us records of mathematical activity as far back as the years 1800–1600 B.C. Writing on clay tablets, they recorded solutions of algebraic problems and compiled tables of squares, cubes, square roots, and cube roots, even some logarithms. They also listed Pythagorean triples, numbers such as 3, 4, 5 and 5, 12, 13, which make up the sides of a right triangle (using of course their own symbols and number system). So the Babylonians knew the Pythagorean Theorem more than a thousand years before the time of Pythagoras! Even earlier than the Babylonians, the so-called Middle Kingdom of Egypt (2000–1800 B.C.) produced some sophisticated calculations of areas and volumes. One of their great achievements was the calculation of the volume of a truncated pyramid with square base and square top. (See Exercise D4 of Section 1.3.)

Beginning in about the sixth century B.C. in Greece, an extraordinary new chapter in the history of mathematics began. For the first time (at least as far as we know) the methods of reasoning and logical deduction, as opposed to trial and error calculations, were employed to arrive at new mathematical truths. Prior to the Greeks, in Egypt and Babylonia for example, geometric results and algebraic formulas were discovered by empirical methods; that is, by trying special cases and then extrapolating to the general case. This empirical method, called *inductive* reasoning, has been used by mathematicians throughout history. It is not merely a perfectly acceptable method of making new discoveries but is really an *indispensable* way to arrive at new results. But the Greeks insisted that any new results had to be *proved* and that meant using the rules of logic. This method is called *deductive* reasoning. In this chapter we introduce some of the important ideas of logic that are frequently used in mathematics.

1.1   STATEMENTS

The Notion of Proof

Contrary to popular opinion, mathematics is not just computations and equations. It might be better described as an attempt to determine which statements are true and which are not. The subject matter may vary from numbers to geometric figures to just about anything, but the form is always the same.

The process of discovery in mathematics is twofold. First comes the formulation of a mathematical statement or conjecture. This formulation often comes after much hard work that usually includes a trial-and-error process, many false starts, and sometimes extensive calculation. The second part of the process is the verification or proof that the statement that we have formulated is true or false. This part too can involve much trial and error and long, hard work. It is this part of the process that we will study in this text.

To begin to prove a mathematical statement it is necessary to begin with certain statements that we accept as given, called **axioms**, and try to logically deduce other statements from them. These deductions are called **propositions**, or, if they're particularly important, **theorems**. The arguments, the logic we use to make the deductions, are called **proofs**.

The beauty of mathematics often comes from the fact that propositions are not always obvious, and can in fact be surprising. Moreover, proofs may not be easy to construct, and may require insight and cleverness. They certainly require a little experience, especially some familiarity with the most commonly used logical methods. This "sophistication" won't come overnight; it's one of the major goals of this book.

Let's begin with an example.

EXAMPLE I   Suppose that the following question were posed to you: Is the square of an even integer itself an even integer? You might start by trying some calculations:

$$2^2 = 4, \qquad 4^2 = 16, \qquad 6^2 = 36, \qquad 8^2 = 64, \qquad 10^2 = 100.$$

Each one of these examples gives a positive answer to the question. Do these answers constitute a proof?

If the problem asked us to show that the squares of *some* even integers are even, we would be done. But that is not the question. There is an implied universality about the question. Is the square of *every* even integer even? Clearly, more must be done than trying a few examples. In fact just working out examples will not suffice, because no matter how many we do, there will always be an infinite number of cases that we haven't done.

So how do we proceed? The first step might be to reword the question as a statement.

"The square of every even integer is even."

Now let's reword the statement using symbols:

"If n is an integer and n is even, then n^2 is even."

In order to begin a proof, we have to ask ourselves: What does it mean for an integer to be even? Well, we know that an integer is even if it is two times an integer. That is the *definition* of even integer.

Now we need to translate this definition into symbols:

"n is an even integer if $n = 2m$ for some integer m."

Now let's think about what we're trying to prove. We want to show that the square of the even integer n is even. So we start by letting n be an even integer and try to show that n^2 is twice an integer. We do the obvious algebra step; we write $n = 2m$ and square both sides.

We get $n^2 = 4m^2 = 2(2m^2)$. Since $2m^2$ is an integer, this shows that n^2 is twice an integer and therefore n^2 is even.

The proof is now complete.

This is an example, admittedly not a complicated one, of the proof of a mathematical statement.

Note that in this last example, we made some assumptions about multiplication of integers; namely, that multiplication satisfies a commutative law and an associative law and that the product of integers is still an integer. These laws make up some of the axioms of the integers and will be discussed in Chapter 5. For the purpose of examples in this and later chapters, we will assume the well-known arithmetic properties of the integers and real numbers. (See Section 5.1 for properties of the integers and Section 7.1 for properties of the real numbers.)

The notion of multiples of an integer is used in some examples. An integer x is called a **multiple** of an integer n if $x = kn$ for some integer k. So the even integers are the multiples of 2. The multiples of 3 consist of the integers 0, ±3, ±6, ±9, and so on.

Statements

In the course of this chapter, we will discuss many of the rules of logic and inference needed in the previous example and others like it in mathematics.

In the previous example, we used the word "statement" several times. In this text, the word will have a specific meaning.

Definition 1.1.1 A **statement** is any declarative sentence that is either true or false.

A statement then will have a truth value. It is either true or false. It cannot be *neither* true nor false and it cannot be *both* true and false.

Some examples of statements are given next:

EXAMPLE 2   John Fitzgerald Kennedy was the 35th president of the United States.

EXAMPLE 3   Marie Curie did not win the Nobel Prize.

EXAMPLE 4 $3 + 1 = 4$.

The fourth example is a mathematical statement, which of course is true. We commonly write mathematical statements with symbols for convenience, although you should think of them not as "formulas," but as full-fledged sentences, with a subject, a verb, and possibly other parts of speech.

In the rest of this chapter, we examine what mathematical statements can look like and some methods that can be used to prove them. For convenience, we often use letters, most often P or Q, to denote statements.

EXAMPLE 5   Some sentences, even some mathematical ones, are not statements. For example, consider a typical sentence from algebra: $x + 1 = 2$. Here, x is a **variable**; it's a symbol that stands for an undetermined number. The sentence is a statement if we specify what number x stands for. It's a true statement if x stands for 1, and it's false for any other x. We could label this sentence $P(x)$ because it depends on the variable x. So $P(1)$ is a true statement and $P(x)$ is a false statement if $x \neq 1$.

Note, however, that if x is not specified, then $P(x)$ is *not* a statement.

We will call any sentence like the one from the previous example an *open sentence*.

Definition 1.1.2 An **open sentence** is any declarative sentence containing one or more variables that is not a statement but becomes a statement when the variables are assigned values.

The values that can be assigned to the variables of an open sentence will depend on the context. They may come from the real numbers as in the example $x + 1 = 2$ or from the complex numbers or even just the positive integers. The values do not even have to be mathematical. For example, the sentence "He was the 16th president of the United States" is an open sentence containing the variable "he" and is therefore a true statement when "he" is assigned the value "Abraham Lincoln" and is false otherwise.

An open sentence is usually written $P(x)$, $P(x, y)$, $P(x, y, z)$, and so on, depending on the number of variables used.

Quantifiers

An open sentence like $x + 1 = 2$ can, as we have seen, be made into a statement by substituting a value for the variable or, in the case of an open sentence with more than one variable, by substituting a value for each of the variables.

Another way an open sentence can be made into a statement is by introducing **quantifiers**. For example, for the open sentence $x + 1 = 2$, we could say: For every real number x, $x + 1 = 2$. This sentence is now a mathematical statement that happens to be false. The quantifier introduced here is the phrase "for every real number x" and is called a **universal** quantifier. Another way to modify $P(x)$ is to write: there is a real number x such that $x + 1 = 2$. Note that this statement is true. The quantifier in this example, "there is a real number x," is called **existential**.

Once a quantifier is applied to a variable, then the variable is called a **bound** variable. In the example "For every real number x, $x + 1 = 2$" of the previous paragraph, then, the variable x is a bound variable. A variable that is not bound is called a **free** variable.

If $P(x)$ is an open sentence, then the statement: "For all x, $P(x)$" means that for *every* assigned value a of the variable x, the statement $P(a)$ is true.

The statement "For some x, $P(x)$" means that for *some* assigned value of the variable x, say $x = a$, the statement $P(a)$ is true. This statement may also be worded: "There exists a value of x such that $P(x)$."

Sometimes, in a statement containing universal quantifiers, the words "for all" or "for every" are not actually in the sentence but are implied by the meaning of the words. Here are some examples.

EXAMPLE 6   If n is an even integer, then n^2 is even.

On the surface, this sentence might seem to be an open sentence rather than a statement since it contains the variable n. However, implicit in the wording is the meaning: for every integer n, if n is even, then n^2 is even. So the variable n has been modified by a universal quantifier and is now a bound variable, making the sentence a statement. As we saw in Example 1, it is actually a true statement.

EXAMPLE 7   A triangle has three sides.

This statement contains a universal quantifier since it is really asserting that every triangle has three sides. Another way to word this statement is "For every plane figure T, if T is a triangle, then T has three sides," or more simply "If T is a triangle, then T has three sides."

EXAMPLE 8   The square of a real number is nonnegative.

Again this statement has a universal quantifier since it is saying that the square of every real number is nonnegative. It could also be written as:

"If x is a real number, then $x^2 \geq 0$."

EXAMPLE 9   All triangles are isosceles.

This statement has a universal quantifier as well. It just happens to be false.

We will sometimes use the symbol $\forall$ to mean "for all" or "for every." Example 8 can be rewritten: "$\forall\ x$, if x is real, then $x^2 \geq 0$."

The following examples give some of the forms that a statement with an existential quantifier can take.

EXAMPLE 10   Some even numbers are multiples of 3.

First note that even though this statement is written in plural form ("some even numbers"), it may be phrased: "There exists an even integer that is a multiple of 3." To prove this statement, one need only find *one* even number that is a multiple of 3. Since 6 is such a number, the proof is complete.

EXAMPLE 11   Some real numbers are irrational.

This statement asserts something about some, but not all, real numbers. It may be reworded as: "There exists a real number x such that x is irrational." It is a true statement provided that there is at least one real number that is not rational. In Section 1.4, we will prove this statement by showing that $\sqrt{2}$ is irrational.

EXAMPLE 12   There is a real number whose square is negative.

This statement also makes an assertion about some real numbers. Note that it is a false statement.

The symbol $\exists$ is used to mean "there exists" or "there is." The symbol $\ni$ is read "such that." Example 11 then can be expressed as: $\exists\ x$, x real, $\ni x$ is irrational.

A statement of course may have more than one quantifier.

EXAMPLE 13   For every real number x, there is an integer n such that $n > x$.

This statement contains both a universal and an existential quantifier.

EXAMPLE 14 The following statement, which is a definition from calculus, also has both a universal and an existential quantifier: A real-valued function $f(x)$ is **bounded** on the closed interval $[a, b]$ if $f(x)$ is defined on $[a, b]$ and there exists a positive real number M such that $|f(x)| \leq M$ for all $x \in [a, b]$.

For example, the function $f(x) = x^2 + 1$ is bounded on $[0, 2]$ because $|f(x)| \leq 5$ for all $x \in [0, 2]$.

The order in which quantifiers appear in a statement is important. If $P(x, y)$ is an open sentence in the variables x and y, then the statement

$$\forall x, \exists y \ni P(x, y)$$

does not always mean the same as the statement

$$\exists y \ni \forall x, P(x, y).$$

To see this, consider the following example.

EXAMPLE 15 The statement "Every real number has a cube root" can be written in the form:

$$\forall x \in \mathbf{R}, \exists y \in \mathbf{R} \ni y^3 = x.$$

This statement is true and is a consequence of the *Intermediate Value Theorem* of Calculus. However, the statement

$$\exists y \in \mathbf{R} \ni \forall x \in \mathbf{R}, y^3 = x$$

means that every real number x is the cube of a single number y and is clearly false.

Negations

Two of the statements just given, "All triangles are isosceles" and "There is a real number whose square is negative," were noted to be false. To *prove* that they are false it is necessary to prove that the negations of these statements are true.

Definition 1.1.3 If P is a statement, the **negation** of P, written $\neg P$ (and read "not P"), is the statement "P is false."

There are several alternative ways to express $\neg P$. For example, "P is not true" and "It is not true that P" are the same as our definition. In addition, it is often possible to express the negation of a sentence more elegantly by using better English style. Note however that the negation of a statement P must be phrased in such a way that exactly one of the statements P or $\neg P$ is true and one is false. Some examples follow:

EXAMPLE 16

P: John Fitzgerald Kennedy was the 35th president of the United States.

$\neg P$: John Fitzgerald Kennedy was not the 35th president of the United States.

Notice (it's obvious!) that exactly one of P and $\neg P$ is true, the other false.

EXAMPLE 17

P: Marie Curie did not win the Nobel Prize.

$\neg P$: Marie Curie won the Nobel Prize.

We really have a double negative here, the negation of a negatively phrased statement, that is better stated positively.

EXAMPLE 18

P: $3 + 1 = 4$.

$\neg P$: $3 + 1 \neq 4$.

As this example shows, symbolic mathematical statements can often be negated symbolically.

We can also talk about the negation of open sentences $P(x)$, $P(x, y)$, $P(x, y, z)$, ... in one or more variables. To use the one-variable case as an example, we will write $\neg P(x)$, read "not $P(x)$," to mean the open sentence in the variable x that becomes the statement $\neg P(a)$ when x is assigned the value a.

The negation of open sentences with more than one variable is defined in a similar fashion. In writing such sentences, we use whatever form good English or mathematical usage would dictate.

EXAMPLE 19

$P(x)$: $x + 1 = 2$.

$\neg P(x)$: $x + 1 \neq 2$.

EXAMPLE 20

$P(x)$: $x > 5$.

$\neg P(x)$: $x \leq 5$.

EXAMPLE 21

$P(n, m)$: $n + m$ is even.

$\neg P(n, m)$: $n + m$ is odd.

In the next example, we consider the negation of a statement with a universal quantifier.

EXAMPLE 22 P: If n is an even integer, then n^2 is even.
$\neg P$:   It is not true that if n is an even integer, then n^2 is even.

Every statement can be negated by simply putting the phrase "it is not true that" in front of it. Usually, though, expressing the negation this way does not convey what the negated statement actually means.

In this example, P contains the universally quantified variable n and can be written:

For all integers n, if n is even, then n^2 is even.

Therefore, P is false if there is *some* even integer whose square is not even. So $\neg P$ must be the statement

For some even integer n, n^2 is not even   or
There exists an even integer n such that n^2 is odd.

Note that in negating P, the universal quantifier is replaced by an existential quantifier.

EXAMPLE 23 P: Every day this week was hot.

The negation of this statement is *not* "Every day this week was not hot," since these two statements do not cover all possibilities. There can be a week in which some days are hot and some are not. To negate P, it is sufficient that there be at least one day that is not hot. So we can write the negation as:

Some day this week was not hot   or
There was a day this week that was not hot.

EXAMPLE 24 P: Every polynomial function is continuous everywhere.
$\neg P$:   There exists a polynomial function that is not continuous somewhere.

There are really two universal quantifiers in statement P. It says that *all* polynomial functions are continuous at *all* real numbers. So the negation should say that *some* polynomial functions are not continuous at *some* real numbers.

EXAMPLE 25   The negation of the statement "All Red Sox players are slow" is "There is a Red Sox player who is not slow," or equivalently, "Not all Red Sox players are slow." An incorrect way to negate this statement is "All Red Sox players are not slow." We make no comment about which of these statements is true.

If we negate a statement with an existential quantifier, then a universal quantifier is required.

EXAMPLE 26   *P:*   There is a real number whose square is negative.

Statement *P* says that we can find a real number whose square is negative. The negation of *P* means that we cannot find such a number. In other words, the negation of *P* is

The square of every real number is not negative.

To negate *P* by saying "There is a real number whose square is nonnegative" would not be correct because this statement and *P* could, theoretically at least, both be true.

EXAMPLE 27   *P:*   Some real-valued functions are not integrable.
                    ¬P: Every real-valued function is integrable.

As Examples 22–27 show, there are thus two basic rules about negating statements with quantifiers.

Rule 1: The negation of the statement "For all *x*, *P(x)*" is the statement "For some *x*, *¬P(x)*."

Rule 2: The negation of the statement "For some *x*, *P(x)*" is the statement "For all *x*, *¬P(x)*."

Of course, if a statement contains both universal and existential quantifiers, then in order to negate the statement, it is necessary to apply both of these rules.

If a statement *S* has the form "For all *x*, $\exists y \ni P(x, y)$" then the negation of *S* is "For some *x*, $\neg(\exists y \ni P(x, y))$" by Rule 1. Since Rule 2 tells us that the negation of "$\exists y \ni P(x, y)$" is "For all *y*, $\neg(P(x, y))$," then the negation of *S* becomes "For some *x*, for all *y*, $\neg(P(x,y))$" or "$\exists x \ni$ for all *y*, $\neg(P(x,y))$."

Similarly, the negation of "$\exists x \ni \forall y, P(x, y)$" is "$\forall x, \exists y \ni \neg(P(x, y))$."

EXAMPLE 28   *P:*   For every real number *x*, there is an integer *n* such that *n > x*.
                    ¬P: There is a real number *x* such that for every integer *n*, *n ≤ x*.

The statement in this last example is known as the Archimedean Principle. See the Historical Comments at the end of Section 1.2.

EXAMPLE 29   *P:*   There is a continuous real-valued function *f(x)* such that *f(x)* is not differentiable at any real number *c*.
                    ¬P: For every continuous real-valued function *f(x)*, there is a real number *c* such that *f(x)* is differentiable at *c*.

Surprisingly, statement P of Example 29 is true. There are continuous functions that are not differentiable at any point! An example, due to K. Weierstrass, is

$$f(x) = \sum_{n=0}^{\infty} \frac{1}{2^n} \cos\left((15)^n \pi x\right).$$

This is a function that is continuous everywhere but differentiable nowhere.

Writing Proofs

As you start to read and write proofs, you will see that they require a certain writing style to ensure clarity and readability. Take, for example, the proof in Example 1, that if n is an even integer, then n^2 is even. Suppose that a proof were written as follows:

$n = \text{even} = 2t.$
$n^2 = 4t^2 = \text{even}.$

This rather brief proof has the correct mathematical steps, but is lacking in explanation and hence in clarity and also suffers from poor notation. One should begin by clearly stating the assumption: "Let n be an integer and suppose that n is even." Then it can be noted that this means that "$n = 2t$ for some integer t." But writing "$n = \text{even}$" is sloppy notation. The word "even" is an adjective and should precede a noun; in this case, the word "number." But even the expression "$n = \text{even number}$" is not appropriate. The phrase "even number" does not belong in an equation. Equations should only contain numbers and symbols.

After writing "$n = 2t$ for some integer t" an explanation should be given for the next step: "Squaring both sides, we get $n^2 = 4t^2 = 2(2t^2)$." Then finally, one should note that "since $2t^2$ is an integer, it follows that n^2 is even."

● HISTORICAL COMMENTS: EARLY GREEK MATHEMATICS

A history of early Greek mathematics was written by **Eudemus** in the fourth century B.C. Although this book is now lost, a summary of it was written by **Proclus** in the fifth century A.D. According to Proclus's work, the earliest known mathematician to use the deductive method was **Thales** of Miletus. Thales founded the earliest Greek school of mathematics and philosophy. Among the results attributed to Thales are: the base angles of an isosceles triangle are congruent; triangles with corresponding angles equal have proportional sides; and an angle inscribed in a semicircle is a right angle. Each of these results was proven by deductive methods.

About sixty to eighty years after Thales, an important school of mathematics and philosophy was founded in southern Italy by **Pythagoras** (ca. 585–497 B.C.). One of the great contributions of the Pythagoreans, as the members of this school were known, is the idea that mathematical entities such as numbers and geometrical figures are in fact abstractions distinct from the material world. To the Pythagoreans the concept of number was the key to understanding the mysteries of the universe. By numbers they always meant whole numbers. Although they worked with ratios of whole numbers, which we call fractions, they did not consider them numbers per se. The Pythagoreans are credited with discovering the Pythagorean Theorem but probably did not give a formal proof of it. And as the records left behind by the Babylonians indicate, the Pythagoreans were not the first to use it.

Exercises 1.1

1. Determine whether each of the following sentences is a statement, an open sentence, or neither.
   (a) The Boston Celtics have won 16 NBA championships.
   (b) The plane is leaving in five minutes.
   (c) Get a note from your doctor.
   (d) Is that the best you can do?
   (e) Excessive exposure to the sun may cause skin cancer.
   (f) $5 + 2 = 6$.
   (g) Someone in this room is the murderer.
   (h) $x^2 + 1 \neq 0$.
   (i) For every real number x, $x^2 + 1 \neq 0$.
   (j) The equation of a circle of radius 1 with center at the origin is $x^2 + y^2 = 1$.
   (k) If n and m are even integers, then nm is even.

2. For each of the following statements, determine if it has any universal or existential quantifiers. If it has universal quantifiers, rewrite it in the form "for all . . . ." If it has existential quantifiers, rewrite it in the form, "there exists . . . such that . . . ." Introduce variables where appropriate.
   (a) The area of a rectangle is its length times its width.
   (b) A triangle may be equilateral.
   (c) $8 - 8 = 0$.
   (d) The sum of an even integer and an odd integer is even.
   (e) For every even integer, there is an odd integer such that the sum of the two is odd.
   (f) A function that is continuous on the closed interval $[a, b]$ is integrable on $[a, b]$.
   (g) A function is continuous on $[a, b]$ whenever it is differentiable on $[a, b]$.
   (h) A real-valued function that is continuous at 0 is not necessarily differentiable at 0.
   (i) All positive real numbers have a square root.
   (j) 1 is the smallest positive integer.

3. Write the negation of each statement in Exercise 2.

4. Write the negation of each of the following statements.
   (a) All triangles are isosceles.
   (b) Some even numbers are multiples of three.
   (c) Every door in the building was locked.
   (d) All new cars have something wrong with them.
   (e) Some angles of a triangle are greater than 90 degrees.
   (f) There are sets that contain infinitely many elements.

5. Write the negation of each of the following statements.
   (a) There is a real number x such that $x^2 + x + 1 = 0$.
   (b) Every real number is less than 100.
   (c) If f is a polynomial function, then f is continuous at 0.
   (d) If f is a polynomial function, then f is continuous everywhere.
   (e) $\forall\, x$, x real, $\exists$ a real number $y \ni y = x^3$.
   (f) There is a real-valued function $f(x)$ such that $f(x)$ is not continuous at any real number x.

6. Consider the following statement P: "The square of an even integer is divisible by 4."
   (a) Write P as a statement in the form, "for all ... , if ... , then ... ."
   (b) Write the negation of P.
   (c) Prove P or $\neg P$. Explain any inductive reasoning you use to conjecture that P is true or that P is false.

7. Consider the following statement P: "The sum of two even integers is divisible by 4."
   (a) Write P as a statement in the form "for all ... , if ... , then ... ."
   (b) Write the negation of P.
   (c) Prove P or $\neg P$. Explain any inductive reasoning you use to conjecture that P is true or that P is false.

8. In Example 14 the definition of a bounded function was given.
   (a) Write the negation of this definition; that is, complete the following statement: "A real-valued function $f(x)$ is *not bounded* on the closed interval $[a, b]$ if ... ."
   (b) Give an example of a bounded function on $[0, 1]$. Justify your answer by determining a value for M.
   (c) Give an example of an unbounded function on $[0, 1]$. Justify your answer.
   (d) Suppose the definition of bounded function were worded this way: "A real-valued function $f(x)$ is said to be *bounded* on the closed interval $[a, b]$ if for all $x \in [a, b]$, there exists a positive real number M such that $|f(x)| \le M$." Does this definition mean the same as the one given in Example 14? If not, explain how they differ. Could this new definition make sense as the definition of a bounded function? Explain.

9. A real-valued function $f(x)$ is said to be *increasing* on the closed interval $[a, b]$ if for all $x_1, x_2 \in [a, b]$, if $x_1 < x_2$, then $f(x_1) < f(x_2)$.
   (a) Write the negation of this definition.
   (b) Give an example of an increasing function on $[0, 1]$.
   (c) Give an example of a function that is not increasing on $[0, 1]$.

10. (a) State a definition for a real-valued function $f(x)$ to be *decreasing* on a closed interval $[a, b]$.
    (b) Give the negation of this definition.
    (c) Give an example of a decreasing function on $[0, 1]$.
    (d) Give an example of a function on $[0, 1]$ that is neither increasing nor decreasing.

11. Prove the following corollary of the Archimedean Principle. (See Example 28 for the statement.) For every positive real number ε, there exists a positive integer N such that $1/n < \varepsilon$ for all $n \geq N$. (Note: this exercise is the basis for the formal proof that the sequence $\{1, \frac{1}{2}, \frac{1}{3}, \frac{1}{4}, \ldots\}$ converges to 0.)

12. Use the Archimedean Principle to prove the following: if x is a real number, then there exists a positive integer n such that $-n < x < n$.

13. Prove that if x is a positive real number, then there exists a positive integer n such that $\dfrac{1}{n} < x < n$.

Discussion and Discovery Exercises

D1. Consider the following question. What positive integers n can be written as the difference of two squares? For example, $5 = 3^2 - 2^2$ and $24 = 5^2 - 1^2$. The following table lists the expression $n = x^2 - y^2$ for varying values of x and y. Since n is positive, we assume that $x > y \geq 0$.

x	y	$x^2 - y^2$	x	y	$x^2 - y^2$	x	y	$x^2 - y^2$
1	0	1	5	4	9	8	0	64
2	0	4	6	0	36	8	1	63
2	1	3	6	1	35	8	2	60
3	0	9	6	2	32	8	3	55
3	1	8	6	3	27	8	4	48
3	2	5	6	4	20	8	5	39
4	0	16	6	5	11	8	6	28
4	1	15	7	0	49	8	7	15
4	2	12	7	1	48	9	0	81
4	3	7	7	2	45	9	1	80
5	0	25	7	3	40	9	2	77
5	1	24	7	4	33	9	3	72
5	2	21	7	5	24	9	4	65
5	3	16	7	6	13	9	5	56

(a) Based on the table, conjecture a theorem that states exactly which positive integers can be written as the difference of two squares.

(b) Try to prove your conjecture.

(c) Notice that some integers appear more than once as the difference of two squares. What accounts for this? Is it possible that an integer will appear infinitely often on this list? Give a reason for your answer.

D2. Reread Example 1. Explain what part of that example involves inductive reasoning and what part involves deductive reasoning. See the introduction to this chapter for a discussion of the difference between inductive and deductive reasoning.

D3. Use inductive reasoning to find a statement about whether or not the sum of two consecutive even integers is divisible by 4. Then prove your statement.

D4. Criticize the following "statement" of the Fundamental Theorem of Calculus: $\int_a^b f(x)\, dx = F(b) - F(a)$.

D5. Criticize the following "proof" of the fact that if n and m are even then $n + m$ is even.

We know that $n = 2t$ and $m = 2t$, so $n + m = 2t + 2t = 4t$. Therefore, $n + m$ is even.

Write out a correct proof.

D6. Is the following a valid proof that every integer multiple of 4 is even? If not, explain what you think is wrong with it and then write your own proof.

Every multiple of 4 has a 4 in it and 4 has a 2 in it.
Therefore, every multiple of 4 has a 2 in it and therefore a multiple of 4 is even.

D7. Consider the following statement P: If n is an integer and n^2 is a multiple of 4, then n is a multiple of 4. The following is a "proof" that P is a false statement:

$6^2 = 36$ and 36 is a multiple of 4 but 6 is not a multiple of 4. Therefore, P is false.

Is this proof valid? Give your reasons and if you think it is not a valid proof, write a correct one.

D8. Consider the following "proof" of the fact that if n is an integer and n^2 is even, then n is even.

By Example 1, the square of an even integer is even. Therefore n can't be odd and hence must be even.

Is this proof correct? Give reasons for your answer. If you think the proof is invalid, write a correct one.

D9. Find a proof of the Pythagorean Theorem and write it out. What assumptions are used in the proof?

1.2 COMPOUND STATEMENTS

Conjunctions and Disjunctions

Statements, especially in mathematics, are often complicated but can be seen as built up from simpler statements. Such compound statements can be constructed in several different ways, at least two of which are very easy.

Definition 1.2.1 Let P and Q be statements.

1. The **conjunction** of P and Q, written $P \wedge Q$ (and read "P and Q"), is the statement "Both P and Q are true."
2. The **disjunction** of P and Q, written $P \vee Q$ (and read "P or Q"), is the statement "P is true or Q is true."

Obviously, $P \wedge Q$ is true if P is true and Q is true. Notice, though, that there are three ways for $P \wedge Q$ to be false: if P is true but Q is false, if P is false but Q is true, and if both P and Q are false.

The disjunction is sometimes referred to as "the inclusive *or*": it's all right for both statements to be true. There is only one way for $P \vee Q$ to be false: both P and Q must be false. That there are three ways for $P \vee Q$ to be true is inherent in the definition.

If $P(x)$ and $Q(x)$ are open sentences in the variable x, we can write $P(x) \wedge Q(x)$, read $P(x)$ and $Q(x)$, to mean the open sentence in the variable x that becomes the statement $P(a) \wedge Q(a)$ when x is assigned the value a.

Similarly, $P(x) \vee Q(x)$, read $P(x)$ or $Q(x)$, means the open sentence in the variable x that becomes the statement $P(a) \vee Q(a)$ when x is assigned the value a.

The conjunction and disjunction of open sentences containing more than one variable are defined in like fashion.

EXAMPLE 1   *P*:   Josef Stalin was the leader of the Soviet Union in 1940.
                  Q: Germany invaded the Soviet Union in 1941.

Here *P* and *Q* are both true, so both $P \wedge Q$ and $P \vee Q$ are true as well.

EXAMPLE 2   *P*: $7 < 5$.
                  Q: $3 - 2 = 1$.

Since *P* is false, $P \wedge Q$ is false, and since *Q* is true, $P \vee Q$ is true.

EXAMPLE 3   *P*:   The Red Sox have won a World Series since World War I ended.
                  Q: F. Scott Fitzgerald wrote "For Whom the Bell Tolls."

Both *P* and *Q* are false, so both $P \wedge Q$ and $P \vee Q$ are also false.

EXAMPLE 4   A statement or open sentence may be compound but not obviously so. For example, if *x* is a real number, then the open sentence $|x| < 3$ can be written without the absolute value sign as $-3 < x < 3$, and this sentence is the conjunction $x < 3$ and $x > -3$.

EXAMPLE 5   The open sentence $|x| > 3$ can be written as the disjunction $x > 3$ or $x < -3$.

Truth Tables

Note that the preceding comments about the truth or falsity of $P \wedge Q$ and $P \vee Q$ apply to *any* statements substituted for *P* and *Q*. For example, if *P* and *Q* are any statements for which *P* is true and *Q* is false, then $P \wedge Q$ is false and $P \vee Q$ is true.

For this reason, it makes sense to consider expressions of the form $P \wedge Q$, $P \vee Q$, or $\neg P$ where *P* and *Q* are variables representing unspecified statements. Such expressions are called **statement forms**. They are not really statements themselves but become statements when the variables *P* and *Q* are replaced by statements.

We can summarize the comments made previously about the truth values of $P \wedge Q$ and $P \vee Q$ by means of **truth tables** for their statement forms.

P	*Q*	$P \wedge Q$
T	T	T
T	F	F
F	T	F
F	F	F

P	*Q*	$P \vee Q$
T	T	T
T	F	T
F	T	T
F	F	F

Note that the first two columns of each table list all of the different combinations of truth values for the variables P and Q. Each combination then determines a corresponding truth value of the statement form. Truth tables for other statement forms can be constructed in this way.

EXAMPLE 6   Let P be a statement form. Then the truth table for $\neg P$ is:

P	$\neg P$
T	F
F	T

EXAMPLE 7   If P and Q are statement forms, the truth table for $P \wedge \neg Q$ is:

P	Q	$P \wedge \neg Q$
T	T	F
T	F	T
F	T	F
F	F	F

Negating Conjunctions and Disjunctions

Sometimes a compound statement formed from two or more statements can be expressed in different ways but mean the same thing. For example, consider the statement

It is not true that today is sunny and warm.

In order for the statement "Today is sunny and warm" to be false, it is only necessary that one of the two conditions fail to happen; namely, it not be sunny or not be warm. Therefore, the statement "It is not true that today is sunny and warm" means the same as

It is not sunny or it is not warm.

If one is true then the other is true and vice versa.

If we let P be the statement "Today is sunny" and Q the statement "Today is warm," then we are saying that if $\neg(P \wedge Q)$ is a true statement, then $(\neg P) \vee (\neg Q)$ is true and conversely if $(\neg P) \vee (\neg Q)$ is true, then $\neg(P \wedge Q)$ is true.

In fact, for any statements P and Q, $\neg(P \wedge Q)$ and $(\neg P) \vee (\neg Q)$ mean the same thing. This follows from the fact that the statement forms $\neg(P \wedge Q)$ and $(\neg P) \vee (\neg Q)$ have the same truth tables.

P	Q	$\neg P$	$\neg Q$	$P \wedge Q$	$\neg(P \wedge Q)$	$\neg P \vee \neg Q$
T	T	F	F	T	F	F
T	F	F	T	F	T	T
F	T	T	F	F	T	T
F	F	T	T	F	T	T

Logically Equivalent Statements

Statements and statement forms like the preceding ones are said to be *logically equivalent*. Here is a formal definition.

> **Definition 1.2.2** We say that two statements are **logically equivalent** or just **equivalent** if they are both true or both false.
>
> We say that two statement forms are **logically equivalent** if the substitution of statements for the variables in the forms always yields logically equivalent statements.

Note: If two statement forms have the same truth tables then they are logically equivalent.

The preceding truth table shows that the statement forms $\neg(P \wedge Q)$ and $(\neg P) \vee (\neg Q)$ are logically equivalent. In other words, given any two statements P and Q, $P \wedge Q$ is false exactly when P *or* Q is false.

Similarly, $P \vee Q$ is false exactly when *both P and Q* are false, so $\neg(P \vee Q)$ is logically equivalent to $\neg P \wedge \neg Q$. A comparison of their truth tables will verify this fact as well.

P	Q	$\neg P$	$\neg Q$	$P \vee Q$	$\neg(P \vee Q)$	$\neg P \wedge \neg Q$
T	T	F	F	T	F	F
T	F	F	T	T	F	F
F	T	T	F	T	F	F
F	F	T	T	F	T	T

EXAMPLE 8

P: The earth is flat.

Q: The earth revolves around the sun.

$(P \wedge Q)$: The earth is flat and revolves around the sun. (False)

$\neg(P \wedge Q)$: The earth is not flat or it does not revolve around the sun. (True)

$P \vee Q$: The earth is flat or it revolves around the sun. (True)

$\neg(P \vee Q)$: The earth is not flat and it does not revolve around the sun. (False)

In Section 1.1, we saw two other examples of logically equivalent statement forms:

(a) $\neg(\forall x \; P(x))$ is logically equivalent to $\exists x \ni (\neg P(x))$.
(b) $\neg(\exists x \ni P(x))$ is logically equivalent to $\forall x \; (\neg P(x))$.

If we combine these two examples with the logical equivalences associated with negating conjunctions and disjunctions, we obtain the following equivalent statement forms:

(c) $\neg(\forall x \; (P(x) \vee Q(x)))$ is equivalent to $\exists x \ni ((\neg P(x)) \wedge (\neg Q(x)))$.
(d) $\neg(\forall x \; (P(x) \wedge Q(x)))$ is equivalent to $\exists x \ni ((\neg P(x)) \vee (\neg Q(x)))$.
(e) $\neg(\exists x \ni (P(x) \vee Q(x)))$ is equivalent to $\forall x \; ((\neg P(x)) \wedge (\neg Q(x)))$.
(f) $\neg(\exists x \ni (P(x) \wedge Q(x)))$ is equivalent to $\forall x \; ((\neg P(x)) \vee (\neg Q(x)))$.

These equivalences are illustrated by the following examples.

EXAMPLE 9 P: Every day this week was sunny or hot.
                   $\neg P$:   Some day this week was not sunny and not hot.

EXAMPLE 10 P: There is a real number x such that $x > 4$ and $x < 10$.
                   $\neg P$:   If x is a real number, then $x \le 4$ or $x \ge 10$.

EXAMPLE 11 P: Every multiple of 6 is even and is not a multiple of 4.
                   $\neg P$:   There is a multiple of 6 that is odd or is a multiple of 4.

Sometimes statements that appear to be logically equivalent are not. For example, suppose that $P(x)$ and $Q(x)$ are open sentences containing the variable x and let S and T be the following statements:

S: For all x, $P(x)$ or $Q(x)$.
T: For all x, $P(x)$ or for all x, $Q(x)$.

Both of these statements have universal quantifiers and involve a disjunction with respect to the statements $P(x)$ and $Q(x)$. At first glance, they appear to be saying the same thing and therefore should be logically equivalent.

But consider the following example: let $P(x)$ be the open sentence "$x > 2$" and $Q(x)$ the open sentence "$x < 5$." Letting x be assigned values from the real numbers, S and T become:

S: For all real numbers x, $x > 2$ or $x < 5$.
T: For all real numbers x, $x > 2$ or for all real numbers x, $x < 5$.

It is clear that S is a true statement but T is a false statement. Therefore S and T are *not* logically equivalent.

As another example, consider Example 9. The statement "Every day this week was sunny or hot" does not mean the same as "Every day this week was sunny or every day this week was hot," because for the first statement to be true, every day of the week had to be sunny or hot, so some could have been sunny and cool and others cloudy but hot. But the second statement is only true if *every* day was sunny or if *every* day was hot.

Now let's consider the following statements, again assuming that $P(x)$ and $Q(x)$ are open sentences containing the variable x.

S: There exists x such that $P(x)$ or $Q(x)$.

T: There exists x such that $P(x)$ or there exists x such that $Q(x)$.

In this case, S and T are logically equivalent. To prove this we must show that S and T are either both true or both false.

First, suppose that S is a true statement. Then there is an assigned value of x, say $x = a$, for which $P(a)$ is a true statement or $Q(a)$ is a true statement. Since there are two possibilities, we'll look at each one separately. Suppose that $P(a)$ is true. It follows that the statement

There exists x such that $P(x)$

is true and therefore the disjunction

There exists x such that $P(x)$ or there exists x such that $Q(x)$

is true also. So statement T is true in this case. On the other hand, if $Q(a)$ is true, then the statement

There exists x such that $Q(x)$

is true, and thus T is true in this case also.

Now suppose that T is true. Then there is an x such that $P(x)$ or there is an x such that $Q(x)$. If there is an x such that $P(x)$, then there is an assigned value of x, say $x = a$, such that $P(a)$ is true. It follows that there is an assigned value of x, namely $x = a$, such that $P(a)$ or $Q(a)$ is true. Thus there exists an x such that $P(x)$ or $Q(x)$, so S is true.

On the other hand, if there is an x such that $Q(x)$, then there is an assigned value of x, say $x = b$, such that $Q(b)$ is true. Hence there is an assigned value of x, namely $x = b$, such that $P(b)$ or $Q(b)$ is true. So there is an x such that $P(x)$ or $Q(x)$ and again S is true.

Therefore if T is true, then S is true. This proves that S and T are logically equivalent.

EXAMPLE 12 The statement

> There was a day this week that was sunny or hot

is logically equivalent to the statement

> There was a day this week that was sunny or there was a day this week that was hot.

EXAMPLE 13 The following statements are equivalent:

> P: There is a real number x such that $x > 2$ or there is a real number x such that $x < 5$.
>
> Q: There is a real number x such that $x > 2$ or $x < 5$.

Our examples of compound statements given in this section so far have been formed from two given statements or statement forms P and Q. We can, of course, construct statements from three or more given statements.

EXAMPLE 14 Consider the statement

> According to today's weather forecast, tomorrow will be cold and cloudy or cold and rainy.

Let P, Q, and R be the statements:

> P: Tomorrow will be cold.
> Q: Tomorrow will be cloudy.
> R: Tomorrow will be rainy.

Then the original statement has the form:

> According to today's weather forecast, $(P \wedge Q) \vee (P \wedge R)$.

If we analyze the meaning of this statement, then we see that it could be rephrased as

> According to today's weather forecast, tomorrow will be cold and it will be cloudy or rainy.

In this form our statement becomes

> According to today's weather forecast, $P \wedge (Q \vee R)$.

EXAMPLE 15 The last example suggests that the statement forms $(P \wedge Q) \vee (P \wedge R)$ and $P \wedge (Q \vee R)$ are logically equivalent. To prove this we will examine their respective truth tables. Since there are three statements involved, the truth tables will have eight rows corresponding to all possible combinations of truth or falsity of P, Q, and R.

P	Q	R	$P \wedge Q$	$P \wedge R$	$(P \wedge Q) \vee (P \wedge R)$
T	T	T	T	T	T
T	T	F	T	F	T
T	F	T	F	T	T
T	F	F	F	F	F
F	T	T	F	F	F
F	T	F	F	F	F
F	F	T	F	F	F
F	F	F	F	F	F

P	Q	R	$Q \vee R$	$P \wedge (Q \vee R)$
T	T	T	T	T
T	T	F	T	T
T	F	T	T	T
T	F	F	F	F
F	T	T	T	F
F	T	F	T	F
F	F	T	T	F
F	F	F	F	F

Since the truth tables of $(P \wedge Q) \vee (P \wedge R)$ and $P \wedge (Q \vee R)$ are the same, they must be logically equivalent.

We now give a mathematical example of the equivalence in the previous example.

EXAMPLE 16 Let x be a real number. Suppose that we wish to write the statement

$$x > -10 \text{ and } |x| > 5$$

in a simpler form without using absolute value signs. The inequality $|x| > 5$ is equivalent to the statement $x > 5$ or $x < -5$.

By the previous example, the statement

$$x > -10 \text{ and } (x > 5 \text{ or } x < -5)$$

is equivalent to the statement

$(x > -10 \text{ and } x > 5) \text{ or } (x > -10 \text{ and } x < -5).$

Because any real number x that is greater than 5 is also greater than -10, the statement $x > -10$ and $x > 5$ is equivalent to $x > 5$. Finally, we can reword our statement as

$x > 5 \text{ or } -10 < x < -5.$

Tautologies and Contradictions

Sometimes a statement form will always be true no matter what statements are substituted for the variables. Such a statement form is called a **tautology**. A simple example is the statement form $P \vee \neg P$. This statement form is true when P is true and it is true when P is false. A statement form is a tautology if each of its truth table values is true.

A statement form that is always false is called a **contradiction**. The statement form $P \wedge \neg P$ is an example. A statement form is a contradiction if each of its truth table values is false.

Note that if S is a tautology then $\neg S$ is a contradiction and if S is a contradiction then $\neg S$ is a tautology. We will see other examples of tautologies and contradictions in this chapter.

●   HISTORICAL COMMENTS: EUCLID'S AXIOMS

In Example 28 of Section 1.1, the following statement was given: For every real number x, there is an integer n such that $n > x$.

This is a true statement. It is in fact a property of the real numbers called the **Archimedean Principle**. An immediate consequence is that there is no "largest" real number. Although this fact may seem obvious from our intuitive grasp of the real number line, a formal proof of the Archimedean Principle depends on an axiom of the real numbers called the Least Upper Bound Axiom. (A discussion of the Least Upper Bound Axiom and a proof of the Archimedean Principle is given in Section 7.2.)

You will often find that statements that seem obvious require not so obvious proofs. The starting point of such proofs are **axioms**, statements that we assume as given. You probably first encountered axioms in your study of plane geometry in high school. And that of course brings us to **Euclid**.

Euclid lived about 300 B.C. and although not much is known about his personal life, he is famous for his text *Elements*. In it Euclid lists five postulates or axioms that form the basis of his study of geometry. In all, the *Elements* consists of 13 books and contains 467 propositions. The proof of each proposition is based on earlier propositions and the postulates. The great achievement of Euclid's *Elements* is the use of logic and deduction to advance the knowledge of mathematics.

The *Elements* is not without its defects, however. Sometimes proofs are incomplete or assumptions are made without proof. For example, in one of his propositions Euclid implicitly uses without proof a statement *equivalent* to the Archimedean Principle. **Archimedes** (287–212 B.C.), perhaps the greatest mathematician of ancient Greece, used it himself without proof (hence the attachment of his name to it) and the Greek mathematician **Eudoxus** (ca. 408–355 B.C.) used it also.

These are the five postulates:

1. A straight line can be drawn from any point to any point.
2. A finite straight line can be produced continuously in a straight line.
3. A circle may be described with any center and distance.
4. All right angles are equal to one another.
5. If a straight line falling on two straight lines makes the interior angles on the same side together less than two right angles, the two straight lines, if produced indefinitely, meet on that side on which the angles are together less than two right angles.

Euclid's fifth postulate, known as the Parallel Postulate, is the most famous of all of the axioms. What strikes one about these postulates is that the first four seem quite straightforward and obvious. The fifth postulate, however, is of a different nature. It reads more like a theorem. One has to read it carefully in order to understand it. Even then it does not seem readily apparent. It seems more like a statement one should try to prove rather than accept as given. In fact many mathematicians from Euclid's time and later felt the same way and numerous attempts were made to prove the fifth postulate. However, all attempts at a proof failed. Usually when someone claimed to have a proof of the fifth postulate, it would turn out that somewhere in the proof, an assumption had been made that was in fact logically equivalent to the fifth postulate. One of the earliest attempts, for instance, was made by the Greek astronomer **Claudius Ptolemy** in about 150 A.D. But Ptolemy assumed the following statement: Given a point P in the plane and a line L that does not intersect P, there is one and only one line that passes through P and is parallel to L. This statement is actually equivalent to Euclid's fifth postulate and is the version of the fifth postulate that actually appears in most high school geometry texts.

The following statements, familiar to students of plane geometry, are also equivalent to the Parallel Postulate:

1. There exist two triangles that are similar but not congruent.
2. There exists a triangle the sum of whose angles is 180 degrees.
3. Through any three points not lying on a straight line there is a circle.

Exercises 1.2

---

1. Let P be a statement form. Prove that P and $\neg(\neg P)$ are logically equivalent.

2. Let P and Q be statement forms. Write the truth tables for the following statement forms.
   (a) $(\neg P) \vee Q$.   (b) $\neg(P \vee Q)$.
   (c) $\neg((\neg P) \wedge Q))$.   (d) $((\neg P) \wedge (\neg Q)) \vee Q$.

3. Prove that the statement forms $\neg((\neg P) \vee Q)$ and $P \wedge (\neg Q)$ are logically equivalent.

4. Write the negation of the following statements.
   (a) August is a hot month and September is sometimes cool.
   (b) Every member of the baseball team is complaining and not hitting.
   (c) Some cars are comfortable and not expensive.
   (d) Math tests are long or difficult.

5. Write the negation of the following statements.
   (a) If x and y are real numbers such that $xy = 0$, then $x = 0$ or $y = 0$.
   (b) For every integer x, x^2 is odd and $x^3 - 1$ is divisible by 4.
   (c) $\forall n$, n an integer, $\exists$ an integer k such that $n = 2k$ or $n = 2k + 1$.
   (d) $\exists$ a rational number $r \ni 1 < r < 2$.
   (e) A real number can be greater than 2 or less than 1.
   (f) Some functions are neither differentiable at 0 nor continuous at 0.

6. Let P be the statement "Every multiple of 6 is even and is not a multiple of 4" of Example 11 in this section.
   (a) Write P in the form, "for all ... , if ... , then. ..." Use variables.
   (b) Write the negation of P. Use variables.
   (c) Prove P or $\neg P$.

7. Repeat Exercise 6 if P is the statement "If the product of two integers is even, then both of the integers are even."

8. Let $n, m \in \mathbf{Z}$. Write the negation of the statement "Exactly one of the integers n or m is odd."

9. Consider the following statement P: "If n is an odd integer, then there exists an integer x such that $n = 4x + 1$ or $n = 4x + 3$."
   (a) Write the negation of P.
   (b) Prove P or $\neg P$.

10. Let P and Q be statement forms.
    (a) Prove that $P \wedge Q$ is logically equivalent to $Q \wedge P$.
    (b) Prove that $P \vee Q$ is logically equivalent to $Q \vee P$.

11. Let $P(x)$ and $Q(x)$ be open sentences containing the variable x. In each part of this problem, determine if the given statements S and T are

logically equivalent. If they are, give a proof. If they are not, give an example of open sentences $P(x)$ and $Q(x)$ for which S and T are not logically equivalent.

(a) S: $\forall x, (P(x) \wedge Q(x))$.

　T: $(\forall x, P(x)) \wedge (\forall x, Q(x))$.

(b) S: $\exists x \ni (P(x) \wedge Q(x))$.

　T: $(\exists x \ni P(x)) \wedge (\exists x \ni Q(x))$.

12. Let P, Q, and R be statement forms. Write the truth tables for the following statement forms.

(a) $(P \wedge \neg Q) \vee (\neg R)$. 　　(b) $(\neg P) \wedge (R \wedge (\neg Q))$.

(Reminder: To do this problem, it is necessary to list all possible combinations of truth values for P, Q, and R. There are eight of them.)

13. Let P, Q, and R be statement forms.

(a) Prove that $(P \vee Q) \vee R$ and $P \vee (Q \vee R)$ are equivalent statement forms.

(b) Prove that $(P \wedge Q) \wedge R$ and $P \wedge (Q \wedge R)$ are equivalent statement forms.

(c) Prove that $P \vee (Q \wedge R)$ and $(P \vee Q) \wedge (P \vee R)$ are equivalent statement forms.

14. Characterize all real numbers x such that $x > 1$ or $|x| < 3$. Express your answer in the simplest possible way without using absolute value signs.

15. Let P, Q, R, and S be statement forms.

(a) Prove that $P \vee ((Q \wedge R) \wedge S))$ and $(P \vee Q) \wedge (P \vee R) \wedge (P \vee S)$ are equivalent statement forms.

(b) Prove that $P \wedge ((Q \vee R) \vee S))$ and $(P \wedge Q) \vee (P \wedge R) \vee (P \wedge S)$ are equivalent statement forms.

16. Let P and Q be statement forms.

(a) Prove that $(P \wedge Q) \vee (P \wedge \neg Q) \vee (\neg P \wedge Q) \vee (\neg P \wedge \neg Q)$ is a tautology.

(b) Prove that $(P \vee Q) \wedge (P \vee \neg Q) \wedge (\neg P \vee Q) \wedge (\neg P \vee \neg Q)$ is a contradiction.

Discussion and Discovery Exercises

D1. Consider the following "proof" that if n or m is an odd integer, then nm is an even integer.

Suppose that n is odd and m is even. Then $m = 2t$ for some integer t. Therefore $nm = n(2t) = 2nt$, which is even. Next suppose that n is even and m is odd. We can write $n = 2s$ for some integer s. Thus $nm = 2sm$, which is also even. In both cases, we get that nm is even. Thus the statement is proved.

Is the proof valid? If you think it is not, explain.

D2. Consider the following two statements:

    P: For every real number x, $x^2 \geq 0$.
    Q: Lyndon Johnson was elected president in 1964.

    Are these statements logically equivalent? Explain.

D3. Using the five statements below as clues, match Sarah, Ann, and Bob with their respective occupations (teacher, entomologist, or poet) and the color of their houses (brown, white, or green). Explain how you arrived at your answer and mention any rules of logic that you use. Assume that no two people have the same occupation or the same color house.
    The first three clues are *true* statements:
      1. Sarah or Ann is the poet.
      2. Sarah's or Bob's house is green.
      3. Ann's house is green or white.
    The next two clues are *false* statements:
      4. The teacher's house is green.
      5. Sarah is the poet or Bob is the entomologist.

D4. Consider the following two statements:

    P: For every even integer n, there is an odd integer m such that $n + m$ is odd.

    Q: There is an odd integer m such that for every even integer n, $n + m$ is odd.

    (a) Do these statements mean the same thing? If not, explain the difference.
    (b) Write the negations of *P* and *Q*.
    (c) Discuss the truth or falsity of statements *P* and *Q*. Give proofs.
    (d) Are these statements logically equivalent? Give reasons for your answer.

D5. Repeat the previous problem for the following statements:

    P: For every real number x between 0 and 1, there is a real number y between 1 and 2 such that $x + y < 2$.

    Q: There is a real number y between 1 and 2 such that for every real number x between 0 and 1, $x + y < 2$.

D6. Euclid's fourth postulate says: "All right angles are equal to one another." Isn't it obvious that all right angles are equal? What do you think Euclid meant by right angle? Why does he consider it necessary to include this postulate?

1.3   IMPLICATIONS

Implication: Definition and Examples

In mathematics, as we have noted before, we are interested in whether a given statement is true or false. Usually such a statement will be proven true (or false) because it (or its negation) can be seen to follow logically from prior statements that we know to be true. Those prior statements may be propositions already proven or they may be axioms. Axioms are statements that we take to be given without proof and serve as the starting point of a particular subject. In Chapter 5, for example, we will list a number of axioms of the integers and then deduce properties of the integers from them.

In the previous two sections, many of the statements we looked at took the form: "If ..., then ..." or "For all ..., if ..., then ...." The "if-then" part of such a statement is called an **implication**. Most theorems will take this form. In Example 1 of Section 1.1, we considered the statement: "If n is an even integer, then n^2 is even." The "if" part of the statement gives the **premise** or **assumption** that is made. In this example, we *assume* that n is an even integer. The "then" part is the conclusion that is asserted to follow from the premise. The statement "n^2 is even" is asserted to be a true statement under the assumption that "n is even." So in order to prove such a statement as "if n is an even integer, then n^2 is even," one starts, as was done in Example 1 of Section 1.1, by assuming that n is an even integer and then proves that n^2 is even.

We begin with a formal definition of implication and then do some examples.

Definition 1.3.1 Let P and Q be statements. The **implication** $P \Rightarrow Q$ (read "P implies Q") is the statement "If P is true, then Q is true."

EXAMPLE 1

P: $3 + 2 = 5$.
Q: $3 + 1 + 1 = 5$.
$P \Rightarrow Q$:   If $3 + 2 = 5$, then $3 + 1 + 1 = 5$.

EXAMPLE 2

P: 4 is an even integer.
Q: 4^2 is an even integer.
$P \Rightarrow Q$:   If 4 is an even integer, then 4^2 is an even integer.

EXAMPLE 3

P: The function $f(x) = x^2$ is differentiable at 0.
Q: The function $f(x) = x^2$ is continuous at 0.
$P \Rightarrow Q$:   If the function $f(x) = x^2$ is differentiable at 0, then it is continuous at 0.

Truth Table for an Implication

Before we consider when an implication might be true, an important distinction must be made. For a mathematician, there is no sense of causality in the statement $P \Rightarrow Q$; we leave this question to philosophers. P might be (apparently) entirely unrelated to Q.

Rather, the statement simply means that in all circumstances under which P is true, Q is also true. Very loosely, whenever P "happens," Q also "happens"; we don't care whether P seems to cause Q or not.

With this in mind, when would $P \Rightarrow Q$ be false? We would need P to "happen" (P true) and Q *not* to "happen" (Q false). This is the *only* case. Seen another way, $P \Rightarrow Q$ can't be false (so must be true) if P is false. Even if P and Q are *both* false the implication $P \Rightarrow Q$ is true. If P is true, though, Q must also be true for $P \Rightarrow Q$ to be true.

In summary, whether $P \Rightarrow Q$ is true or false depends only on the truth or falsity of P and of Q, so there are four cases, just as with conjunction and disjunction. The following table gives the truth values of the statement form $P \Rightarrow Q$.

P	Q	$P \Rightarrow Q$
T	T	T
T	F	F
F	T	T
F	F	T

EXAMPLE 4 P: $3 + 2 = 5$.
Q: $3 + 1 + 1 = 5$.

In this case, both P and Q are true, so $P \Rightarrow Q$ is true as well. Intuitively, P "forces" Q to be true, since $2 = 1 + 1$, but the truth of the two statements is sufficient.

EXAMPLE 5 P: Gerald Ford was vice president under Jimmy Carter.
Q: $2 < 7$.

Here, P is false and Q is true, so $P \Rightarrow Q$ is true. Notice that if R is the statement $7 < 2$, $P \Rightarrow R$ is also true; a false statement implies anything!

EXAMPLE 6 P: The American Revolution ended in 1781.
Q: George Washington served three terms as president.

P is true and Q is false; however related the statements might be, the statement $P \Rightarrow Q$ is false.

EXAMPLE 7   *P*:   4 is an even integer.
                 Q: 4^2 is an even integer.

Since *P* and *Q* are both true, $P \Rightarrow Q$ is true. Example 1 of Section 1.1 does imply that there is a causality between statements *P* and *Q* but the mere fact that *P* and *Q* are both true statements is sufficient to establish that $P \Rightarrow Q$ is true.

EXAMPLE 8   *P*:   The function $f(x) = x^2$ is differentiable at 0.
                 Q: The function $f(x) = x^2$ is continuous at 0.

Both *P* and *Q* are true statements verified in calculus. So the implication $P \Rightarrow Q$ is true. There is a causality between *P* and *Q*, since a theorem from calculus shows that differentiability implies continuity. But again this causality, although it is certainly helpful to use, is not needed to establish the truth of $P \Rightarrow Q$.

EXAMPLE 9   *P*:   The function $f(x) = x^2$ is continuous at 0.
                 Q: The function $f(x) = x^2$ is differentiable at 0.

Here there is no causality between *P* and *Q* since continuity does not in general imply differentiability. Nevertheless, the implication $P \Rightarrow Q$ is true since *P* and *Q* are both true.

Proving Statements Containing Implications

Most often we will be interested in establishing the truth of (proving) statements of the form $\forall x, P(x) \Rightarrow Q(x)$ where $P(x)$ and $Q(x)$ are open sentences.

Since $P(x)$ and $Q(x)$ are not statements, the expression $P(x) \Rightarrow Q(x)$ is not a statement either. $P(x) \Rightarrow Q(x)$ is the open sentence in the variable *x* that becomes the statement $P(a) \Rightarrow Q(a)$ when *x* is assigned the value *a*. But recall that the expression $\forall x, P(x) \Rightarrow Q(x)$ is a statement because the variable *x* has been quantified.

(For simplicity of notation, we assume there is only one variable involved but in fact the same principles apply no matter how many variables are being used.)

Since, for an assigned value *a* of *x*, the statement $P(a) \Rightarrow Q(a)$ will always be true if $P(a)$ is false, we need not consider this case. Rather, we can assume that, if the variable *x* is assigned a value *a*, then $P(a)$ is true and proceed from there to prove $Q(a)$. $P(a)$ is called the **hypothesis**, and $Q(a)$ the **conclusion**. We use a letter, in this case *a*, to denote the assigned value of the variable *x* rather than give a specific value to *x* like 0 or 2, since we must prove that $P(a) \Rightarrow Q(a)$ for every possible assigned value of *x*. Here are some examples.

EXAMPLE 10   The statement "The square of every even integer is even" can be written "If n is an even integer, then n^2 is even." This statement contains a universal quantifier. Recall from Example 6 of Section 1.1 that the sentence can be reworded "For every integer n, if n is even, then n^2 is even." This statement has the form $\forall$ integers n, $P(n) \Rightarrow Q(n)$ where $P(n)$ is the open sentence "n is even" and $Q(n)$ is the open sentence "n^2 is even."

The proof of this statement would start as follows: Let a be an integer and suppose a is even. The statement "a is even" is the hypothesis. We are thus starting out by asserting the truth of the hypothesis $P(a)$. The conclusion is the statement "a^2 is even." The remainder of the proof involves verifying that the statement $Q(a)$ is now true. The details of this proof were done in Example 1 of Section 1.1.

When we start this proof by saying "Let a be an integer and suppose a is even," we are assigning the variable n an even integer value but without specifying exactly what that value is. The letter a thus stands for *any* even integer and once it is proven that a^2 is even, the truth of the statement $P(a) \Rightarrow Q(a)$ is established for *every* integer a.

Note: In practice, it is usually simpler to use the same letter for the variable and its assigned value. So we could use just the letter n in the proof of Example 10 without confusion. But remember, once we say "Let n be an integer," then $P(n)$ is a statement.

EXAMPLE 11   Prove or disprove the statement S: "If n and m are odd integers, then $n + m$ is even."

This statement can be reworded

For all integers n and m, $P \Rightarrow Q$

where P is the open sentence "n and m are odd integers" and Q is the open sentence: "$n + m$ is even."

Note that statement S contains two variables. We should more properly write $P(n, m)$ and $Q(n, m)$, since P and Q depend on the variables n and m, but for the sake of simplicity of notation, we will simply write P and Q.

Because the truth or falsity of S is not immediately clear, we might begin by making a conjecture as to whether or not it is true. A good way to start is to try a few examples. Our hypothesis is that n and m are odd integers, so we'll try some odd integers and see if their sums are even or odd:

$$3 + 5 = 8, \quad 11 + 17 = 28, \quad 59 + 43 = 102.$$

In each of these cases, the sum is even. These examples do not prove that the statement is true since the statement contains a universal quantifier. We must prove that for *all* odd integers n and m, the sum $n + m$ is even. But it certainly seems reasonable to conjecture that S is true.

Now let's prove it!

To begin our proof, let n and m be odd integers. This means that we can write n as $2t + 1$ and m as $2s + 1$, where t and s are integers. Note that if we write $n = 2t + 1$, then we cannot write $m = 2t + 1$ since n and m are not necessarily the same integer. We must use another letter besides t and so we *arbitrarily* chose the letter s. Thus

$$n + m = (2t + 1) + (2s + 1) = 2t + 2s + 2 = 2(t + s + 1)$$

and since $t + s + 1$ is an integer, $n + m$ is even. The proof is complete.

Note that we started our proof by letting n and m be odd integers. The variables n and m stand for *any* pair of odd integers, not specific ones. So the fact that $n + m$ is even has been established for *all* pairs of odd integers n and m. This method of proof is necessary since the statement has a universal quantifier.

Negating an Implication: Counterexamples

Sometimes a statement containing an implication may be false. In that case, it becomes necessary to state its negation in a coherent fashion in order to prove that it is false.

EXAMPLE 12   Consider the statement "If a real-valued function f is continuous at 0, then f is differentiable at 0." This statement from calculus is false. To see how to write its negation, we break it down into its simple components. Let $P(f)$ be the open sentence "f is continuous at 0" and $Q(f)$ the open sentence "f is differentiable at 0." (Note that the variable in this example is f.) So our statement is "For every real-valued function f, $P(f) \Rightarrow Q(f)$." As we saw in Section 1.1, the negation must be the statement "There is some real-valued function f such that $\neg(P(f) \Rightarrow Q(f))$."

As we have seen, if P and Q are statements, then the only way for $P \Rightarrow Q$ to be false is that P be true and Q be false. Thus the negation must take the form "There is an assigned value of the variable f, say $f = g$, such that $P(g)$ is true and $Q(g)$ is false." More simply, using just the letter f, "There is a real-valued function f that is continuous at 0 but not differentiable at 0." Since the absolute value function f defined by $f(x) = |x|$ is such a function, the proof is complete. This function is called a **counterexample** because it serves to disprove a statement with a universal quantifier.

Given the previous example, it should not be surprising that if P and Q are statement forms, then $\neg(P \Rightarrow Q)$ is logically equivalent to $P \wedge \neg Q$. We leave it as an exercise to show that these statements have the same truth tables.

It follows then that if $P(x)$ and $Q(x)$ are open sentences, the negation of the statement "For all x, $P(x) \Rightarrow Q(x)$" is the statement "There exists x such that $P(x) \wedge \neg Q(x)$," or in other words, "For some x, $P(x)$ is true and $Q(x)$ is false."

The value assigned to the variable x that makes $P(x)$ true and $Q(x)$ false is called a **counterexample** to the statement "For all x, $P(x) \Rightarrow Q(x)$."

Note that the negation of an implication is not an implication!

EXAMPLE 13   Consider the statement "The sum of two perfect squares is a perfect square." To prove or disprove this statement, let's first write it as a statement that has a universal quantifier and contains an implication: "For every pair of integers n and m, if n and m are perfect squares, then $n + m$ is a perfect square." Symbolically, we can write this as "$\forall$ integers n and m, $P \Rightarrow Q$" where P is the open sentence "n and m are perfect squares" and Q is the open sentence "$n + m$ is a perfect square." The negation of this statement would be "There exist integers n and m such that $P \wedge \neg Q$." In other words, "There exist integers n and m such that n and m are perfect squares but $n + m$ is not a perfect square."

Since 4 and 9 are perfect squares, but $4 + 9 = 13$ is not a perfect square, the negation must be true and the original statement false. Again we have used a counterexample to disprove a statement with a universal quantifier.

Necessary and Sufficient Conditions

Given statements P and Q, the implication $P \Rightarrow Q$ means, as we have seen, that if P is true, then Q is true. We say then that P is a **sufficient condition** for Q. In other words, in order for Q to be true, it is *sufficient* that P be true.

Also, if $P \Rightarrow Q$ is a true statement, we say that Q is a **necessary condition** for P, meaning that Q *must* be true in order for P to be true. In other words, if Q is false, then P is false. The statement $\neg Q \Rightarrow \neg P$ is logically equivalent to $P \Rightarrow Q$. This equivalence will be discussed in the next section.

Note however that if $P \Rightarrow Q$ is a true statement, then it is not necessary that P be true in order for Q to be true. Similarly, Q is not a sufficient condition for P. In other words, even if Q is true, P may be false.

EXAMPLE 14   Let x be a real number. Let P be the statement "$x > 5$" and Q the statement "$x > 0$." It is clear from properties of inequalities of real numbers that $P \Rightarrow Q$ is true. So $x > 5$ is a sufficient condition for $x > 0$. But it is not a necessary condition. x need not be greater than 5 for x to be greater than 0. On the other hand, $x > 0$ is a necessary condition for $x > 5$; that is, if $x > 0$ is false, then $x > 5$ is false. We can also easily see that $x > 0$ is not a sufficient condition for $x > 5$.

EXAMPLE 15   Let f be a real-valued function. It follows from Example 12 that f being continuous at 0 is a necessary but not sufficient condition for f to be differentiable at 0. On the other hand, f being differentiable at 0 is a sufficient but not necessary condition for f to be continuous at 0.

● HISTORICAL COMMENTS: LOBACHEVSKIAN GEOMETRY

In Section 1.2 we discussed the attempts to prove Euclid's fifth postulate and how all such attempts failed. Then, in the 19th century, three mathematicians, **Nicolai Lobachevsky** (1793–1856), **Johann Bolyai** (1802–1860), and **Carl Friedrich Gauss** (1777–1855) arrived at the same conclusion independently: that Euclid's fifth postulate was indeed an independent axiom and could not be deduced from the other axioms. This assertion followed from the realization that by altering Euclid's fifth postulate, a new consistent geometry could be created. Their revised postulate was the following: given a point P and a line L that does not contain P, there is *more* than one line that passes through P and is parallel to L (that is, does not intersect L). From this postulate, together with the other Euclidean axioms, a new non-Euclidean geometry was born. This geometry has come to be known as Lobachevskian geometry.

Lobachevskian geometry has some interesting and surprising theorems. In this geometry, the sum of the angles of any triangle is *less* than 180 degrees and if two triangles are similar, then they *must* be congruent.

A model for Lobachevskian geometry is the interior of a circle where "lines" are defined to be chords of the circle with the endpoints excluded. In the accompanying figure, we see that we can draw infinitely many lines through the point P that do not intersect the line L.

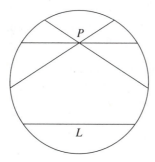

Exercises 1.3

---

1. Express the following statements in the form "for all . . . , if . . . then . . ." using symbols to represent variables. Then write their negations, again using symbols.
   (a) Every hexagon has 6 sides.
   (b) An integer is odd or even.
   (c) A function that is differentiable at 0 is continuous at 0.
   (d) All positive real numbers have a square root.

2. Repeat Exercise 1 for the following statements.
   (a) All angles of an equilateral triangle are equal.

(b) 1 is the smallest positive integer.

(c) When the product of two integers is even, then both integers are even.

(d) Between any two real numbers there is a rational number.

3. Let P and Q be statements.
   (a) Prove that $\neg(P \Rightarrow Q)$ is logically equivalent to $P \wedge \neg Q$.
   (b) Prove that $\neg(P \Rightarrow Q)$ is *not* logically equivalent to $\neg P \Rightarrow \neg Q$.
   (c) Give an example of statements P and Q such that $\neg P \Rightarrow \neg Q$ is true and $\neg(P \Rightarrow Q)$ is false.
   (d) Explain why it is impossible to give an example of statements P and Q such that $\neg(P \Rightarrow Q)$ is true and $\neg P \Rightarrow \neg Q$ is false.

4. Consider the following statement: "The product of an even integer with any integer is always even."
   (a) Rewrite the statement in the form "for all ... , if ... , then ..." using symbols to represent variables.
   (b) Write the negation of the statement, again using symbols.
   (c) Prove the statement if you think it is true or disprove it if you think it is false.

5. Repeat Exercise 4 for the statement "The cube of an odd integer is odd."

6. Repeat Exercise 4 for the statement "The sum of the squares of two consecutive integers is odd."

7. Repeat Exercise 4 for the statement "The sum of the squares of three consecutive integers is even."

8. Repeat Exercise 4 for the statement: "If the square root of a positive integer is an even integer, then the integer itself is even."

9. Prove that the statement form $(P \Rightarrow Q) \vee (Q \Rightarrow P)$ is a tautology.

10. Let P, Q, and R be statements. Prove that if the statement $P \Rightarrow Q$ is true and the statement $Q \Rightarrow R$ is true, then the statement $P \Rightarrow R$ is true.

11. Let $P(x)$, $Q(x)$, and $R(x)$ be open sentences containing the variable x. Prove that if the statement $(\forall x(P(x) \Rightarrow Q(x))) \wedge (\forall x(Q(x) \Rightarrow R(x)))$ is true, then the statement $\forall x(P(x) \Rightarrow R(x))$ is true.

12. Let P, Q, and R be statement forms. Prove that the statement form $P \Rightarrow (Q \vee R)$ is logically equivalent to the statement form $(P \wedge \neg Q) \Rightarrow R$.

13. Let P, Q, and R be statement forms. Is the statement form $P \Rightarrow (Q \vee R)$ logically equivalent to the statement form $(P \Rightarrow Q) \vee (P \Rightarrow R)$? Justify your answer.

14. Let P, Q, and R be statement forms. Prove that the statement form $P \Rightarrow (Q \wedge R)$ is logically equivalent to the statement form $(P \Rightarrow Q) \wedge (P \Rightarrow R)$.

15. Let x and y be real numbers. Consider the statement S: "If $xy = 0$, then $x = 0$ or $y = 0$."
    (a) Show that S can be expressed in the form $P \Rightarrow (Q \lor R)$ for appropriate statements P, Q, and R.
    (b) Rewrite statement S using its logically equivalent form $(P \land \neg Q) \Rightarrow R$. (See Exercise 12.)
    (c) Using the revised form of statement S from part (b) and some properties of the real numbers, prove statement S.

16. Let $P(x)$, $Q(x)$, and $R(x)$ be open sentences containing the variable x. If the statement $(\exists\, x \ni (P(x) \Rightarrow Q(x))) \land (\exists\, x \ni (Q(x) \Rightarrow R(x)))$ is true, does it follow that the statement $\exists\, x \ni (P(x) \Rightarrow R(x))$ is true also? If you think it does, give a proof. If you think it does not, give an example of open sentences $P(x)$, $Q(x)$, and $R(x)$ for which the statement $(\exists\, x \ni (P(x) \Rightarrow Q(x))) \land (\exists\, x \ni (Q(x) \Rightarrow R(x)))$ is true and the statement $\exists\, x \ni (P(x) \Rightarrow R(x))$ is false.

17. Let n be an integer. Let P be the statement "n is a multiple of 4" and Q be the statement "n^2 is a multiple of 4." Which of the following are true statements? Prove your answers.
    (a) P is a sufficient condition for Q.
    (b) P is a necessary condition for Q.
    (c) Q is a sufficient condition for P.
    (d) Q is a necessary condition for P.

18. Let $n, m \in \mathbf{Z}$. Repeat Exercise 17 where P is the statement "n and m are odd integers" and Q is the statement "$n + m$ is even." See Example 11 of this section.

Discussion and Discovery Exercises

D1. In Example 11, the phrase "we arbitrarily chose the letter s" is used. Write a short explanation of what the word "arbitrarily" means in this context.

D2. Write out the truth table for $\neg P \Rightarrow \neg Q$ and show that it is logically equivalent to $P \lor \neg Q$. Write out an informal explanation of why this equivalence makes sense.

D3. (a) Prove that if n is a positive integer, then $1 + 2 + 2^2 + 2^3 + \dots + 2^n = 2^{n+1} - 1$.
    (b) Suppose that you place a piece of paper on a table, then place 2 pieces on top of it, then 4 pieces on top of that, then 8 pieces, then 16 pieces, and so on, doubling the number of pieces each time. Assuming that you do this doubling 50 times, how many pieces of paper are in the pile?

(c) Assuming that each piece is 1/200th of an inch thick, estimate how tall the pile is.

(d) Use a calculator to give a good approximation of the size of the pile. Express your answer in inches, feet, or miles, whatever seems most appropriate. How good was your estimate?

(e) How does this problem fit the theme of this chapter? Explain.

D4. The ancient Egyptians were able to find a formula for the volume of a truncated pyramid with square base and square top. Although we don't know how they arrived at the formula, it is an ingenious application of inductive reasoning and perhaps of deductive reasoning as well, if in fact they were able to prove their result. If the base has side b, the top has side a, and the vertical height is h, find the formula for the volume V of the truncated pyramid. Justify your answer. (Recall that the volume of a full pyramid is one-third the area of the base times the height.)

1.4 CONTRAPOSITIVE AND CONVERSE

Contrapositive

An alternative way to prove statements of the form $P \Rightarrow Q$ is to verify the statement $\neg Q \Rightarrow \neg P$. This would mean showing that whenever Q is false then P is false. Then it would follow that if P is true, Q could not be false (because Q false implies P false) and thus Q is true. Another way to see this is to note that the statements $P \Rightarrow Q$ and $\neg Q \Rightarrow \neg P$ are logically equivalent. We leave it as an exercise to verify that they have the same truth tables.

Besides the formalism of showing that the truth tables are the same, it is important to *understand in an intuitive way* why these statements are logically equivalent. It should make sense that they mean the same thing. Let's take a nonmathematical example.

EXAMPLE I Let P be the statement "John lives in Boston" and Q be the statement "John lives in Massachusetts." The implication $P \Rightarrow Q$ is the statement "If John lives in Boston, then John lives in Massachusetts." This is a true statement. (We'll assume we mean Boston, Massachusetts.) The implication $\neg Q \Rightarrow \neg P$ is the statement "If John does not live in Massachusetts, then John does not live in Boston." It means the same as $P \Rightarrow Q$.

> **Definition 1.4.1** The statement $\neg Q \Rightarrow \neg P$ is called the **contrapositive** of the statement $P \Rightarrow Q$.

EXAMPLE 2   Implication: If it rained today, then the game was cancelled.
                   Contrapositive: If the game was not cancelled, then it did not rain
                                     today.

EXAMPLE 3   Implication: If $x + 1 > 5$, then $x > 4$.
                   Contrapositive: If $x \leq 4$, then $x + 1 \leq 5$.

The following example illustrates why it is often better to prove an implication by proving its contrapositive.

EXAMPLE 4   Prove that if n is an integer and n^2 is even, then n is an even integer.
      We'll start by trying to prove this statement directly. Let n be an integer such that n^2 is even. We wish to show that n is even. Because we are assuming that n^2 is even, we can write n^2 as twice an integer. This means we can write $n^2 = 2t$ for some integer t.
      Because we want to show that n is twice an integer, the obvious thing to do is to take the square root of both sides of this equation. This gives $n = \sqrt{2t}$. What conclusion can we draw from this equation? Unfortunately, not much at all because the square root of an integer is not necessarily an integer itself. Thus this approach to proving our statement has hit a dead end.
      Now we'll try proving the contrapositive of our statement instead. The first step is to write the contrapositive in a coherent fashion.
      The statement we are trying to prove is an implication $P \Rightarrow Q$ where, under the assumption that n is an integer, P is the statement "n^2 is even" and Q is the statement "n is even." Thus $\neg P$ is the statement "n^2 is not even" or "n^2 is odd" and $\neg Q$ the statement "n is not even" or "n is odd."
      Therefore, the contrapositive of $P \Rightarrow Q$ is the statement "If n is odd, then n^2 is odd." It is this statement that we will now try to prove.
      If n is odd, then we can write $n = 2t + 1$ for some integer t. Squaring both sides gives us

$$n^2 = (2t + 1)^2 = 4t^2 + 4t + 1 = 2(2t^2 + 2t) + 1$$

and because $2t^2 + 2t$ is an integer, it follows that n^2 is odd. The proof of the contrapositive is now complete.
      In proving $\neg Q \Rightarrow \neg P$, we also have proved $P \Rightarrow Q$, because these two statements are logically equivalent. We have now completed the proof of the statement: "If n is an integer and n^2 is even, then n is even."

Converse

Recall that in Example 1 of Section 1.1, we proved that if n is even, then n^2 is even. Now in this last example we have proven that if n^2 is even, then n is even. In symbolic language, if P is the statement "n is even" and Q is the statement "n^2 is even" (the reverse of the notation used in the previous example), then Example 1 of Section 1.1 proves that $P \Rightarrow Q$ and Example 2 above proves that $Q \Rightarrow P$.

> **Definition 1.4.2** The statement $Q \Rightarrow P$ is called the **converse** of the statement $P \Rightarrow Q$.

EXAMPLE 5   Implication: If it rained today, then the game was cancelled.
                  Converse: If the game was cancelled, then it rained today.

EXAMPLE 6   Implication: If $x + 1 > 5$, then $x > 4$.
                  Converse: If $x > 4$, then $x + 1 > 5$.

It is important to note that an implication and its converse are NOT necessarily logically equivalent. The truth of one does not imply the truth of the other. In Example 5, even if the implication is true, the converse could be false. The game could be cancelled for a different reason. On the other hand, the statement "If n is even, then n^2 is even" and its converse "If n^2 is even, then n is even" are both true. However, one does not follow logically from the other. Two different mathematical proofs are required.

EXAMPLE 7   In Example 11 of Section 1.3, we proved that if n and m are odd integers, then $n + m$ is even. The converse of this statement is: if n and m are integers and $n + m$ is even, then n and m are odd. But this converse is false because $4 + 2$ is even but neither 4 nor 2 is odd.

Biconditional

When a statement $P \Rightarrow Q$ and its converse $Q \Rightarrow P$ are both true, then P and Q are either both true or both false. Hence P and Q are logically equivalent. Using the terminology introduced at the end of Section 1.3, we can say that P is a necessary and sufficient condition for Q.

> **Definition 1.4.3** Let P and Q be statements. The statement $P \Leftrightarrow Q$ (or P iff Q, read P if and only if Q) is the statement $(P \Rightarrow Q) \wedge (Q \Rightarrow P)$. The symbol $\Leftrightarrow$ is called the **biconditional**.

Combining Example 1 of Section 1.1 and Example 4 in this section, we have the following result, which we label as a theorem.

Theorem 1.4.4   Let n be an integer. Then n is even if and only if n^2 is even.

This theorem may also be worded this way:

Theorem 1.4.5   Let n be an integer. Then the following are equivalent statements:

1. n is even.
2. n^2 is even.

Notice that proving an equivalence requires two proofs!

Sometimes a mathematical statement or proposition may seem like a new result when in fact it merely follows logically from an earlier proposition or theorem. For example, if P and Q are statement forms, then $\neg P \Leftrightarrow \neg Q$ is logically equivalent to $P \Leftrightarrow Q$. The proof of this fact can easily be seen by comparing their truth tables. The details are left as an exercise. The following theorem is a consequence of this equivalence.

Theorem 1.4.6   Let n be an integer. Then n is odd if and only if n^2 is odd.

PROOF: Given an integer n, let P be the statement "n is odd" and Q be the statement "n^2 is odd." The theorem has the form $\forall n, P \Leftrightarrow Q$. Because $P \Leftrightarrow Q$ is logically equivalent to $\neg P \Leftrightarrow \neg Q$, it suffices to prove $\forall n, \neg P \Leftrightarrow \neg Q$. But $\neg P \Leftrightarrow \neg Q$ is the statement "n is even if and only if n^2 is even," proved in Theorem 1.4.4. The proof is now complete. ●

Note: The symbol ● used at the end of the last proof will be used throughout the text to signify that a proof has been completed.

Proof by Contradiction

Recall that in order to prove an implication $P \Rightarrow Q$, we can prove its contrapositive $\neg Q \Rightarrow \neg P$ instead. Since P is our hypothesis or premise and Q is the conclusion that we wish to deduce from P, in effect what we are doing is assuming that our conclusion Q is false and proving that our premise P is false, thus getting a contradiction. The contradiction is that P and $\neg P$ cannot both be true. This approach is often called **proof by contradiction**. It can work in other ways as well. In order to prove a statement S, for example, we could assume that S is false and deduce a statement that we know is false; a statement like "$0 = 1$" or "the square of a real number is negative." The justification of this method of proof is given by the following theorem, followed by two examples.

Theorem 1.4.7   Let S be a statement and let C be a false statement. Then the statement $\neg S \Rightarrow C$ is logically equivalent to S.

PROOF: If S is true, then $\neg S$ is false and therefore the implication $\neg S \Rightarrow C$

is true. On the other hand, if S is false, then $\neg S$ is true and since C is false, the implication $\neg S \Rightarrow C$ is false. Thus the statements S and $\neg S \Rightarrow C$ are either both true or both false. Hence they are logically equivalent. ●

EXAMPLE 8 Prove that there are no integers x and y such that $x^2 = 4y + 2$.

To prove this by contradiction, we let S be the statement "There are no integers x and y such that $x^2 = 4y + 2$." This statement can be reworded "For all integers x and y, $x^2 \neq 4y + 2$."

Now we assume that S is false; then there exist integers x and y such that $x^2 = 4y + 2$.

Since $x^2 = 2(2y + 1)$, x^2 is even. Hence by Theorem 1.4.4, x is even also. So we can write $x = 2n$ for some integer n. Therefore $x^2 = 4n^2 = 4y + 2$.

Subtracting and factoring, we get: $4(n^2 - y) = 2$ or $n^2 - y = \frac{1}{2}$. Because n^2 and y are integers, their difference must be an integer also. Now we have a contradiction, namely the statement that $\frac{1}{2}$ is an integer. Since the assumption that S is false leads to a false statement, Theorem 1.4.6 implies that S must be true. (*Note:* the fact that $\frac{1}{2}$ is not an integer can be deduced from the axioms for the integers discussed in Chapter 5.) Our proof by contradiction is now complete.

EXAMPLE 9 We will prove by contradiction that $\sqrt{2}$ is not a rational number. The assertion that $\sqrt{2}$ is not a rational number can be expressed: if r is a rational number, then $r^2 \neq 2$. Suppose, on the contrary, that $\sqrt{2}$ is rational. Then there is a rational number r such that $r^2 = 2$. Thus we can write $\sqrt{2} = a/b$ where a and b are integers and a/b is reduced to lowest terms. This means that a and b have no common factors other than 1 and -1. (Note that we are assuming a fact about rational numbers, namely that any rational number can be expressed in such a reduced form.) Squaring both sides of the above equation and multiplying through by b^2 gives $a^2 = 2b^2$. It follows that a^2 is even and so by Theorem 1.4.4, a is even. So $a = 2c$, for some integer c. We get $a^2 = 4c^2 = 2b^2$ or $2c^2 = b^2$, which implies that b^2 is even and thus, again by Theorem 1.4.4, that b is even. This is a contradiction, since we assumed that a/b was reduced to lowest terms, and therefore a and b can have no common factors other than 1. Thus our assumption that $\sqrt{2}$ is a rational number must be false and we can conclude that $\sqrt{2}$ is not a rational number.

● HISTORICAL COMMENTS: RIEMANNIAN GEOMETRY

In proposing an alternative to Euclid's fifth postulate, Lobachevski, Bolyai, and Gauss assumed that given a point P and a line L that does not intersect P, there is *more* than one line that passes through P and is parallel to L. Now suppose that the Parallel Postulate were replaced by the following axiom: given a point P and a line L that does not intersect P, there is *no* line that passes through P and is parallel to L.

Mathematicians like Lobachevski and Bolyai believed that this revision of the Parallel Postulate was not valid because they could derive a contradiction from it. But in doing so they assumed that Euclid's second postulate "a finite straight line can be produced continuously in a straight line" meant that a line has infinite length. In fact, Euclid's second postulate only implies that a straight line has no boundary. It does not mean that a line must be of infinite length.

In 1854, in a celebrated lecture given to the faculty at the University of Göttingen in Gemany, **Bernhard Riemann** (1826–1866) proposed an ambitious program of study that led to a revolution in geometry. As one facet of this program, Riemann pointed out the distinction between a line being without boundary and a line having infinite length. By allowing a "line" to have no boundary but at the same time have finite length, Riemann was able to describe another non-Euclidean geometry, different from Lobachevski's. In this geometry, Euclid's fifth postulate is replaced by the axiom previously mentioned: through a given point P not on a line L there is no line that is parallel to L. In other words, all lines intersect!

In what has come to be called Riemannian geometry, it can be shown that if perpendicular lines are drawn on the same side of a given line L, then these lines will all intersect in a common point Q and the lengths of these perpendiculars from Q to L will all have the same length α. Moreover, all lines in this geometry have finite length 4α.

One model of Riemannian geometry is the surface of a sphere. On a sphere the shortest distance between two points is the arc of a great circle. By a great circle we mean a circle whose center is the center of the sphere. The great circles then are the "lines" of the geometry of the sphere. Note that these lines are finite in length, their lengths equal the circumference of the sphere, but have no boundary; that is, they have no beginning or end.

Note also that the larger the sphere is, the larger are the lengths of the lines in this geometry. At the same time, the larger the sphere the more the geometry resembles Euclidean geometry. Imagine the surface of the earth as a sphere, even though it is not exactly that shape. To observers on the earth, the surface appears flat and straight lines seem to extend indefinitely and be infinite in length. But we know that any line drawn on the earth will eventually circle the earth and come back on itself. By contrast, if the spherical surface is a ball of radius 3 inches, then any line on the ball more than an inch long will definitely look curved. The difference between the surface of the earth and the ball is "curvature." *Curvature* is a mathematical term that measures how much a surface bends or curves. The curvature of the plane is 0 and the curvature of a sphere is $1/r$ where r is its radius. So the greater r is, the closer the curvature is to 0 and the more closely the geometry appears to be Euclidean.

Exercises 1.4

---

1. Let P and Q be statement forms. Prove that $\neg Q \Rightarrow \neg P$ is logically equivalent to $P \Rightarrow Q$.

2. Let P and Q be statement forms.
   (a) Prove that $(P \wedge \neg Q) \Rightarrow Q$ is logically equivalent to $P \Rightarrow Q$.
   (b) Explain why this logical equivalence makes sense.

3. Let P and Q be statement forms.
   (a) Prove that $(P \wedge \neg Q) \Rightarrow \neg P)$ is logically equivalent to $P \Rightarrow Q$.
   (b) Explain why this logical equivalence makes sense.

4. Write the contrapositive and converse of the following statements:
   (a) If the power goes off, the food will spoil.
   (b) If the light is on, then the door is not locked.
   (c) If it is 9 A.M., then we started the test.

5. Let x be a real number. Write the contrapositive and converse of the statement "If $x^2 - x - 2 = 0$, then $x = -1$ or $x = 2$."

6. Let n and m be integers.
   (a) Prove the statement "If n and m are even, then $n + m$ is even."
   (b) State the contrapositive.
   (c) Prove or disprove the converse.

7. Let f be a real-valued function. Write the contrapositive and converse of the statement "If f is differentiable at 0, then f is continuous at 0."

8. Professor Know-It-All tells Jonathan: "If you get at least a B on the final exam, then you will pass the course." Jonathan passes the course. What can he conclude?
   (a) He got at least a B on the final.
   (b) He cannot conclude anything.
   Give reasons for your answer.

9. The professor tells Karen: "It is necessary that you get at least a B on the final in order to pass the course." Karen gets a B. What can she conclude?
   (a) She passed the course.

(b) She can't conclude anything.
Give reasons for your answer.

10. The professor tells Michael: "If you get at least a C on the final exam, then you will pass the course." Michael finds out that he got a D on the final. What can he conclude?
(a) He'd better start looking for a summer school course.
(b) There's still hope.
Give reasons for your answer.

11. Let P and Q be statement forms.
(a) Write out the truth table for $P \Leftrightarrow Q$.
(b) Prove that $\neg P \Leftrightarrow \neg Q$ is logically equivalent to $P \Leftrightarrow Q$.

12. Let $P(n, m)$ be the open sentence "n and m are odd integers" and $Q(n, m)$ be the open sentence "nm is an odd integer." Determine the truth or falsity of the statements that follow. Justify your answers by writing out complete proofs. If you feel a statement is false, state its negation and then prove it. Explain any logical principles you use in your proofs.
(a) For all integers n and m, $P(n, m) \Rightarrow Q(n, m)$.
(b) For all integers n and m, $Q(n, m) \Rightarrow P(n, m)$.
(c) For all integers n and m, $P(n, m) \Leftrightarrow Q(n, m)$.

13. Explain how the statement "For all integers n and m, n or m is even $\Leftrightarrow$ nm is even" follows logically from Exercise 12.

14. Repeat Exercise 12 for the open sentences $P(n, m)$: "n or m is an odd integer" and $Q(n, m)$: "$n + m$ is an odd integer."

15. Let $P(n)$ be the open sentence "n is even" and $Q(n)$ be the open sentence "The positive square root of n is an even integer." Prove or disprove the following statements:
(a) For all integers n, $P(n) \Rightarrow Q(n)$.
(b) For all integers n, $Q(n) \Rightarrow P(n)$.
(c) For all integers n, $P(n) \Leftrightarrow Q(n)$.

16. Let x and y be real numbers. Let S be the statement "If $xy = 0$, then $x = 0$ or $y = 0$."
(a) Write the contrapositive of statement S.
(b) Write the converse of statement S. Prove or disprove.

17. Prove that the equation $x^2 = 4y + 3$ has no integer solutions.

18. Let a, b, and c be integers such that $a^2 + b^2 = c^2$. Prove that a or b is even.

19. Explain how Theorem 1.4.7 is used in the proof in Example 9. More specifically, what is S and what is C?

20. Let n be an integer.
(a) Prove that n is even if and only if n^3 is even.

(b) Prove that n is odd if and only if n^3 is odd.

21. Prove that $\sqrt[3]{2}$ is irrational.

22. Let P and Q be statement forms. Prove that the statement forms $P \Leftrightarrow Q$ and $(P \wedge Q) \vee (\neg P \wedge \neg Q)$ are logically equivalent.

23. Write the negation of the following statement: "For every integer n, n is even if and only if n^4 is even."

Discussion and Discovery Exercises

D1. Let P, Q, and R be statements.
   (a) Suppose you are asked to prove that $P \wedge Q \Rightarrow R$ is a true statement and you prove that if R is false, then P is false. Is that a valid proof? Explain.
   (b) Suppose you are asked to prove that $P \vee Q \Rightarrow R$ is a true statement and you prove that if R is false, then P is false or Q is false. Is that a valid proof? Explain.

D2. Let P and Q be statements. Suppose that whenever P is true, then Q is false and whenever Q is true, then P is false. Are P and $\neg Q$ logically equivalent? Explain.

D3. Let P be the statement "If n and m are integers and $n + m$ is even, then n and m are odd." In Example 7, it is proved that P is false. Explain the logic behind this proof. In particular, explain why using the integers 4 and 2 is sufficient. In the course of your explanation, write the negation of statement P. Explain when it is appropriate to use examples in a proof and when it is not.

D4. Criticize the following "proof" that if n and m are odd integers, then $n + m$ is even.

   Suppose that n or m is even; say n is even and m is odd. Then $n = 2x$ and $m = 2y + 1$ for some integers x and y. Then $n + m = 2x + 2y + 1 = 2(x + y) + 1$ is odd. Therefore, if n and m are odd, $n + m$ must be even.

D5. Bill, Francie, Matt, and Liz were relaxing in their living room. Each of them was doing one of the following activities: watching a football game, practicing the piano, playing a video game, reading a detective novel. Assume that two of them could be doing the same thing. Suppose that:
   1. Someone is reading a detective novel.
   2. Only one person is playing the piano.
   3. Bill or Francie is watching a football game.
   4. The statement "Bill is playing a video game or Francie is reading a detective novel" is false.

5. Matt is playing a video game or practicing the piano.
6. If Bill is reading a detective novel, then Matt is watching a football game.
7. If Bill is watching a football game, then Liz is playing a video game.

Match the person with his or her activity. Explain how you arrived at your answer and mention any rules of logic that you use.

D6. Look up the definition of *linear independence* of vectors in $\mathbf{R}^n$ in a linear algebra book. Write out the definition clearly, explaining any notation or terminology. Analyze the definition in terms of quantifiers and implications. Use symbols where appropriate.

D7. The following problem shows how working out examples and looking for patterns can lead to a general solution.

A kindly prison warden decided to free his 25 prisoners for good behavior. The prisoners were locked in separate cells, numbered 1 through 25. Each cell had a lock that opened when the key was turned once and locked when the key was turned again. The next turn would open the lock, the next would close it, and so on. One night, when the prisoners were sleeping, the warden quietly turned each lock once, opening all of the cells. Then he began to worry that he had freed too many prisoners so he went back and turned every other lock beginning with cell 2, thus locking those cells. Thinking that there might be still too many prisoners freed he gave every third lock a turn beginning with cell 3. Then he turned every fourth lock, then every fifth lock, then every sixth lock, and so on all the way to the 25th lock. (Of course he only turned one lock for every 13th lock, one lock for every 14th lock, and so on.)
(a) Which cells ended up unlocked? Explain how you arrived at your answer.
(b) Do the numbers suggest any patterns?
(c) If the prison had 50 cells, which cells would be unlocked?
(d) State a general result if there are n cells. State a theorem about integers that you would need in order to prove this general result and then prove it.

D8. Let T be a right triangle of sides a, b, and c. The Pythagorean Theorem states that if T is a right triangle and c is the hypotenuse, then $a^2 + b^2 = c^2$. There is a generalization of the Pythagorean Theorem called the Law of Cosines.
(a) State and prove the Law of Cosines.
(b) State the converse of the Pythagorean Theorem.
(c) If this converse is true, give a proof. If it is false, give a counterexample.

D9. In Euclidean geometry, the sum of the angles of any triangle is equal to 180 degrees and as we noted in Section 1.3, in Lobachevskian geometry, the sum of the angles of any triangle is *less* than 180 degrees. Using the sphere model, conjecture a statement about the sum of the angles of a triangle in Riemannian geometry. Can you give an example of a triangle on the sphere in which the sum of the angles is not 180 degrees?

D10. Consider the following statement: "Euclid was the first non-Euclidean geometer." In light of the discussions in this chapter, does this statement make sense? Explain.

SETS

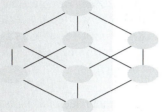

2

2.1　SETS AND SUBSETS

The Notion of a Set

We now begin the study of a fundamental idea, that of a **set** or collection of objects. This apparently simple notion is important in every field of mathematics. But we will not give a formal definition of set. Instead, the word will be a basic, undefined term, part of our intuitively understood vocabulary. (Try to define "set" yourself. You will wind up using words such as "collection" or "group," which are no more defined than "set" is.) We must start somewhere, or else every definition we give will lead to another, and another—an unending process. We choose "set" as a beginning.

The objects in a set are called **elements**. We use the notation $a \in A$ to express that the object a is an element of the set A, and often simply say "a is in A." You should realize that this idea of membership in a set is also undefined, part of the concept of set.

There are several different ways to denote a set, to convey which objects are elements and which are not.

EXAMPLE I　The elements may simply be listed. For example, $\{1, 2, 3, 4, 5\}$ is the set consisting of the first five positive integers. Putting the elements of a set in brackets $\{\ \}$ is standard notation.

EXAMPLE 2　If there are too many elements to list (infinitely many, perhaps!), we may just indicate a pattern for the elements. For example, $\{1, 2, 3, \ldots 10\}$ would denote the set of all positive integers from 1 to 10 and $\{1, 2, 3, \ldots\}$ would denote the set of all positive integers.

Special notation is sometimes used for very common sets. We use $\mathbf{R}$ for the set of real numbers, and $\mathbf{Z}$ (from the German "Zahlen," meaning "numbers") for the set of all integers, which might also be written $\{\ldots -3, -2, -1, 0, 1, 2, 3, \ldots\}$. The set consisting of just the positive integers is denoted $\mathbf{Z}^+$. The set of rational numbers, those of the form m/n, where m and n are integers with $n \neq 0$, is denoted $\mathbf{Q}$ (for "quotient").

EXAMPLE 3   Let A be the set of all integers that have 0 as one of their decimal digits. Now of course A has infinitely many elements, so we cannot list them, and determining an obvious pattern to indicate the elements may not be easy. Instead, we take our cue from the way we have described A: as the set of integers that satisfy a certain property or statement. We write

$A = \{n \in \mathbf{Z} \mid$ one of the decimal digits of n is 0$\}$.
The symbol $\mid$ is read "such that."

When a set is written in the form used in Example 3, it should consist of an expression before the $\mid$ symbol and an expression after the $\mid$ symbol. The expression before the $\mid$ symbol should introduce the variable(s) used to represent the elements of the set and the expression after the $\mid$ symbol should be an open sentence, say $P(x)$ if there is one variable x, which definitively describes the elements of the set. For example, $\{x \in \mathbf{Z} \mid x$ is even$\}$ is the set of even integers and here $P(x)$ is the open sentence "x is even."

In many discussions, we will be dealing only with elements of a certain, fixed set, say U. Such a set U provides a context, or a "universal set" for the discussion. For the set A of the previous example, the universal set could be taken to be the set of integers.

EXAMPLE 4 $\{1, 2, 3, 4, 5\}$ may also be written $\{n \in \mathbf{Z} \mid 1 \leq n \leq 5\}$. Here $U = \mathbf{Z}$.

EXAMPLE 5 $\{x \in \mathbf{R} \mid -1 \leq x \leq 1\}$ is the set of all real numbers between -1 and 1. In this case, $U = \mathbf{R}$.

In general, if S is a given set, $A = \{x \in S \mid P(x)\}$ denotes the set of all elements x in S such that the open sentence $P(x)$ is a true statement. We say that A is the **truth set** of the open sentence $P(x)$.

In many cases, the set S may be clear from the context, and we would just write $A = \{x \mid P(x)\}$. In any case, the $\mid$ symbol should always be followed by an open sentence that describes the elements of the set in some meaningful fashion.

EXAMPLE 6 Suppose A is the set of all integer multiples of 4. One possible description of A is

$\{x \in \mathbf{Z} \mid x = 4t \text{ for some integer } t\}$. Here $S = \mathbf{Z}$.

Another possible description of A is

$\{4t \mid t \in \mathbf{Z}\}$.

The sentence $t \in \mathbf{Z}$ after the $\mid$ symbol indicates that t takes on all integer values, so that as t runs through the integers, $4t$ runs through the multiples of 4.

An *incorrect* way to write A in this example would be

$\{t \in \mathbf{Z} \mid 4t\}$.

First of all, the expression $4t$ after the $\mid$ symbol is not an open sentence. Moreover, writing the expression $t \in \mathbf{Z}$ before the $\mid$ symbol is wrong because it is not the element t that is in A but rather $4t$.

EXAMPLE 7 Example 6 is one example of an important type of set of integers; namely, a set consisting of all the multiples of a given integer. If $n \in \mathbf{Z}$, we will denote by $n\mathbf{Z}$ the set of all multiples of n. So $2\mathbf{Z}$ is the set of even integers, $3\mathbf{Z}$ the set of multiples of 3, and so on. In set notation,

$n\mathbf{Z} = \{x \in \mathbf{Z} \mid x = nt \text{ for some integer } t\} = \{nt \mid t \in \mathbf{Z}\}$.

EXAMPLE 8 Let S be the set of integer multiples of π. We can describe S by listing elements; that is, $S = \{0, \pm\pi, \pm 2\pi, \pm 3\pi, \ldots\}$. Note, however, that, except for 0, the elements of S are not integers but real numbers, so the universal set is $\mathbf{R}$. It would be incorrect to write $S = \{n \in \mathbf{Z} \mid n\pi\}$ but rather we should write $S = \{x \in \mathbf{R} \mid x = n\pi \text{ for some integer } n\}$ or simply $S = \{n\pi \mid n \in \mathbf{Z}\}$.

EXAMPLE 9 The property to be satisfied by the elements may be compound; consider $\{x \in \mathbf{R} \mid x > 0 \text{ and } x^2 \in \mathbf{Z}^+\}$. This is the set of positive square roots of positive integers: $\{1, \sqrt{2}, \sqrt{3}, 2, \sqrt{5}, \ldots\}$.

Many of our examples of sets come from the real numbers. In one-variable calculus, for instance, the functions one looks at are defined on sets of real numbers. One particular class of such sets are **intervals**. For example, the set in Example 5 is an interval. The different types of intervals and the standard notations used to represent them are the following: if a and b are real numbers with $a < b$, then

1. $[a, b] = \{x \in \mathbf{R} \mid a \le x \le b\}$, called the **closed interval** from a to b.
2. $(a, b) = \{x \in \mathbf{R} \mid a < x < b\}$, called the **open interval** from a to b.
3. $(a, b] = \{x \in \mathbf{R} \mid a < x \le b\}$.
4. $[a, b) = \{x \in \mathbf{R} \mid a \le x < b\}$.
5. $[a, \infty) = \{x \in \mathbf{R} \mid x \ge a\}$.
6. $(a, \infty) = \{x \in \mathbf{R} \mid x > a\}$.
7. $(-\infty, a] = \{x \in \mathbf{R} \mid x \le a\}$.
8. $(-\infty, a) = \{x \in \mathbf{R} \mid x < a\}$.
9. $(-\infty, \infty) = \mathbf{R}$.

The first four types of intervals are called **bounded** intervals and the last five are called **unbounded** intervals. Types 3 and 4 are called **half-open, half-closed** intervals. Note that the symbol ∞ is not a real number but is only used as a convenient notation for unbounded intervals.

Some of the sets we've seen are **infinite**, that is, contain infinitely many elements. The others are said to be **finite**, and we denote the number of elements in such a set, called the **cardinality** of A, by $|A|$. For example, $|\{1, 5, 7, 8\}| = 4$. A more detailed discussion of infinite sets and cardinality is given in Chapter 6.

Subsets

Usually the elements of a set belong to another, bigger set. For instance, all of the elements of the set of integers $\mathbf{Z}$ belong to the set $\mathbf{R}$ of real numbers. (Can you give an example of a set that contains all the real numbers and is bigger than the set of real numbers?)

> **Definition 2.1.1** Let A and B be sets. We say that A is a **subset** of B, and write $A \subseteq B$, if every element of A is also an element of B.
>
> If A is a subset of B and $A \ne B$, we write $A \subset B$ and say that A is a **proper** subset of B.

Note: If A is a subset of a set B, we will usually write $A \subseteq B$ even if $A \ne B$ unless we wish to make particular note of the fact that A is a proper subset of B.

EXAMPLE 10 $\{1, 2, 3\} \subseteq \{0, 1, 2, 3, 4\}$.

EXAMPLE 11 $\{x \in \mathbf{R} \mid x > 0\} \subseteq \{x \in \mathbf{R} \mid x \ge 0\}$. This is just the statement that positive real numbers are nonnegative. In interval notation, we could write $(0, \infty) \subseteq [0, \infty)$ or $(0, \infty) \subset [0, \infty)$.

EXAMPLE 12 $\mathbf{Z}^+ \subseteq \mathbf{Z}$, $\mathbf{Z} \subseteq \mathbf{Q}$, and $\mathbf{Q} \subseteq \mathbf{R}$.

Note: The expression $A \subseteq B$ is a statement "A is a subset of B." It has a subject, a verb, and an object. It is also a statement with an implied universal quantifier. It may be expressed in the form "Every element of A is an element of B" or "$\forall x \in U$, where U is the universal set, if $x \in A$, then $x \in B$." We can write it as "$\forall x \in U, P(x) \Rightarrow Q(x)$" where $P(x)$ is the open sentence "$x \in A$" and $Q(x)$ the open sentence "$x \in B$."

Therefore, to prove the statement $A \subseteq B$, it is usually necessary to take an *arbitrary* element of A and show that it is an element of B. This means letting the variable x be assigned a value, say $x = a$, and then proving the implication "$a \in A$" $\Rightarrow$ "$a \in B$." So we start the proof by saying: let $a \in A$, and then we prove that $a \in B$. By using a variable, in this case a, which signifies an arbitrary but unspecified element of A, we will have proved that *every* element of A is an element of B.

EXAMPLE 13   Prove that $A \subseteq B$ where $A = \{n \in \mathbf{Z} \mid n$ is a multiple of 4$\}$ and $B = \{n \in \mathbf{Z} \mid n$ is even$\}$. Here the universal set is $\mathbf{Z}$. The statement $A \subseteq B$ is the statement "For every integer n, if n is a multiple of 4, then n is even."

To prove this, we must take an arbitrary element of A and show that it is in B. So let $n \in A$. Then $n = 4t$ for some t in $\mathbf{Z}$. It follows that $n = 2(2t)$ and hence n is even. Therefore $n \in B$. We have now established that for every $n \in \mathbf{Z}$, if $n \in A$, then $n \in B$. We can conclude that $A \subseteq B$.

In many proofs we'll encounter in this text, it will be necessary to prove that a set A is a subset of a set B.

EXAMPLE 14   Consider the statement "If x is a real number and $x > 10$, then $x > 5$." This statement can be worded "For all $x \in \mathbf{R}, P(x) \Rightarrow Q(x)$" where $P(x)$ is the open sentence "$x > 10$" and $Q(x)$ is the open sentence "$x > 5$."

If we let $A = \{x \in \mathbf{R} \mid x > 10\}$ and $B = \{x \in \mathbf{R} \mid x > 5\}$, then the statement "For all $x \in \mathbf{R}, P(x) \Rightarrow Q(x)$" is equivalent to the statement "$A \subseteq B$."

Next, we might ask: how do we prove the statement $A \subseteq B$ in this example? First, the proof depends on knowing certain facts about the real numbers; namely, that $10 > 5$ and that inequalities satisfy a so-called transitive property: if $a > b$ and $b > c$, then $a > c$. These properties of the real numbers are discussed in Chapter 7. For purposes of this proof, we will assume that these properties have already been established.

Now let $x \in A$. We have $x > 10$ and since $10 > 5$, by the transitive property it follows that $x > 5$. Hence $x \in B$. We have now proven that $A \subseteq B$.

The following proposition gives a result that holds for all sets and is a further illustration of showing one set as a subset of another.

Proposition 2.1.2   Let A, B, and C be sets, and suppose $A \subseteq B$ and $B \subseteq C$. Then $A \subseteq C$.

PROOF: To prove that $A \subseteq C$, it is necessary to show that any element of A is an element of C. So we let $a \in A$. Because $A \subseteq B$, we have $a \in B$. But then since $B \subseteq C$, $a \in C$. We can now conclude that $A \subseteq C$. ●

Sometimes two sets may be the same even though they are described differently. The following example is an illustration.

EXAMPLE 15   Let $A = \{n \in \mathbf{Z} \mid n \text{ is even}\}$ and $B = \{n \in \mathbf{Z} \mid n^2 \text{ is even}\}$. It was proven in Theorem 1.4.4 that n is even if and only if n^2 is even. Therefore, any element of A must be in B and any element of B must be in A. Therefore the two sets are really the same.

> **Definition 2.1.3** We say two sets A and B are **equal**, written $A = B$, if they have the same elements.

Note: Two sets A and B are equal if and only if every element of A is an element of B and every element of B is an element of A. In other words, $A = B \Leftrightarrow (A \subseteq B \wedge B \subseteq A)$. So proving that two sets are equal often requires two proofs.

Sometimes the statement $A \subseteq B$ is false. Let's analyze how to prove such a statement is false.

Recall that the statement $A \subseteq B$ contains a universal quantifier. It means that for *all* a, if $a \in A$, then $a \in B$. As previously noted, $A \subseteq B$ is the statement "$\forall x, P(x) \Rightarrow Q(x)$" where $P(x)$ is the open sentence "$x \in A$" and $Q(x)$ is the open sentence "$x \in B$." So the negation of $A \subseteq B$ must be the statement "$\exists$ a value of the variable x, say $x = a$, such that the statement $P(a) \Rightarrow Q(a)$ is false." Now the only way an implication $P(a) \Rightarrow Q(a)$ can be false is if $P(a)$ is true and $Q(a)$ is false; that is, if $a \in A$ but $a \notin B$. (As you might expect, $\notin$ means "is not an element of.")

To summarize: $A \subseteq B$ is false if $\exists a \in A$ such that $a \notin B$.

EXAMPLE 16   The set $A = \{1, 2, 3, 4, 5\}$ is not a subset of $B = \{1, 2, 3, 4, 6\}$, since $5 \in A$ but $5 \notin B$.

EXAMPLE 17   Consider the statement

All triangles in the Euclidean plane are isosceles.

This statement is equivalent to the statement $A \subseteq B$ where A is the set of all triangles in the Euclidean plane and B is the set of all isosceles triangles in the Euclidean plane.

The statement $A \subseteq B$ is false. To prove it is false, we need merely find an element of A that is not in B. Since the right triangle with sides of length 3, 4, and 5 is such an element, the proof is complete.

EXAMPLE 18   Let our universal set U be the set of all real-valued functions of one variable. Let $[a, b]$ be a closed interval on the real line; that is, $[a, b] = \{x \in \mathbf{R} \mid a \leq x \leq b\}$. Let $A = \{f \mid f \text{ is continuous on } [a, b]\}$ and $B = \{f \mid f \text{ is integrable on }$

$[a, b]\}$. It is an important theorem of calculus that every continuous function on $[a, b]$ is integrable on $[a, b]$. In other words, $A \subseteq B$. The converse of this statement is $B \subseteq A$. This converse, however, is false since there are integrable functions that are not continuous. The function f defined by $f(x) = 0$ if $x = a$ and $f(x) = 1$ if $a < x \leq b$ is one example of an element of B that is not in A.

Complements

As the previous examples suggest, we often want to determine if a given element is *not* in some set. We introduce the following definition.

> **Definition 2.1.4** Let A, B be sets. The **complement of A in B**, denoted $B - A$, is $\{b \in B \mid b \notin A\}$.

EXAMPLE 19 $\{1, 2, 3, 4, 5\} - \{1, 3\} = \{2, 4, 5\}$.

EXAMPLE 20 $\{1, 2, 3, 4, 5\} - \{1, 7, 8\} = \{2, 3, 4, 5\}$. Notice that there is no need for either set to be a subset of the other.

For convenience, if U is a universal set, we will write $U - A = \overline{A}$, called simply the **complement** of A.

For example, if our universal set is $\mathbf{Z}$, so that we are only considering integers, then $\overline{\mathbf{Z}^+} = \{\ldots -3, -2, -1, 0\}$, the nonpositive integers.

Sometimes we encounter the set that contains *no elements*. For example, if $U = \mathbf{R}$ and $A = \{x \mid x^2 < 0\}$, then A has no elements since for every real number x, $x^2 \geq 0$. We call a set with no elements the **empty set**, and we denote it $\varnothing$.

Note that if U is a universal set, then $\overline{U} = \varnothing$ and $\overline{\varnothing} = U$.

The following result shows how the statement $A \subseteq B$ is affected by taking complements. Note how the proof just involves applying a rule of logic that we discussed in Chapter 1.

Theorem 2.1.5 Let A and B be sets, contained in some universal set U. Then $A \subseteq B$ if and only if $\overline{B} \subseteq \overline{A}$.

PROOF: Let $x \in U$. Let $P(x)$ be the open sentence "$x \in A$" and $Q(x)$ be the open sentence "$x \in B$." Then $\neg P(x)$ is the open sentence "$x \notin A$" or equivalently "$x \in \overline{A}$" and $\neg Q(x)$ is the open sentence "$x \notin B$" or equivalently "$x \in \overline{B}$."

$A \subseteq B$ is the statement "$\forall x, P(x) \Rightarrow Q(x)$" and $\overline{B} \subseteq \overline{A}$ is the statement "$\forall x, \neg Q(x) \Rightarrow \neg P(x)$." Since for any value a of x, $P(a) \Rightarrow Q(a)$ is logically equivalent to its contrapositive $\neg Q(a) \Rightarrow \neg P(a)$, it follows that $A \subseteq B$ if and only if $\overline{B} \subseteq \overline{A}$. ●

EXAMPLE 21   Let $A = \{f \mid f \text{ is continuous on } [a, b]\}$ and $B = \{f \mid f \text{ is integrable on } [a, b]\}$ as in Example 18. As we saw in that example, the true statement from calculus "Every continuous function on $[a, b]$ is integrable on $[a, b]$" is equivalent to the statement $A \subseteq B$. The statement $\overline{B} \subseteq \overline{A}$ is the statement "If f is not integrable on $[a, b]$, then f is not continuous on $[a, b]$." This latter statement is also true and requires no proof since it is logically equivalent to the statement $A \subseteq B$.

●   MATHEMATICAL PERSPECTIVE: INTERVALS

Earlier in this section we discussed the different types of intervals on the real line. Intuitively, an interval is a set of real numbers without "holes" in it. This would mean that if x and y are two numbers in the interval then all numbers between x and y are in the interval. A formal definition of interval would be the following:

> **Definition 2.1.6** A subset I of the set of real numbers **R** is called an **interval** if $I \neq \varnothing$, I contains more than one element, and for every x and y in I such that $x < y$, $[x, y] \subseteq I$.

It is not hard to show that any of the nine types of intervals previously mentioned are in fact intervals by this definition. (See Exercise 24 following.)

The difference between an open interval (a, b) and the closed interval $[a, b]$ is more profound than the simple fact that the closed interval contains the endpoints and the open interval does not. Many mathematical results that are true for closed intervals are not true for open intervals and vice versa. For example, the Max-Min Theorem of calculus states that if a real valued function f is continuous on the closed interval $[a, b]$, then f has a maximum and a minimum value on $[a, b]$. This theorem is false if the closed interval $[a, b]$ is replaced by the open interval (a, b). The function f defined by $f(x) = x^2$, for instance, is continuous on the open interval $(0, 1)$ but has no maximum or minimum value on this interval.

Another example is the theorem from calculus that says that if a real-valued function f is continuous on the closed interval $[a, b]$, then f is bounded on $[a, b]$. (Recall from Example 14 of Section 1.1 that a real-valued function f is bounded on a closed interval $[a, b]$ if there exists a positive real number M such that $|f(x)| \leq M$ for all $x \in [a, b]$.) As in the Max-Min Theorem, this theorem is false for open intervals. For example, the function $f(x) = 1/x$ is continuous on the open interval $(0, 1)$ but is unbounded on this interval.

Both this last theorem and the Max-Min Theorem are usually not proved in a beginning calculus course because the proofs depend on a property of the real numbers called the Least Upper Bound Axiom. This axiom is discussed in Chapter 7.

Exercises 2.1

---

1. Describe the following finite sets by listing their elements.
   (a) The set of all positive integers between 15 and 21 inclusive
   (b) The set consisting of the 8 smallest squares of positive integers
   (c) The set of all integers of absolute value less than 1
   (d) The set consisting of the areas of circles of radius 1, 2, and 3
   (e) $\{x \in \mathbf{Z} \mid x^2 < 13\}$
   (f) The set of all U.S. presidents who were born in this century
   (g) The set of the first ten prime numbers that are the sums of two squares. (A positive integer p is a **prime** number if $p > 1$ and the only positive integers that are factors of p are 1 and p.)
   (h) The set of the first 20 positive integers that are the sums of two odd prime numbers
   (i) The set of all remainders when a positive integer is divided by 7

2. Describe the following sets by listing enough elements to indicate a pattern for all of the elements of the set.
   (a) $\{x \in \mathbf{Z} \mid x > 10\}$
   (b) $\{x \in \mathbf{Z} \mid x^2 > 4\}$
   (c) The set of all integer multiples of 4
   (d) The set of all positive integers that have a remainder of 3 when divided by 4
   (e) The set of all remainders when a positive integer is divided by 93
   (f) The set of all integers that are sums of consecutive odd integers starting at 1
   (g) The set of all numbers x for which $\sin x = 0$
   (h) The set of all numbers x for which $\cos x = 0$

3. (a) Based on your answer to 1(g), make a conjecture about all prime numbers that are the sums of two squares.
   (b) Based on your answer to 1(h), make a conjecture about all positive integers that are the sums of two odd prime numbers.
   (c) Based on your answer to 1(i) and 2(e), make a conjecture about the set of all remainders when a positive integer is divided by a positive integer n.

4. Describe the following sets by writing them in the form $\{x \in S \mid \ldots\}$ for the appropriate set S.
   (a) $\{0, \pm 1, \pm 2, \pm 3, \pm 4, \pm 5, \pm 6\}$
   (b) The closed interval $[1, 2]$

(c) The set of all rational numbers between 0 and 1 inclusive
(d) The set of all integer multiples of 4
(e) $\{3, -1, 7, -5, 11, -9, \ldots\}$
(f) The set of all positive integers that have a remainder of 3 when divided by 4
(g) The set of all integer multiples of $\pi/2$
(h) The set of all real numbers x for which $\cos x = 0$
(i) The set of all real numbers whose distance from the origin is greater than 5

5. Let A be the set of all integer multiples of 5: $A = 5\mathbf{Z}$.
   (a) Express A in the form $\{x \in S \mid \ldots\}$ for the appropriate set S.
   (b) Prove that if $x, y \in A$, then $x + y \in A$ and $xy \in A$.

6. Let $n \in \mathbf{Z}$ and let $A = n\mathbf{Z}$. Prove that if $x, y \in A$, then $x + y \in A$ and $xy \in A$.

7. Let $P(x)$ be the open sentence "$x < 5$" and $Q(x)$ be the open sentence "$x \geq -1$." Express the following sets in interval notation if possible. If the set cannot be expressed in interval notation, give a reason.
   (a) $\{x \in \mathbf{R} \mid P(x) \wedge Q(x)\}$.
   (b) $\{x \in \mathbf{R} \mid P(x) \vee Q(x)\}$.
   (c) $\{x \in \mathbf{R} \mid \neg P(x)\}$.
   (d) $\{x \in \mathbf{R} \mid \neg Q(x)\}$.
   (e) $\{x \in \mathbf{R} \mid P(x) \wedge \neg Q(x)\}$.
   (f) $\{x \in \mathbf{R} \mid \neg P(x) \wedge Q(x)\}$.
   (g) $\{x \in \mathbf{R} \mid \neg P(x) \wedge \neg Q(x)\}$.
   (h) $\{x \in \mathbf{R} \mid \neg P(x) \vee \neg Q(x)\}$.

8. For the following open sentences $P(x)$, express the set $\{x \in \mathbf{R} \mid P(x)\}$ in interval notation if possible. If the set cannot be expressed in interval notation, give a reason.
   (a) $|x| < 4$ (b) $|x| \leq 7$
   (c) $|x| > 2$ (d) $|x - 1| < 3$
   (e) $|x - 7| > 9$

9. Let A be the set of all integer multiples of 9 and B be the set of all integer multiples of 3.
   (a) Express the statement $A \subseteq B$ as a statement of the form "For all ... if ... then ..." but without using the symbols A and B.
   (b) Express A and B in the form $\{x \in S \mid \ldots\}$ for the appropriate set S.
   (c) Prove $A \subseteq B$.
   (d) Write the *negation* of the statement in part (a).
   (e) Write the *contrapositive* of the implication in part (a). Is it true or false? Give a reason for your answer.
   (f) Write the *converse* of the implication in part (a). Then prove it or disprove it.

10. Let n and m be integers. Let $A = n\mathbf{Z}$ and $B = m\mathbf{Z}$.
    (a) Prove that if n is a multiple of m, then $A \subseteq B$.
    (b) Write the *converse* of the statement in part (a). Prove or disprove.

11. Let U be the set of all real-valued functions of one variable. Let $[a, b]$ be a closed interval on the real line. Repeat Exercise 9 for the sets A and B where A is the set of all functions that are differentiable on $[a, b]$ and B is the set of all functions that are continuous on $[a, b]$. You may use any relevant theorem from calculus to do this problem.

12. Let $A = \{n \in \mathbf{Z} \mid n \text{ is a multiple of } 4\}$ and $B = \{n \in \mathbf{Z} \mid n^2 \text{ is a multiple of 4}\}$.
    (a) Prove that $A \subseteq B$.
    (b) Is $B \subseteq A$? Prove or disprove.

13. Let $A = \{n \in \mathbf{Z} \mid n + 3 \text{ is odd}\}$. Prove that A is equal to the set of all even integers.

14. Let $A = \{n \in \mathbf{Z} \mid n = 8t + 7 \text{ for some } t \in \mathbf{Z}\}$ and $B = \{n \in \mathbf{Z} \mid n = 4t + 3 \text{ for some } t \in \mathbf{Z}\}$.
    (a) List 5 elements of A and 5 elements of B.
    (b) Is $A \subseteq B$? Prove or disprove.
    (c) Is $B \subseteq A$? Prove or disprove.

15. Let $A = \{n \in \mathbf{Z} \mid n = 4t + 1 \text{ for some } t \in \mathbf{Z}\}$ and $B = \{n \in \mathbf{Z} \mid n = 4t + 9 \text{ for some } t \in \mathbf{Z}\}$. Prove that $A = B$.

16. Let $A = \{n \in \mathbf{Z} \mid n = 3t + 1 \text{ for some } t \in \mathbf{Z}\}$ and $B = \{n \in \mathbf{Z} \mid n = 3t + 2 \text{ for some } t \in \mathbf{Z}\}$. Prove that A and B have no elements in common.

17. Find sets A and B so that the following statements can be expressed in the form $A \subseteq B$. Also write the universal set U.
    (a) Every hexagon has 6 sides.
    (b) Every polynomial is differentiable on $\mathbf{R}$.
    (c) All prime numbers greater than 2 are odd.
    (d) Every integer that is a multiple of 6 is even.

18. Without using set notation, write the statement $\overline{B} \subseteq \overline{A}$, where A and B are the sets defined in
    (a) Exercise 9
    (b) Exercise 11
    (c) Exercise 17(a)
    (d) Exercise 17(b)
    (e) Exercise 17(c)
    (f) Exercise 17(d)

19. Compute the complements of the following sets.
    (a) $A = \{x \in \mathbf{Z} \mid x > 5\}$, $U = \mathbf{Z}$.
    (b) $A = \{x \in \mathbf{Z} \mid 1 \leq x \leq 10\}$, $U = \mathbf{Z}$.

(c) $A = \{x \in \mathbf{Z} \mid 1 \le x \le 10\}$, $U = \mathbf{R}$.
(d) $A = \{x \in \mathbf{R} \mid -2 < x \le 4\}$, $U = \mathbf{R}$.
(e) $A = \{x \in \mathbf{R} \mid x < 2 \text{ or } x > 20\}$, $U = \mathbf{R}$.

20. For the following sets A and B, compute $A - B$ and $B - A$.
(a) $A = \{1, 2, 3, 4, 5, 6, 7, 8, 9, 10\}$, $B = \{6, 7, 9, 11, 12\}$.
(b) $A = \{1, 2, 3, \ldots, 100\}$, $B = \{1, 3, 5, \ldots, 99\}$.
(c) $A = \{0, 1, 2, 3, \ldots\}$, $B = \{0, 1, 2, 3\}$.
(d) $A = \{0, 1, 2, 3, \ldots\}$, $B = \{2, 4, 6, \ldots\}$.
(e) $A = \{x \in \mathbf{R} \mid x > 4\}$, $B = \{x \in \mathbf{R} \mid x < 10\}$.
(f) $A = \{x \in \mathbf{R} \mid x > 4\}$, $B = \{x \in \mathbf{R} \mid x > 10\}$.

21. Let A and B be sets contained in a universal set U. Which one(s) of the following statements are equivalent to the statement $A \subseteq B$? Give reasons for your answers.
(a) $\forall x \in U$, $x \in A$ and $x \in B$.
(b) $\forall x \in U$, if $x \notin B$, then $x \notin A$.
(c) $\exists x \in U \ni x \in A$ and $x \in B$.
(d) $\forall x \in U$, if $x \in A$, then $x \in B$.
(e) $\forall x \in U$, if $x \notin A$, then $x \notin B$.

22. It has been noted in this section that two sets A and B are equal if and only if $A \subseteq B$ and $B \subseteq A$. Using this characterization of equal sets, complete the following statement "Two sets A and B are *not* equal if and only if. . . ."

23. Use Exercise 11 of Section 1.3 to prove Proposition 2.1.2.

24. Prove that the open interval (a, b) is in fact an interval by Definition 2.1.6.

25. (a) Give an example to show that the complement of an interval need not be an interval.
(b) Give an example of an interval whose complement is also an interval.

Discussion and Discovery Exercises

D1. Let A and B be sets. Criticize the following "proof" that $A \subseteq B$ or $B \subseteq A$.

Let P be the statement "$x \in A$" and Q be the statement "$x \in B$." Then $A \subseteq B$ is the statement $P \Rightarrow Q$ and $B \subseteq A$ is the statement $Q \Rightarrow P$. Exercise 9 of Section 1.3 asserts that the statement $(P \Rightarrow Q) \vee (Q \Rightarrow P)$ is always true. Hence the statement $A \subseteq B$ or $B \subseteq A$ is always true.

D2. Let $A = \{x \in \mathbf{R} \mid x < 1\}$ and $B = \{x \in \mathbf{R} \mid x < 0\}$. Criticize the following "proof" that $B \subseteq A$.

$x \in B \qquad x < 0 < 1 \qquad x \in A$
Therefore, $B \subseteq A$.

D3. Does the statement "*A* is not subset of *B*" mean that if $x \in A$, then $x \notin B$? Give reasons for your answer.

D4. Show how you might "discover" the truth of Theorem 2.1.5 before actually giving a formal proof of it.

D5. Explain why the function $f(x) = x^2$ does not have a maximum or a minimum value on the open interval $(0, 1)$.

D6. Give an example of a function that does not have a maximum or a minimum value on the closed interval $[0, 1]$. Why does your example not contradict the Max-Min Theorem?

2.2   COMBINING SETS

Unions and Intersections

It is possible to create new sets out of two or more given sets. The two most common ways of doing this are given in the following definition.

> **Definition 2.2.1** Let *A* and *B* be sets. The **union** of *A* and *B*, denoted $A \cup B$, is $\{x \mid x \in A \vee x \in B\}$.
>   The **intersection** of *A* and *B*, denoted $A \cap B$, is $\{x \mid x \in A \wedge x \in B\}$.

For an element *x* to be in the union of two sets, *x* must be in one *or* the other of the two sets, although it could be in both. To be in the intersection of two sets, *x* must be in *both* sets.

EXAMPLE 1 $\{1, 2, 3, 4, 5\} \cup \{6, 7\} = \{1, 2, 3, 4, 5, 6, 7\}$.

EXAMPLE 2 $\{1, 2, 3\} \cup \{2, 3, 4\} = \{1, 2, 3, 4\}$ and $\{1, 2, 3\} \cap \{2, 3, 4\} = \{2, 3\}$. Notice that we cannot compute $|A \cup B|$ by simply adding $|A|$ and $|B|$.

EXAMPLE 3 $[0, 3] \cup [2, 5] = [0, 5]$ and $[0, 3] \cap [2, 5] = [2, 3]$ but note that $[0, 3] \cup [4, 5]$ is not an interval.

EXAMPLE 4   We encounter something of a puzzle if we try to compute, for example, $\{1, 2\} \cap \{3, 4\}$ or $[0, 3] \cap [4, 5]$. There are no elements that are in both of these sets; the intersection, therefore, must contain *no* elements! So the intersection is the empty set.
    We say two sets are **disjoint** if their intersection is the empty set.

In order to visualize the union and intersection of sets, we often use the interiors of circles to represent a set. This type of representation is called a **Venn diagram**.

Next we see the Venn diagrams for the union and intersection of two sets.

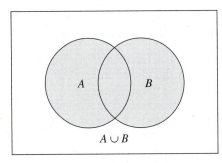
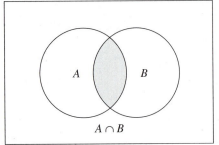

It is clear from this picture that the order in which we write the union and intersection does not matter. In other words, for any sets A and B, $A \cup B = B \cup A$ and $A \cap B = B \cap A$. Of course, a picture is not a proof because a set is not necessarily the interior of a circle. But the picture helps us to understand why these properties are true.

If we have three sets A, B, and C, the following diagrams represent their union and intersection.

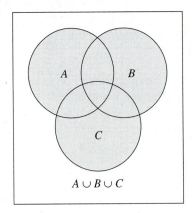
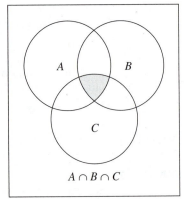

It is clear from these pictures that $(A \cup B) \cup C = A \cup (B \cup C)$ and $(A \cap B) \cap C = A \cap (B \cap C)$.

We list these properties of sets as well as some additional ones in the following proposition.

Proposition 2.2.2   The following statements are true for all sets A, B, and C.

1. $A \cup B = B \cup A$.
2. $A \cap B = B \cap A$.

3. $(A \cup B) \cup C = A \cup (B \cup C)$.
4. $(A \cap B) \cap C = A \cap (B \cap C)$.
5. $A \subseteq A \cup B$.
6. $A \cap B \subseteq A$.
7. $\varnothing \subseteq A$; that is, the empty set is a subset of every set.
8. $A \cup \varnothing = A$.
9. $A \cap \varnothing = \varnothing$.

PROOF: Properties 1 through 4 are clear from the appropriate Venn diagrams above. Similarly, Venn diagrams can be drawn to illustrate Properties 5 and 6. (See Exercise 10.)

For 7, a diagram is no help because we cannot draw the empty set. (There's nothing there!)

To prove that $\varnothing \subseteq A$, it is necessary to prove the statement $\forall x \in U$, $P(x) \Rightarrow Q(x)$, where $P(x)$ is the open sentence "$x \in \varnothing$" and $Q(x)$ is the open sentence "$x \in A$." But for any choice of the variable x, the hypothesis $P(x)$ is false since $\varnothing$ contains no elements. Therefore, the implication $P(x) \Rightarrow Q(x)$ is always true. (Why?) Hence $\varnothing \subseteq A$ for every set A.

We leave the proofs of 8 and 9 to the exercises. ●

Note that Properties 3 and 4 allow use to write $A \cup B \cup C$ and $A \cap B \cap C$ without ambiguity.

Note also that we did not give formal proofs of Properties 1 through 6 of Proposition 2.2.2, since the Venn diagrams made the properties seem obvious. That will not always be the case, so it is necessary to know how to write out formal proofs of properties of sets. Usually such proofs require proving either that two sets are equal or that one set is a subset of another. Recall that proving two sets are equal means giving *two* proofs; namely, that each of the two sets is a subset of the other. Here are two examples. You will see more in the exercises.

EXAMPLE 5 Let A and B be subsets of a universal set U. We will prove that $A - B = A \cap \overline{B}$. To prove this statement, we need to prove that $A - B$ is a subset of $A \cap \overline{B}$ and that $A \cap \overline{B}$ is a subset of $A - B$.

First, we prove that $A - B$ is a subset of $A \cap \overline{B}$. To do this, we must demonstrate that any element of $A - B$ is an element of $A \cap \overline{B}$. Let $x \in A - B$. Then $x \in A$ and $x \notin B$. So $x \in A$ and $x \in \overline{B}$, which implies that $x \in A \cap \overline{B}$. Therefore, $A - B \subseteq A \cap \overline{B}$.

Now, we show that $A \cap \overline{B}$ is a subset of $A - B$. Let $x \in A \cap \overline{B}$. Then $x \in A$ and $x \in \overline{B}$, implying that $x \in A$ and $x \notin B$. Thus $x \in A - B$. This proves that $A \cap \overline{B} \subseteq A - B$.

We can now conclude that $A \cap \overline{B} = A - B$.

EXAMPLE 6 Let A and B be subsets of a universal set U. We will prove that $A \subseteq B \Leftrightarrow A \cup B = B$.

This example is different from the previous one in the sense that it is necessary to prove two sets equal but within a biconditional; that is, under the assumption that $A \subseteq B$, we must prove that the sets $A \cup B$ and B are equal. Then assuming that these two sets are equal, we need to prove that A is a subset of B. The proof goes as follows.

PROOF: Suppose that $A \subseteq B$. We will prove that $A \cup B$ is a subset of B and B is a subset of $A \cup B$. Let $x \in A \cup B$. Then $x \in A$ or $x \in B$. If $x \in A$, then $x \in B$ because $A \subseteq B$. So in either case, $x \in B$. This proves that $A \cup B \subseteq B$. The fact that $B \subseteq A \cup B$ is a consequence of Proposition 2.2.2. Hence $A \cup B = B$.

Now suppose that $A \cup B = B$. We want to prove that $A \subseteq B$. Let $x \in A$. Then $x \in A \cup B$ by Proposition 2.2.2. But we are assuming that $A \cup B = B$, and so $x \in B$. It follows that $A \subseteq B$. ●

The following two properties of union and intersection are a bit more complicated to prove than the properties given in Proposition 2.2.2. We draw the Venn diagrams first to illustrate why the theorem is true and then get to the formal proofs.

Theorem 2.2.3   Let A, B, and C be sets. Then

1. $A \cap (B \cup C) = (A \cap B) \cup (A \cap C)$.
2. $A \cup (B \cap C) = (A \cup B) \cap (A \cup C)$.

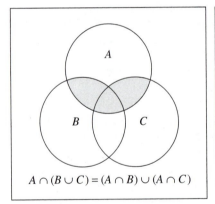

$A \cap (B \cup C) = (A \cap B) \cup (A \cap C)$

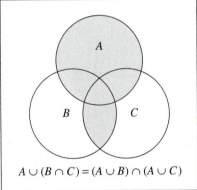

$A \cup (B \cap C) = (A \cup B) \cap (A \cup C)$

PROOF: We give two proofs of 1 and leave the proof of 2 as an exercise.

(First Proof) Let $x \in A \cap (B \cup C)$. Then $x \in A$ and $x \in B \cup C$, so $x \in B$ or $x \in C$.

If $x \in B$, then since we know that $x \in A$, we have $x \in A \cap B$. But $A \cap B \subseteq (A \cap B) \cup (A \cap C)$ by Property 5 of Proposition 2.2.2, so $x \in (A \cap B) \cup (A \cap C)$.

Similarly, if $x \in C$, we have that $x \in A \cap C$ and thus that $x \in (A \cap B) \cup (A \cap C)$. We have therefore shown that $A \cap (B \cup C) \subseteq (A \cap B) \cup (A \cap C)$.

To establish the reverse, let $x \in (A \cap B) \cup (A \cap C)$. Then $x \in A \cap B$ or $x \in A \cap C$.

In the former case, we have $x \in A$ and $x \in B$, but since $B \subseteq B \cup C$ (why?), $x \in B \cup C$, so $x \in A \cap (B \cup C)$.

A similar argument works in the latter case, so in either case we have $x \in A \cap (B \cup C)$.

Therefore, $(A \cap B) \cup (A \cap C) \subseteq A \cap (B \cup C)$.

We have now proven that $A \cap (B \cup C) = (A \cap B) \cup (A \cap C)$.

(Second Proof) This proof relies on Example 15 of Section 1.2.

Let $x \in U$. Let P, Q, and R be the following statements: P: "$x \in A$," Q: "$x \in B$," and R: "$x \in C$."

Then $P \wedge (Q \vee R)$ is the statement "$x \in A$ and either $x \in B$ or $x \in C$," which is equivalent to the statement "$x \in A \cap (B \cup C)$." Similarly, $P \wedge Q$ is the statement "$x \in A \cap B$," $P \wedge R$ is the statement "$x \in A \cap C$," and $(P \wedge Q) \vee (P \wedge R)$ is the statement "$x \in (A \cap B) \cup (A \cap C)$."

Since $P \wedge (Q \vee R)$ is logically equivalent to $(P \wedge Q) \vee (P \wedge R)$ by Example 15 of Section 1.2, it follows that $x \in A \cap (B \cup C)$ if and only if $x \in (A \cap B) \cup (A \cap C)$. Thus $A \cap (B \cup C) = (A \cap B) \cup (A \cap C)$. ●

DeMorgan's Laws

The following example shows what happens when we take complements of unions and intersections.

EXAMPLE 7   Let $U = \{1, 2, 3, 4, 5, 6, 7, 8, 9, 10\}$ be our universal set. Let $A = \{2, 3, 4, 5, 6\}$ and $B = \{1, 4, 7, 10\}$.

Then $\overline{A} = \{1, 7, 8, 9, 10\}$ and $\overline{B} = \{2, 3, 5, 6, 8, 9\}$.

Now $A \cup B = \{1, 2, 3, 4, 5, 6, 7, 10\}$ and so $\overline{A \cup B} = \{8, 9\}$.

Note that $\overline{A \cup B}$ is just the *intersection* of $\overline{A}$ and $\overline{B}$. Similarly, since $A \cap B = \{4\}$, we see that $\overline{A \cap B} = \overline{A} \cup \overline{B}$.

Now we prove that this example can be generalized to all sets. The results are known as **DeMorgan's Laws**.

Theorem 2.2.4   Let A and B be sets contained in some universal set U. Then

1. $\overline{A \cup B} = \overline{A} \cap \overline{B}$.
2. $\overline{A \cap B} = \overline{A} \cup \overline{B}$.

PROOF: We give two proofs of 1 and leave the proof of 2 as an exercise.

(First Proof) We need to show two things: $\overline{A \cup B} \subseteq \overline{A} \cap \overline{B}$ and $\overline{A} \cap \overline{B} \subseteq \overline{A \cup B}$.

To show that $\overline{A \cup B} \subseteq \overline{A} \cap \overline{B}$, we start by letting $x \in \overline{A \cup B}$. This means that x is not in $A \cup B$ and so x cannot be in either A or B. Thus x is not in A *and* x is not in B. Hence $x \in \overline{A}$ and $x \in \overline{B}$ or equivalently $x \in \overline{A} \cap \overline{B}$. This proves that $\overline{A \cup B} \subseteq \overline{A} \cap \overline{B}$.

Conversely, let $x \in \overline{A} \cap \overline{B}$. Then $x \in \overline{A}$ and $x \in \overline{B}$ or equivalently $x \notin A$ and $x \notin B$. So x is not in either A or B, or in other words $x \notin A \cup B$. Hence $x \in \overline{A \cup B}$. Thus $\overline{A} \cap \overline{B} \subseteq \overline{A \cup B}$. It follows that $\overline{A \cup B} = \overline{A} \cap \overline{B}$.

(Second Proof) Let $x \in U$. Let P be the statement "$x \in A$" and Q be the statement "$x \in B$."

Then $P \vee Q$ is the statement "$x \in A$ or $x \in B$," which means "$x \in A \cup B$." So $\neg(P \vee Q)$ is the statement "$x \notin A \cup B$" or "$x \in \overline{A \cup B}$."

Also, $\neg P$ is the statement "$x \notin A$" or equivalently "$x \in \overline{A}$" and $\neg Q$ is the statement "$x \notin B$" or equivalently "$x \in \overline{B}$." It follows that $\neg P \wedge \neg Q$ is the statement "$x \in \overline{A}$ and $x \in \overline{B}$" or equivalently "$x \in \overline{A} \cap \overline{B}$."

We saw in Section 1.2 that $\neg(P \vee Q)$ and $\neg P \wedge \neg Q$ are logically equivalent statement forms. It follows that $x \in \overline{A \cup B}$ if and only if $x \in \overline{A} \cap \overline{B}$.

Therefore, $\overline{A \cup B} = \overline{A} \cap \overline{B}$. ●

Cartesian Products

We sometimes think of the union of two sets as their "sum," as if we had "added" them, although, as we saw in Example 2, not even the numbers of elements in the sets are really added. Nevertheless, you might wonder if there can be a similar analog to multiplication. We now define a set that is the "product" of two sets in a very satisfactory way.

Definition 2.2.5 Let A and B be sets. The **Cartesian product** of A and B, written $A \times B$, is $\{(a, b) \mid a \in A$ and $b \in B\}$.

Informally, $A \times B$ is a set of pairs of objects. The first object is chosen from A and the second from B. Notice that we distinguish the order (these are commonly called **ordered pairs**).

EXAMPLE 8 Let $A = \{1, 2\}$ and $B = \{3, 4, 5\}$. Then $A \times B = \{(1, 3), (1, 4), (1, 5), (2, 3), (2, 4), (2, 5)\}$.

It's worth making a few comments about this simple example. First, notice how important order is: $(1, 3) \in A \times B$ but $(3, 1) \notin A \times B$. Even more basic, notice that $(1, 3)$ is *one* element of $A \times B$, which is made up of two objects. In fact, $|A \times B| = 6$; there are six ordered pairs. Here by the way is a reason why $A \times B$ is called a product: if A and B are finite sets, $|A \times B| = |A||B|$. You will be asked to prove this in Chapter 5. (See Exercise 34 of Section 5.2.)

EXAMPLE 9 There is no reason why A and B must be different sets. For example, $\mathbf{R} \times \mathbf{R} = \{(x, y) \mid x, y \in \mathbf{R}\}$. This should be a very familiar set to you: it's usually visualized as the Cartesian plane in analytic geometry and calculus. Again, be aware of the importance of order in these sets; you know that $(1, 2)$ is not the same point in the plane as $(2, 1)$.

HISTORICAL COMMENTS: IRRATIONAL NUMBERS

As the theory behind the concepts of calculus developed in the 19th century, it became clear to mathematicians that a rigorous definition of the real number system was necessary. The difficult part of any definition of the real numbers was defining the concept of irrational number. (At this point, the properties of the integers and rational numbers were taken for granted.)

In Section 1.4, we proved that $\sqrt{2}$ is irrational, a proof that is attributed to the Pythagoreans. The discovery that there could be a mathematical quantity that is not the ratio of two whole numbers was startling to the Pythagoreans. But it was not something that could be ignored, because the diagonal of a unit square has length $\sqrt{2}$. Such quantities were called *incommensurable ratios* as opposed to fractions, which were called *commensurable ratios*. Although irrationals were used freely in calculations, they were not fully accepted as numbers even as late as the 17th century. It was not until the 19th century, most notably with the work of **Karl Weierstrass** (1815–1897), **Georg Cantor** (1845–1918), and **Richard Dedekind** (1831–1916), that a formal definition of irrational number was given.

Proving that a given real number is irrational can be a difficult problem in general. For example, it is considerably harder to prove that π and e are irrational than it is to prove that $\sqrt{2}$ is irrational and, although it is conjectured that $\pi + e$ is irrational, no proof has yet been found. The irrationality of π was first proved in 1761 by **J. H. Lambert** (1728–1777) by means of continued fractions. There are many other proofs today. A general theorem about the trigonometric functions is the following, which we state without proof. We leave the corollaries as exercises.

Theorem 2.2.6 If r is a nonzero rational number, then $\cos r$ is irrational.

Corollary 2.2.7 π is irrational.

Corollary 2.2.8 If r is a nonzero rational number, then $\sin r$, $\tan r$, $\sec r$, $\csc r$, and $\cot r$ are all irrational.

The proof that e is irrational is easier than the one for π so we give the proof next.

Theorem 2.2.9 e is irrational.

PROOF: As with most of these irrationality proofs, we assume that e is rational and derive a contradiction. First of all, recall that e can be represented by the infinite series

$$e = 1 + \frac{1}{1!} + \frac{1}{2!} + \frac{1}{3!} + \ldots .$$

This is just the MacLaurin series for e^x with $x = 1$.

Suppose now that $e = n/m$ where n and m are positive integers. Notice that the expression

$$m! \left(e - 1 - \frac{1}{1!} - \frac{1}{2!} - \frac{1}{3!} - \cdots - \frac{1}{m!} \right)$$

is an integer. (See Exercise 33 following.)

Using the infinite series just given, we get that

$$m! \left(\frac{1}{(m+1)!} + \frac{1}{(m+2)!} + \frac{1}{(m+3)!} + \cdots \right) = t \in \mathbf{Z} \text{ for some } t.$$

Multiplying through by $m!$ on the left-hand side gives

$$t = \frac{1}{m+1} + \frac{1}{(m+1)(m+2)} + \frac{1}{(m+1)(m+2)(m+3)} + \cdots$$

$$= \sum_{k=1}^{\infty} \frac{1}{(m+1)(m+2)\ldots(m+k)}.$$

For each positive integer k,

$$\frac{1}{(m+1)(m+2)\ldots(m+k)} < \frac{1}{(m+1)^k}$$

and the geometric series $\sum_{k=1}^{\infty} 1/(m+1)^k$ coverages to $1/m$. (See Exercise 34 following.)

We now get the contradiction that t is a positive integer and $t < 1$. Therefore, e must be an irrational number. ●

The following theorem is a generalization of Theorem 2.2.9 but more difficult to prove. We state it without proof.

Theorem 2.2.10   If r is a nonzero rational number, then e^r is irrational.

An excellent source for problems of this type, including a proof of Theorem 2.2.10, is *Irrational Numbers* by Ivan Niven [16].

Exercises 2.2

---

1. Let $A = \{1, 2, 3, 4, 5\}$, $B = \{0, 1, 4, 8\}$, and $C = \{2, 5, 7, 9, 11, 13, 17\}$. Compute each of the following.
   (a) $A \cup B$.
   (b) $A \cap B$.
   (c) $A \cap C$.
   (d) $A - B$.
   (e) $A - (B \cup C)$.
   (f) $(A - B) \cup (A - C)$.
   (g) $A \cap (B \cup C)$.
   (h) $(A \cap B) \cup (A \cap C)$.
   (i) $A \cup (B \cap C)$.
   (j) $(A \cup B) \cap (A \cup C)$.

2. Let A, B, and C be the sets given in the previous exercise and let the universal set $U = \{0, 1, 2, 3, 4, \ldots, 20\}$. Compute the following.
   (a) $\overline{A}$ (b) $\overline{B}$ (c) $\overline{A \cap B}$
   (d) $\overline{A} \cup \overline{B}$ (e) $\overline{A \cup B}$ (f) $\overline{A} \cap \overline{B}$

3. (a) Let A, B, and C be the sets given in Exercise 1. Compute $|A \cup B|$, $|A \cup C|$, and $|B \cup C|$.
   (b) Let A and B be arbitrary finite sets. Based on your answers to part (a), deduce a formula for $|A \cup B|$ and then prove it.

4. Let $U = \mathbf{R}$ and let A, B, and C be the following intervals: $A = [1, 4]$, $B = (0, 2)$, and $C = [2, \infty)$. Compute the following. Express your answer as an interval or a union of intervals whenever possible.
   (a) $A \cup B$ (b) $A \cap B$ (c) $A \cup C$
   (d) $A \cap C$ (e) $B \cup C$ (f) $B \cap C$

5. Let A, B, and C be the sets given in Exercise 4. Compute the following.
   (a) $\overline{A}$ (b) $\overline{B}$ (c) $A - B$
   (d) $B - A$ (e) $A - C$ (f) $B - C$

6. Express each of the following sets as an interval or a union of intervals:
   (a) $\{x \in \mathbf{R} \mid |x| > 6\}$
   (b) $\{x \in \mathbf{R} \mid |x - 3| > 10\}$
   (c) $\{x \in \mathbf{R} \mid 0 < |x| < 1\}$
   (d) $\{x \in \mathbf{R} \mid 0 < |x - 1| < 2\}$

7. Let A and B be open, bounded intervals such that $A \cap B \neq \emptyset$.
   (a) Prove that $A \cup B$ is an open interval.
   (b) Prove that $A \cap B$ is an open interval.

8. Let A and B be closed, bounded intervals such that $A \cap B \neq \emptyset$.
   (a) Is $A \cup B$ a closed interval? Prove your answer.
   (b) Is $A \cap B$ a closed interval? Prove your answer.

9. Let A be the set of all integer multiples of 5 and B the set of all even integers. Compute
   (a) $A \cup B$ (b) $A \cap B$
   (c) $A - B$ (d) $B - A$

10. Draw Venn diagrams to illustrate Properties 5 and 6 of Proposition 2.2.2. Then give formal proofs of these properties.

11. Prove Properties 8 and 9 of Proposition 2.2.2.

12. Draw Venn diagrams to illustrate Examples 5 and 6 of this section.

13. Let $A = \{x \in \mathbf{R} \mid x \leq 10\}$ and $B = \{x \in \mathbf{R} \mid x > 5\}$. Prove that
    (a) $A \cap B = (5, 10]$.
    (b) $A \cup B = \mathbf{R}$.
    (c) $A - B = (-\infty, 5]$.

14. Let A and B be sets. Prove the following:
    (a) $A \cap (A \cup B) = A$.   (b) $A \cup (A \cap B) = A$.

15. Let A, B, and C be sets.
    (a) Prove that $A \subseteq (B \cap C) \Rightarrow (A \subseteq B) \wedge (A \subseteq C)$.
    (b) State the contrapositive of part (a).
    (c) State the converse of part (a). Prove or disprove.

16. Let A, B, and C be sets.
    (a) Prove that $(A \subseteq C) \wedge (B \subseteq C) \Rightarrow A \cup B \subseteq C$.
    (b) State the contrapositive of part (a).
    (c) State the converse of part (a). Prove or disprove.

17. Let A, B, and C be sets.
    (a) Prove or disprove: If $A \subseteq B \cup C$, then $A \subseteq B$ or $A \subseteq C$.
    (b) State the contrapositive of part (a).
    (c) State the converse of part (a). Prove or disprove.

18. Let A, B, and C be sets. Prove that if A is not a subset of C, then A is not a subset of B or B is not a subset of C.

19. Give two proofs of part 2 of Theorem 2.2.3 similar to the two proofs of part 1 of that theorem.

20. Draw Venn diagrams to illustrate the two parts of Theorem 2.2.4.

21. Give two proofs of part 2 of Theorem 2.2.4 similar to the two proofs of part 1 of that theorem.

22. Let A and B be subsets of a universal set U. Prove the following:
    (a) $(A \cup B) \cap \overline{A} = B - A$.
    (b) $(A \cup B) - (A \cap B) = (A - B) \cup (B - A)$.

23. Let A and B be subsets of a universal set U. Prove that $A \subseteq B \Leftrightarrow A \cap B = A$.

24. Let A and B be subsets of a universal set U.
    (a) Prove that $A = A - B \Leftrightarrow A \cap B = \emptyset$.
    (b) Prove that $A \cup B = U \Leftrightarrow \overline{A} \subseteq B$.

25. Let A, B, and C be sets. Prove the following:
    (a) $A - (B \cup C) = (A - B) \cap (A - C)$.
    (b) $A - (B \cap C) = (A - B) \cup (A - C)$.

26. Characterize the following expressions as either:
    (i) a statement,
    (ii) not a statement but an expression that makes sense mathematically,
    (iii) an expression that makes no sense mathematically.
    Assume that A and B are sets and P and Q are statements.

(a) $B \subseteq A$.   (b) $B \cap A$.
(c) $A \Rightarrow B$.   (d) $P \cup Q$.
(e) $P \Rightarrow (A = B)$.   (f) $A + B$.
(g) $(B - A) \Rightarrow Q$.   (h) $\forall x \in P, x \in Q$.
(i) $P \subseteq Q$.   (j) $A \wedge B$.
(k) $P = Q$.

27. Write out the elements of $A \times B$ where $A = \{-1, 0\}$ and $B = \{7, 12, 19, 21\}$.

28. Let $\mathbf{Q}$ and $\mathbf{I}$ be the sets of rational and irrational numbers respectively.
    (a) Prove that if $r \in \mathbf{Q}$ and $x \in \mathbf{I}$, then $r + x \in \mathbf{I}$.
    (b) Prove that if $r \in \mathbf{Q}$, $r \neq 0$, and $x \in \mathbf{I}$, then $rx \in \mathbf{I}$.

29. (a) Is the sum of two irrational numbers always irrational? If true, give a proof and if false, give a counterexample.
    (b) Is the product of two irrational numbers always irrational? If true, give a proof and if false, give a counterexample.

30. Prove Corollary 2.2.7.

31. Prove Corollary 2.2.8.

32. State the converse of Theorem 2.2.6. Prove or disprove.

33. Assuming that $e = n/m$ as in the proof of Theorem 2.2.9, prove that the expression

$$m! \left(e - 1 - \frac{1}{1!} - \frac{1}{2!} - \frac{1}{3!} - \cdots - \frac{1}{m!} \right)$$

is an integer.

34. Prove that the geometric series

$$\sum_{k=1}^{\infty} \frac{1}{(m + 1)^k}$$

converges to $1/m$.

35. Prove that if r is a positive rational number and $r \neq 1$, then $\ln r$ is irrational. (*Note*: ln denotes the natural logarithm function.)

Discussion and Discovery Exercises

D1. Let A and B be sets. Criticize the following "proof" of Example 6; that $A \subseteq B \Leftrightarrow A \cup B = B$.

Let $x \in A \subseteq B$. Then $x \in A$ and B so $x \in A \cup B$ and $A \cup B = B$.
Let $x \in A \cup B = B$. Then $x \in A$ or $x \in B$ and $x \in B$. Therefore $A \subseteq B$.

2.3   COLLECTIONS OF SETS

In this section we consider sets whose elements are themselves sets. For example, we could consider the collection of all sets of integers that contain the integer 1. Such a collection, let's call it S, could be expressed in the following way: $S = \{A \subseteq \mathbf{Z} \mid 1 \in A\}$. S is a set whose elements are sets, specifically, sets of integers. The set $\{1, 2, 3\}$ is an element of S, as is the set $\mathbf{O}$ of odd integers and the set $A = \{n \in \mathbf{Z} \mid n < 10\}$. The set $B = \{n \in \mathbf{Z} \mid n \geq 5\}$, on the other hand, is not in S. Note also that the integer 1 is not in S, but $\{1\}$ is in S.

Power Set

An important example of a set of sets is the **power set** of a set A. It is defined as follows:

> **Definition 2.3.1** Let A be a set. The **power set** of A, written $\mathbf{P}(A)$, is $\{X \mid X \subseteq A\}$.

The power set of A, then, is the set of all subsets of A. The elements of $\mathbf{P}(A)$ are themselves sets of elements of A.

EXAMPLE 1   Let $A = \{1, 2, 3\}$. Then, for example, $\{1, 2\} \subseteq A$, so $\{1, 2\} \in \mathbf{P}(A)$. Notice that $1 \notin \mathbf{P}(A)$. However, $\{1\}$ is a subset of A, so $\{1\} \in \mathbf{P}(A)$.

By the same token, $A \in \mathbf{P}(A)$ and at the other extreme, $\varnothing \in \mathbf{P}(A)$; that is, $\varnothing \subseteq A$. This was proven in Proposition 2.2.2.

For completeness, here's a listing of the whole power set of A:

$$\mathbf{P}(A) = \{\varnothing, \{1\}, \{2\}, \{3\}, \{1, 2\}, \{1, 3\}, \{2, 3\}, \{1, 2, 3\}\}.$$

Indexing Sets

Before introducing other examples of sets of sets, we need to introduce some notation.

Suppose that we have a finite number of sets, $A_1, A_2, \ldots A_n$, all contained in some universal set U. The collection of such sets would be written $\{A_1, A_2, \ldots A_n\}$. The subscripts that are used to distinguish the sets in this collection form what is called the **indexing set** for the collection. If we denote this indexing set by I, then $I = \{1, 2, \ldots n\}$ and this collection can be written $\{A_i \mid i \in I\}$.

Suppose that we wanted to talk about the union of the sets in this collection. This would be the set $A_1 \cup A_2 \cup \ldots \cup A_n = \{x \in U \mid x \in A_i$ for some $i = 1, 2, \ldots n\}$. A shorthand notation for this union is $\bigcup_{i=1}^{n} A_i$.

Similarly, the intersection of the sets in this collection is written $\bigcap_{i=1}^{n} A_i = \{x \in U \mid x \in A_i$ for all $i = 1, 2, \ldots n\}$.

EXAMPLE 2 Let n be a fixed positive integer and let $A_i = \{i, i + 1\}$ for $i = 1, 2, 3, \ldots, n$. Then $\{A_i \mid i \in I\} = \{\{1, 2\}, \{2, 3\}, \ldots, \{n, n + 1\}\}$. Notice that this collection is itself a finite set, with n elements.

We leave as an exercise the computation of $\bigcup_{i=1}^{n} A_i$ and $\bigcap_{i=1}^{n} A_i$.

Now suppose that we have an infinite collection of sets $\{A_1, A_2, A_3, \ldots\}$ where the indexing set is the set of all positive integers. Then for the union and intersection we write

$$\bigcup_{i=1}^{\infty} A_i = \{x \in U \mid x \in A_i \text{ for some } i \in \mathbf{Z}^+\}$$

and

$$\bigcap_{i=1}^{\infty} A_i = \{x \in U \mid x \in A_i \text{ for all } i \in \mathbf{Z}^+\}.$$

EXAMPLE 3 For each positive integer i, let $A_i = (-i, i) = \{x \in \mathbf{R} \mid -i < x < i\}$. Then the set $S = \{A_1, A_2, A_3, \ldots\}$ is a collection of open intervals on the real line. The union of the elements in S is $\bigcup_{i=1}^{\infty} (-i, i) = \mathbf{R}$ and the intersection is $\bigcap_{i=1}^{\infty} (-i, i) = (-1, 1)$. The fact that the union is $\mathbf{R}$ is a consequence of the Archimedean Principle discussed in Section 1.2. We leave the details as an exercise.

Notice that the sets A_i in the previous example have a very interesting property: $A_1 \subseteq A_2 \subseteq A_3 \subseteq \ldots$. That is, for all $i, j \in \mathbf{Z}^+$, if $i \leq j$, then $A_i \subseteq A_j$. When this happens, the collection is said to be **increasing**, or to be an **ascending chain**. (We can similarly define **decreasing** collections or **descending chains**.)

EXAMPLE 4 Let $A_i = \{i, i + 1, i + 2, \ldots\}$. Then $\bigcup_{i=1}^{\infty} A_i = \mathbf{Z}^+$ and $\bigcap_{i=1}^{\infty} A_i = \varnothing$.

To prove the assertion about the union, first note that since each $A_i \subseteq \mathbf{Z}^+$, it follows that $\bigcup_{i=1}^{\infty} A_i \subseteq \mathbf{Z}^+$. (Prove this!) Conversely, if $n \in \mathbf{Z}^+$, then $n \in A_n$. This implies that $n \in \bigcup_{i=1}^{\infty} A_i$. Thus $\bigcup_{i=1}^{\infty} A_i = \mathbf{Z}^+$.

To prove that $\bigcap\limits_{i=1}^{\infty} A_i = \varnothing$, we need merely show that, for any $n \in \mathbf{Z}^+$, n is not in every A_i or equivalently, $n \notin A_i$ for some i. But $n \notin A_{n+1}$. So the proof is complete.

Is the collection in this example increasing, decreasing, or neither?

Not all infinite collections of sets can be indexed by the positive integers. Here is an example.

EXAMPLE 5   If r is a real number, let $A_r = (r, \infty) = \{x \in \mathbf{R} \mid x > r\}$. Here the indexing set is the set $\mathbf{R}$ of real numbers. This collection is decreasing: if $r \le s$, then $(r, \infty) \supseteq (s, \infty)$. (As you might expect, $A \supseteq B$ just means $B \subseteq A$.)

If we want to express the union and intersection of the sets in this collection, the notation of Example 3 will not work. In this case, we express the union as $\bigcup\limits_{r \in \mathbf{R}} A_r$ and the intersection as $\bigcap\limits_{r \in \mathbf{R}} A_r$.

In general, we have the following notation.

> **Definition 2.3.2** Let I be a set, and $\{A_i \mid i \in I\}$ a collection of sets.
>
> The **union** of the collection, denoted $\bigcup\limits_{i \in I} A_i$, is $\{a \mid a \in A_i,$ for some $i \in I\}$.
>
> The **intersection** of the collection, denoted $\bigcup\limits_{i \in I} A_i$, is $\{a \mid a \in A_i,$ for all $i \in I\}$.

Sometimes we may wish to discuss a collection of sets without actually specifying a particular indexing set. We may simply say: let S be a collection of sets, and to denote the union and intersection we write $\bigcup\limits_{A \in S} A$ and $\bigcap\limits_{A \in S} A$.

An example of this notation appears next in the definition of **partition**, a concept that will be important for us when we discuss equivalence relations in Chapter 4.

Partitions

> **Definition 2.3.3** Let A be a set. A **partition** of A is a subset $\mathscr{P}$ of $\mathbf{P}(A)$ such that
>
> 1. if $X \in \mathscr{P}$ then $X \ne \varnothing$.
> 2. $\bigcup\limits_{X \in \mathscr{P}} X = A$.
> 3. if $X, Y \in \mathscr{P}$ and $X \ne Y$, then $X \cap Y = \varnothing$.

That is, a partition of a set divides the set into different nonempty subsets so that every element of the set is in one of the subsets and no element is in more than one.

EXAMPLE 6 Let $A = \{1, 2, \ldots 10\}$. Then $\mathscr{P} = \{\{1, 4, 9\}, \{2, 3, 5\}, \{6\}, \{7, 8, 10\}\}$ is a partition of A but $\{\{1, 4, 9\}, \{2, 3, 5\}, \{6\}, \{7, 8, 9, 10\}\}$ is not.

EXAMPLE 7 For $n \in \mathbf{Z}$, let $[n, n + 1) = \{x \in \mathbf{R} \mid n \leq x < n + 1\}$. Then $\mathscr{P} = \{[n, n + 1) \mid n \in \mathbf{Z}\}$ is a partition of $\mathbf{R}$.

The Pigeonhole Principle

Recall that if A is a finite set, then $|A|$, the cardinality of A, is the number of elements in A. The following result, that the cardinality of the union of two finite disjoint sets is the sum of the cardinalities of the two sets, may seem obvious, as it should, but a rigorous proof requires a more formal definition of what it means for a set to be finite than we have presented here, so we simply state the result.

Theorem 2.3.4 Let A and B be finite disjoint sets. Then $|A \cup B| = |A| + |B|$.

The following corollary is an extension of Theorem 2.3.4 to any finite collection, $A_1, A_2, \ldots A_n$, of finite mutually disjoint sets. By mutually disjoint, we mean that $A_i \cap A_j = \varnothing$ if $i \neq j$. The proof requires the Principle of Mathematical Induction and is given as an exercise in Chapter 5.

Corollary 2.3.5 Let $A_1, A_2, \ldots A_n$ be a collection of finite mutually disjoint sets. Then

$$\left| \bigcup_{i=1}^{n} A_i \right| = \sum_{i=1}^{n} |A_i|.$$

Corollary 2.3.5 can be used to give the following generalization of Theorem 2.3.4. We leave the proof as an exercise.

Corollary 2.3.6 Let A and B be finite sets. Then $|A \cup B| = |A| + |B| - |A \cap B|$.

We can now discuss the **Pigeonhole Principle**. In its simplest form, the Pigeonhole Principle says that if n objects are placed in k containers and $n > k$, then at least one container will have more than one object in it. A set theory version of the Pigeonhole Principle is the following: Let A_1, $A_2, \ldots A_n$ be a collection of finite mutually disjoint sets. Let $A = \bigcup_{i=1}^{n} A_i$. If $|A| = k$ and $k > n$, then, for some i, $|A_i| \geq 2$.

EXAMPLE 8   We will show that in any list of four positive integers, at least two will have the same remainder when divided by 3.

There are three possible remainders when a number is divided by 3; namely, 0, 1, or 2. Since there are four numbers in our list and only three remainders, at least two of the numbers in the list will have the same remainder.

A generalization of the Pigeonhole Principle is the following theorem, which is a consequence of Theorem 2.3.4. The proof is left as an exercise.

Theorem 2.3.7   Let $A_1, A_2, \ldots A_n$ be a collection of finite mutually disjoint sets. Let $A = \bigcup_{i=1}^{n} A_i$. Suppose that $|A| > nr$ for some positive integer r. Then, for some i, $|A_i| \geq r + 1$.

EXAMPLE 9   Suppose that a certain small town has more than 30,000 people on its voting list. Each voter is classified as Democrat, Republican, or Independent. Prove that at least one of these classifications must contain more than 10,000 people.

If we let A_1 be the set of Democrats on the list, A_2 the set of Republicans, A_3 the set of Independents, and $A = A_1 \cup A_2 \cup A_3$, then $|A| > (3)(10,000)$. So by Theorem 2.3.7, for some $i = 1, 2, 3$, $|A_i| \geq 10,001$.

⬤ MATHEMATICAL PERSPECTIVE: AN UNUSUAL SET

In the previous section, we mentioned the mathematician Georg Cantor in connection with a rigorous definition of the rational numbers. Cantor is also known for defining a set of real numbers with some interesting properties. This set has become known as the **Cantor set**.

To define the Cantor set, we first define a descending chain of sets as follows:

Let A_0 be the closed interval [0, 1].

We form the set A_1 by removing from [0, 1] its middle third; namely, the open interval (1/3, 2/3). Thus $A_1 = [0, 1/3] \cup [2/3, 1]$.

We define A_2 by removing the middle thirds from each of the closed intervals that make up A_1. So $A_2 = [0, 1/9] \cup [2/9, 1/3] \cup [2/3, 7/9] \cup [8/9, 1]$.

We continue in this way to form sets A_n by removing the middle thirds from each of the closed intervals that make up the previous set A_{n-1}.

Note that each A_n is the union of 2^n closed intervals, each of which has length $1/3^n$, and that the sets A_n form a descending chain.

Finally, we define the Cantor set C by $C = \bigcap_{n=1}^{\infty} A_n$.

The Cantor set is difficult to visualize and impossible to draw with complete accuracy because its construction involves an infinite process. Even determining what real numbers are in the set C is not so easy. Certainly, at

least, the endpoints of every closed interval in each of the sets A_n are in C:
0, 1, 1/3, 2/3, 1/9, 2/9, 7/9, 8/9, ...

A good way of determining whether or not a given real number in the interval [0, 1] is in C is to write the number in *ternary* form. The *decimal* representation of a number x in [0, 1] is of course a representation of the form

$$x = 0.a_1a_2a_3\ldots = \frac{a_1}{10} + \frac{a_2}{10^2} + \frac{a_3}{10^3} + \cdots$$

where the numbers $a_1, a_2, a_3, \ldots$ are integers between 0 and 9 inclusive.

The ternary expansion of x has the form

$$x = 0.b_1b_2b_3\ldots = \frac{b_1}{3} + \frac{b_2}{3^2} + \frac{b_3}{3^3} + \cdots$$

where $b_1, b_2, b_3, \ldots$ are from among the integers 0, 1, and 2. This representation is unique except for numbers such as 1/3 that can be represented as 0.100000 . . . ending in a string of 0's and also by 0.022222 . . . ending in a string of 2's. We will adopt the convention that the representation using a string of 2's will be used instead of the one using a string of 0's.

As an example, we will compute the ternary representation of the number 5/7. By simple calculation, one can show that: 2/3 < 5/7 < 1, 6/9 < 5/7 < 7/9, 19/27 < 5/7 < 20/27, 57/81 < 5/7 < 58/81, 173/243 < 5/7 < 174/243, . . . and so on. Thus 2/3 + 0/9 + 1/27 + 0/81 + 2/243 < 5/7 < 2/3 + 0/9 + 1/27 + 1/81. So we get 5/7 = 0.201021201021 . . . in ternary form.

Now how do we determine whether 5/7 is in C? Note that because 2/3 < 5/7 < 1, 5/7 is in A_1. Because 6/9 < 5/7 < 7/9, 5/7 is in A_2. But the middle third of the interval [6/9, 7/9] is (19/27, 20/27) and 5/7 is in that interval. Therefore 5/7 is *not* in A_3 and therefore is *not* in C.

The fact that 5/7 is in the interval (19/27, 20/27) is reflected in the fact that the ternary expansion of 5/7 has a 1 in the third place. So in fact a number x will *not* be in C if a 1 appears at any point in its ternary expansion. Even more strongly, $x \in C$ if and only if the ternary expansion of x consists of just 0's and 2's.

Let $x = 0.020202\ldots$ in ternary form where the alternating 0's and 2's continue indefinitely. Then x is in the Cantor set C. To see what number this ternary expansion represents, we write

$$x = \frac{0}{3} + \frac{2}{3^2} + \frac{0}{3^3} + \frac{2}{3^4} + \cdots.$$

Now this is just the geometric series $\sum_{n=1}^{\infty} 2/9^n$, which converges to 1/4. So 1/4 is in C.

One might ask just how "large" is the Cantor set. It is certainly an infinite set and also a bounded set because it is contained in the interval [0, 1]. We can measure the size of any subset of [0, 1] that is an interval or is a union of a finite number of disjoint intervals by just adding the lengths of each subinterval. For example, the set $S = [1/5, 3/7] \cup [5/8, 9/11]$ has size 1649/3080.

Even more strongly, the size of each A_n is $2^n/3^n = (2/3)^n$. Because C is the intersection of all A_n's, it must be contained in each one. Hence C has "size" less than $(2/3)^n$ for all positive integers n. But $(2/3)^n$ approaches 0 as n approaches ∞. So if the Cantor set has size in any meaningful way, it must have size 0.

On the other hand, it is possible to show that, in a sense that will be made precise in Chapter 6, C has "more" elements than the set $\mathbf{Q}$ of all rational numbers. This is one of the paradoxical features of the Cantor set. In one sense it seems small and in another sense it appears quite large.

Exercises 2.3

---

1. Compute $\mathbf{P}(A)$, where
   (a) $A = \{4, 7, 10\}$.
   (b) $A = \{0, 2, 4, 6\}$.

2. Write four elements of $\mathbf{P}(\mathbf{Z})$.

3. Compute $\mathbf{P}(\mathbf{P}(A))$, where $A = \{0, 1\}$.

4. Let $A = \{0, 1, 2, 3, 4, 5\}$.
   (a) Write 4 subsets of A.
   (b) Write 3 elements of $\mathbf{P}(A)$.
   (c) Write 3 subsets of $\mathbf{P}(A)$.
   (d) Write 3 partitions of A.

5. Let $A = \{0, 1, 2\}$. Characterize the following statements as true or false.
   (a) $\{0\} \subseteq \mathbf{P}(A)$.
   (b) $\{1, 2\} \in \mathbf{P}(A)$.
   (c) $\{\{0, 1\}, \{1\}\} \subseteq \mathbf{P}(A)$.
   (d) $\varnothing \in \mathbf{P}(A)$.
   (e) $\varnothing \subseteq \mathbf{P}(A)$.
   (f) $\{\varnothing\} \in \mathbf{P}(A)$.
   (g) $\{\varnothing\} \subseteq \mathbf{P}(A)$.

6. Let A and B be sets contained in a universal set U.
   (a) Prove that if $A \subseteq B$, then $\mathbf{P}(A) \subseteq \mathbf{P}(B)$.
   (b) State the converse of part (a). Prove or disprove.

7. Let n be a fixed positive integer and let $A_i = \{i, i + 1\}$ for $i = 1, 2, 3, \ldots n$ as in Example 2. Compute $\bigcup_{i=1}^{n} A_i$ and $\bigcap_{i=1}^{n} A_i$. Prove your answer.

8. Let $A_i = (-i, i) = \{x \in \mathbf{R} \mid -i < x < i\}$ as in Example 3. Prove that $\bigcup_{i=1}^{\infty} (-i, i) = \mathbf{R}$ and $\bigcap_{i=1}^{\infty} (-i, i) = (-1, 1)$. (See Exercise 12 of Section 1.1.)

9. Let S be a set and let $\{A_i \mid i \in I\}$ be a collection of subsets of S. Prove that $\bigcup_{i \in I} A_i \subseteq S$.

10. Let $A_i = \{1, 2, 3, \ldots i\}$ for $i \in \mathbf{Z}^+$. Compute $\bigcup_{i=1}^{\infty} A_i$ and $\bigcap_{i=1}^{\infty} A_i$. Prove your answer.

11. Let $A_i = [i, i + 1) = \{x \in \mathbf{R} \mid i \leq x < i + 1\}$ for $i \in \mathbf{Z}^+$. Compute $\bigcup_{i=1}^{\infty} A_i$ and $\bigcap_{i=1}^{\infty} A_i$. Prove your answer.

12. Let $A_i = (1/i, i] = \{x \in \mathbf{R} \mid 1/i < x \leq i\}$ for $i \geq 2$. Compute $\bigcup_{i=2}^{\infty} A_i$ and $\bigcap_{i=2}^{\infty} A_i$. Prove your answer.

13. Let $A_i = [1, 1 + 1/i]$ for $i \in \mathbf{Z}^+$. Compute $\bigcup_{i=1}^{\infty} A_i$ and $\bigcap_{i=1}^{\infty} A_i$.

14. Let $A_i = (1, 1 + 1/i)$ for $i \in \mathbf{Z}^+$. Compute $\bigcup_{i=1}^{\infty} A_i$ and $\bigcap_{i=1}^{\infty} A_i$.

15. Which of the collections of sets in Exercises 10–14 are increasing chains and which are decreasing chains?

16. Let $\{A_1, A_2, A_3, \ldots\}$ be an infinite collection of sets indexed by the positive integers. Assume each A_i is contained in a universal set U and that $\overline{A_i}$ is the complement of A_i in U. Prove that

    (a) $\overline{\bigcup_{i=1}^{\infty} A_i} = \bigcap_{i=1}^{\infty} \overline{A_i}$.

    (b) $\overline{\bigcap_{i=1}^{\infty} A_i} = \bigcup_{i=1}^{\infty} \overline{A_i}$.

17. (a) Prove that if the collection $\{A_i \mid i \in \mathbf{Z}^+\}$ is decreasing, then $\bigcup_{i \in \mathbf{Z}^+} A_i = A_1$.

    (b) Give an example of a decreasing chain $\{A_i \mid i \in \mathbf{Z}^+\}$ with $\bigcap_{i \in \mathbf{Z}^+} A_i \neq \emptyset$.

18. (a) Prove that if the collection $\{A_i \mid i \in \mathbf{Z}^+\}$ is increasing, then $\bigcap_{i \in \mathbf{Z}^+} A_i = A_1$.

(b) Give an example of an increasing chain $\{A_i \mid i \in \mathbf{Z}^+\}$ with $A_i \subseteq \mathbf{Z}^+$ and $\bigcup_{i \in \mathbf{Z}^+} A_i \neq \mathbf{Z}^+$.

19. Prove that $\bigcup_{i \in I} \{i\} = I$.

20. Write two partitions each of $\mathbf{Z}$ and $\mathbf{R}$.

21. Let A be a set and let S be a subset of $\mathbf{P}(A)$. Complete the following statement: "S is not a partition of A if . . . ."

22. Let $\mathbf{E}$ and $\mathbf{O}$ denote the sets of even and odd integers respectively. Prove that $\mathscr{P} = \{\mathbf{E}, \mathbf{O}\}$ is a partition of $\mathbf{Z}$.

23. Let $A = 3\mathbf{Z}$, $B = \{n \in \mathbf{Z} \mid n = 3t + 1 \text{ for some } t \in \mathbf{Z}\}$ and $C = \{n \in \mathbf{Z} \mid n = 3t + 2 \text{ for some } t \in \mathbf{Z}\}$. Prove that $\{A, B, C\}$ is a partition of $\mathbf{Z}$.

24. Prove Corollary 2.3.6.

25. Prove Theorem 2.3.7.

26. Use the Pigeonhole Principle to prove the following: Let S be a square whose sides have length 1. Let P_1, P_2, P_3, P_4, P_5 be five points that lie inside S. Prove that the distance between at least two of the points is less than $\sqrt{2}/2$.

27. The following numbers are given in ternary form. Find the real number that each represents. Which ones are in the Cantor set C?
    (a) $x = 0.002002002 \ldots$ (with 002 repeated indefinitely).
    (b) $x = 0.011011011 \ldots$ (with 011 repeated indefinitely).
    (c) $x = 0.0010000000 \ldots$ (with 0's repeated indefinitely).

28. Determine whether the following numbers are in C:
    (a) 3/5
    (b) 37/50
    (c) 1/40
    Hint: 1/40 = 2/(3^4 − 1).

Discussion and Discovery Exercises

D1. Even for sets with a small number of elements, the power set can be fairly large. (See Exercise 1(b).) Conjecture the number of elements in the power set of a set with n elements. Try to prove your conjecture.

D2. Let A and B be sets contained in a universal set U.
    (a) What can be said about the power set of $A \cup B$ in relation to $\mathbf{P}(A) \cup \mathbf{P}(B)$? For example, are they equal sets? State and prove some results along these lines.
    (b) Compare the power set of $A \cap B$ with $\mathbf{P}(A) \cap \mathbf{P}(B)$. State and prove any results you can.

FUNCTIONS

3.1 DEFINITION AND BASIC PROPERTIES

The idea of a function is everywhere in such mathematical subjects as trigonometry, analytic geometry, and, of course, calculus. These courses deal almost exclusively with real numbers only, so you are no doubt used to this type of expression:

$$f(x) = x^2 \quad \text{or} \quad f(x) = \sqrt{x} \quad \text{or} \quad f(x) = \cos x.$$

But we will be dealing with many other sets, and so we will need to generalize what we mean by the word "function."

You probably think of a function as a formula that allows you to take a number and compute a value; in the first example above, you just multiply the number by itself. But what's important is not that you do (or even could do) some algebraic manipulation, but rather that you can *find* the value; better yet, that there simply *is* a single value associated with, or assigned to, a given object. In this spirit, we define a function between arbitrary sets.

Definition 3.1.1 Let A and B be nonempty sets. A **function** f from A to B is a rule that assigns to each element in the set A one and only one element b in the set B.

We call A the **domain** of f and B the **codomain**; we write $f: A \to B$, and for each $a \in A$, we write $f(a) = b$ if b is assigned to a.

The idea of a function can be suggested by the following simple diagram:

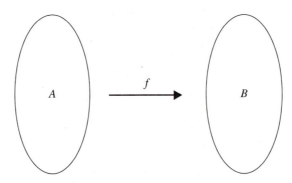

How we find $f(a)$ is not important; what matters is that it exists. Sometimes we say that f maps a to $f(a)$ or a is mapped to $f(a)$ by f.

Note that in defining a function f, we may simply state the domain and codomain of f and specify what $f(x)$ is for any x in the domain. This specification may be done by one or more formulas or by some verbal description.

From now on, throughout the text, whenever we discuss a function f: $A \rightarrow B$, it will be understood, if not explicitly stated, that A and B are nonempty sets.

EXAMPLE 1   Let $A = \{1, 6, 12\}$, $B = \{0, 2, 12\}$. Define $f: A \rightarrow B$ by $f(1) = 2$, $f(6) = 0$, and $f(12) = 12$. Note that there need not be any specific "formula" for f, but since the domain has only three elements it is easy to describe the assignment of an element of B to each element of A. Of course, there are other functions from A to B.

EXAMPLE 2   With A and B as in the previous example, let $g: A \rightarrow B$ be defined by $g(1) = 0$, $g(6) = 0$, and $g(12) = 0$. Notice that distinct elements of the domain may be assigned the same element of the codomain. In this example, g maps every element of A to 0.

EXAMPLE 3   For any nonempty set A, we can define a very simple function $i_A: A \rightarrow A$, called the **identity function** on A, by $i_A(a) = a \ \forall a \in A$; that is, i_A maps every element of A to itself.

EXAMPLE 4   Let $a, b \in \mathbf{R}$, and define $f: \mathbf{R} \rightarrow \mathbf{R}$ by $f(x) = ax + b$. This is called a **linear function** (you may have recognized the form from analytic geometry); it is *constant* (that is, $f(x) = b \ \forall x \in \mathbf{R}$) if $a = 0$ and *nonconstant* otherwise.

EXAMPLE 5   Let $f: \mathbf{R} \rightarrow \mathbf{R}$ be defined by $f(x) = x^2$. The domain of f is, of course, $\mathbf{R}$, and the codomain of f is also $\mathbf{R}$. But not every real number is an assigned value by means of this function; only the nonnegative ones are. That is to say, the codomain is in some sense "too big." For this reason, we single out a subset of the codomain, by making the following definition.

Image of a Function

Definition 3.1.2 Let $f: A \to B$ be a function. The **image** of the function f is Im $f = \{y \in B \mid y = f(x) \text{ for some } x \in A\}$.
  More generally, if $X \subseteq A$, we define

$$f(X) = \{y \in B \mid y = f(x) \text{ for some } x \in X\}.$$

  $f(X)$ is called the **image of X** under f. (Note that the image of f is just $f(A)$.)
  The **graph of f** is the set $\{(a, b) \in A \times B \mid b = f(a)\}$.

Note: A consequence of this definition is that for any function f, $f(\emptyset) = \emptyset$. Do you see why?

As above, we can suggest the definition of Im f with a simple diagram:

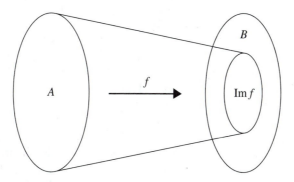

Similarly, for subsets $X \subseteq A$, the definition of $f(X)$ leads us to this diagram:

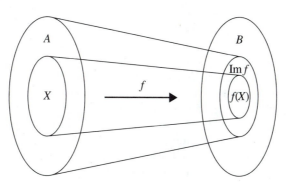

The image of a function f, then, is always a subset of the codomain. In fact, if we have a function $f: A \to B$, it is always possible to consider a very much related function $g: A \to \text{Im } f$, which we would define by $g(x) = f(x)$, $\forall x \in A$. A subtle point here is that if $\text{Im } f \neq B$, then, despite how much they look alike, f and g really are different functions because they have different codomains.

So, it is important to define just what we mean by two functions being equal.

> **Definition 3.1.3** Two functions f and g are **equal** if they have the same domain and the same codomain and if $f(a) = g(a)$ for all a in the domain.

Let's examine these ideas with a few more examples.

EXAMPLE 6 Let $A = \{1, 6, 12\}$, $B = \{0, 2, 12\}$, and let f be defined as in Example 1. Then the image of f is $\{0, 2, 12\}$, the same as the codomain. If $X = \{1, 12\}$, then $f(X) = \{2, 12\}$.

EXAMPLE 7 Let g be as defined in Example 2. Then $\text{Im } g = \{0\}$. Here the image is *not* the same as the codomain.

EXAMPLE 8 Consider the ordinary **sine function** from trigonometry, which we can express as $\sin: \mathbf{R} \to \mathbf{R}$. The codomain of this function is $\mathbf{R}$, but we know that all sine values are between -1 and 1 and that every real number between -1 and 1 is the sine of some number; that is, $\text{Im sin} = [-1, 1]$. In fact, $\sin([-\pi, \pi]) = [-1, 1]$.

EXAMPLE 9 Define $f: \mathbf{R} \to \mathbf{Z}$ by the following rule: $f(x)$ is the greatest integer less than or equal to x. For obvious reasons, f is called the **greatest integer function**, and we write $f(x) = [x]$, so that, for example, $[6.5] = 6$, $[\pi] = 3$, $[-1.2] = -2$, and $[2] = 2$.

In this case, the image of f is the entire codomain, since any integer is the greatest integer less than or equal to, among other things, itself.

Note that the image of $\mathbf{Z}$ under f is also the codomain $\mathbf{Z}$.

We now consider some examples in which we give formal proofs that the image of a given function is a certain set. Proofs like these are not always easy. First of all, it is important to understand exactly what it means for an element of the codomain of a function to be in the image of that function. Specifically, if $f: A \to B$ is a function and $y \in B$, then $y \in \text{Im } f$ if and only if there exists an element x in A such that $f(x) = y$. In some cases we will actually solve for x in terms of y. In other cases it is difficult, if not impossible, to solve for x and we merely show that x exists.

EXAMPLE 10 Let $f: \mathbf{R} \to \mathbf{R}$ be defined by $f(x) = x^2$. We claim that Im $f = S$ where $S = \{y \in \mathbf{R} \mid y \geq 0\}$.

Clearly, Im $f \subseteq S$, because the square of any real number is nonnegative. Conversely, suppose that $y \in S$. Letting $x = \sqrt{y}$, it follows that $f(x) = x^2 = (\sqrt{y})^2 = y$, so that $y \in$ Im f. Therefore, $S \subseteq$ Im f and we have Im $f = S$.

In Example 10, we assumed some well-known properties of the real numbers; namely, that the square of any real number is nonnegative and that every nonnegative real number has a square root. We will continue to assume these and other properties of the real numbers throughout this chapter. For proofs, see Chapter 7.

EXAMPLE 11 Let $f: \mathbf{R} \to \mathbf{R}$ be defined by $f(x) = 3x + 7$. You may recognize this function as a linear function whose graph is a straight line of slope 3. You could conclude from the graph that the image of f is $\mathbf{R}$. Here is a proof.

Clearly, Im $f \subseteq \mathbf{R}$. Let $y \in \mathbf{R}$. We must find an x in $\mathbf{R}$ such that $f(x) = y$. It is easy to find x. Just take the equation $y = 3x + 7$ and solve for x in terms of y. We get $x = (y - 7)/3$. We leave it to the reader to check that $f(x) = y$. It now follows that Im $f = \mathbf{R}$.

EXAMPLE 12 Let $f(x) = 3x + 7$ as in the previous example. We will compute $f(X)$ where X is the closed interval $[0, 1]$. A look at the graph of f will lead us to conclude that $f(X) = [7, 10]$.

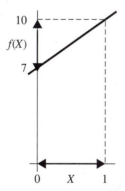

We now give a rigorous proof.

As in the previous example, we are proving that two sets are equal. So we will prove that each set is a subset of the other.

First, let $y \in f(X)$. Then $y = f(a)$ for some $a \in [0, 1]$. Because $0 \leq a \leq 1$, it follows from properties of inequalities that $7 \leq 3a + 7 \leq 10$. Hence $y \in [7, 10]$. Therefore, $f(X) \subseteq [7, 10]$.

Now let $y \in [7, 10]$. We know from the previous example that $f(x) = y$ where $x = (y - 7)/3$. Thus to show that $y \in f(X)$, it is sufficient to show that $x \in [0, 1]$. But since $7 \le y \le 10$, it follows easily that $0 \le (y - 7)/3 \le 1$. Hence $[7, 10] \subseteq f(X)$. We have now proved that $f(X) = [7, 10]$.

EXAMPLE 13   Let $f: \mathbf{Z} \to \mathbf{Z}$ be defined by

$$f(n) = \begin{cases} n + 1, & \text{if } n \text{ is even} \\ n - 3, & \text{if } n \text{ is odd.} \end{cases}$$

We will compute $f(\mathbf{E})$ where $\mathbf{E}$ is the set of even integers. From the definition of f, it seems reasonable to expect that $f(\mathbf{E}) = \mathbf{O}$ where $\mathbf{O}$ is the set of odd integers. We will now prove it. Remember: To prove two sets equal, we usually prove that each is a subset of the other.

First, let $n \in \mathbf{E}$. Then $f(n) = n + 1$ and since n is even, $n + 1$ is odd. Thus $f(n) \in \mathbf{O}$, proving that $f(\mathbf{E}) \subseteq \mathbf{O}$.

Now let $m \in \mathbf{O}$. To prove that $m \in f(\mathbf{E})$, we must find an $n \in \mathbf{E}$ such that $f(n) = m$. Well, clearly $m - 1$ is even and $f(m - 1) = m - 1 + 1 = m$. This proves that $m \in f(\mathbf{E})$ and hence $\mathbf{O} \subseteq f(\mathbf{E})$. It follows that $f(\mathbf{E}) = \mathbf{O}$.

EXAMPLE 14   Let f be the function of the previous example. In this example, we will compute $f(X)$, where

$$X = \{n \in \mathbf{Z} \mid n = 4t + 1 \text{ for some } t \in \mathbf{Z}\}.$$

A good way to start is to try some values. Some elements in X are: -3, 1, 5, and 9, and $f(-3) = -6$, $f(1) = -2$, $f(5) = 2$, $f(9) = 6$. Each one of these values is two more than a multiple of 4, so we can reasonably conjecture that

$$f(X) = \{n \in \mathbf{Z} \mid n = 4t + 2 \text{ for some } t \in \mathbf{Z}\}.$$

Now let's prove it.

We'll denote the set $\{n \in \mathbf{Z} \mid n = 4t + 2 \text{ for some } t \in \mathbf{Z}\}$ by S and prove that $f(X) = S$. Let $n \in X$. Then $n = 4t + 1$ for some integer t. So n is odd and

$$f(n) = f(4t + 1) = 4t + 1 - 3 = 4t - 2 = 4(t - 1) + 2.$$

We see that $f(n)$ is in S since it is 2 more than a multiple of 4. Thus $f(X) \subseteq S$.

Let $m \in S$. Then $m = 4t + 2$ for some t in $\mathbf{Z}$. We want to find an element $n \in X$ such that $f(n) = m$. Since $m + 3$ is odd and $f(m + 3) = m + 3 - 3 = m$, it makes sense to let $n = m + 3$. To complete the proof, we must show that $m + 3 \in X$. But

$$m + 3 = 4t + 2 + 3 = 4t + 5 = 4(t + 1) + 1 \in X.$$

So $S \subseteq f(X)$ and we conclude that $f(X) = S$.

For some functions, computing their images can be difficult and the methods we used for the previous examples will not suffice. Sometimes the methods of calculus can be used to great advantage. In the next two examples, we will use the Intermediate Value Theorem. We state it here without proof, because a formal proof belongs in an introductory text in real analysis or advanced calculus.

Theorem 3.1.4 **Intermediate Value Theorem** Let f be a function whose domain and codomain are subsets of $\mathbf{R}$. Assume that f is continuous on the closed interval $[a, b]$. If y is any number between $f(a)$ and $f(b)$, then there is a real number x in $[a, b]$ such that $f(x) = y$.

EXAMPLE 15 Consider the function $f: \mathbf{R} \to \mathbf{R}$ defined by $f(x) = x^3 + 4x + 1$. The first derivative of $f(x)$ is the function $f'(x) = 3x^2 + 4$, which is always positive. Thus $f(x)$ is an increasing function and has no relative max or min points. We also see that $\lim_{x \to \infty} f(x) = \infty$ and $\lim_{x \to -\infty} f(x) = -\infty$. Since f is continuous, the Intermediate Value Theorem tells us that any real number is in the image of f. We can see this more clearly by looking at the graph of f:

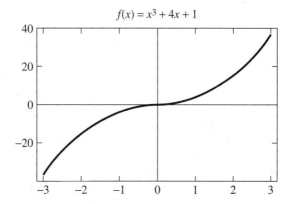

$$f(x) = x^3 + 4x + 1$$

Here is a formal proof that Im $f = \mathbf{R}$. Since Im $f \subseteq \mathbf{R}$, it suffices to show that if $y \in \mathbf{R}$, then $\exists x \in \mathbf{R} \ni f(x) = y$.

Let $y \in \mathbf{R}$. The fact that $\lim_{x \to \infty} f(x) = \infty$ essentially means that the function f will eventually surpass any given number if x is taken large enough; in particular, it will eventually surpass y. Thus there is a real number $M > 0$ such that $f(x) > y$ whenever $x > M$. Similarly, since $\lim_{x \to -\infty} f(x) = -\infty$, there is a real number $N < 0$ such that $f(x) < y$ whenever $x < N$. So there exist real numbers x_1 and x_2 where $x_1 < x_2$ such that $f(x_1) < y < f(x_2)$. Since f is continuous everywhere, it is in particular continuous on the interval $[x_1, x_2]$. Then by the Intermediate Value Theorem there exists $x \in [x_1, x_2]$ such that $f(x) = y$.

EXAMPLE 16 Let f be the function of the previous example. We will now compute $f(X)$ where X is the closed interval $[0, 1]$. Since $f(0) = 1$ and $f(1) = 6$, it follows from the Intermediate Value Theorem and the fact that f is increasing that $f(X) = [1, 6]$. We leave the formal proof as an exercise.

EXAMPLE 17   Let $f: \mathbf{R} \to \mathbf{R}$ be defined by $f(x) = x^3 - 3x^2 + 1$. Let's compute $f(X)$ where X is the closed interval $[1, 3]$. Since $f(1) = -1$ and $f(3) = 1$, it may seem that $f(X) = [-1, 1]$. However, f is not an increasing or a decreasing function on $[1, 3]$. In fact f has a local minimum at $x = 2$ and $f(2) = -3$. We claim that $f(X) = [-3, 1]$.

We first draw the graph of f.

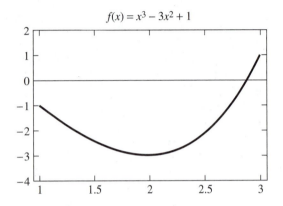

$$f(x) = x^3 - 3x^2 + 1$$

It is easy to see from this graph why $f(X) = [-3, 1]$. We now give a formal proof. First note that f is decreasing on $[0, 2]$ and increasing on $[2, \infty]$. Let $x \in [1, 3]$. If $1 \le x \le 2$, then $-3 = f(2) \le f(x) \le f(1) = -1$. So $f(x) \in [-3, 1]$. If $2 \le x \le 3$, then $-3 = f(2) \le f(x) \le f(3) = 1$. Again we have that $f(x) \in [-3, 1]$. This proves that $f(X) \subseteq [-3, 1]$. Now let $y \in [-3, 1]$. Since y is between $f(2)$ and $f(3)$ and f is continuous, by the Intermediate Value Theorem there exists $x \in [2, 3]$ such that $f(x) = y$. Hence $y \in f([2, 3])$ and since $[2, 3] \subseteq [1, 3]$, we get that $y \in f([1, 3])$. Therefore, $[-3, 1] \subseteq f([1, 3])$. Hence $f(X) = [-3, 1]$.

If $f: A \to B$ is a function and if X and Y are subsets of A such that X is a subset of Y, it seems reasonable that $f(X)$ will be a subset of $f(Y)$. We prove that result now.

Proposition 3.1.5   Let A and B be sets and let $f: A \to B$ be a function. Let X and Y be subsets of A. If $X \subseteq Y$, then $f(X) \subseteq f(Y)$.

PROOF: To prove that $f(X) \subseteq f(Y)$, we must take an arbitrary element in $f(X)$ and prove that it is in $f(Y)$. So we let $y \in f(X)$. Then $y = f(x)$ for some $x \in X$. Since $X \subseteq Y$, it follows that $x \in Y$ and therefore $y \in f(Y)$. This proves that $f(X) \subseteq f(Y)$. ●

Let's take a moment to analyze the proof of Proposition 3.1.5. The statement of the proposition is of the form $P \Rightarrow Q$ where P is the statement

$X \subseteq Y$ and Q is the statement $f(X) \subseteq f(Y)$. It is statement Q that we want to prove, so we ignore statement P for the moment. To prove statement Q, we have to prove that one set is a subset of another set. The usual way of doing this is to take an arbitrary element of the first set and prove that it is in the second set. That is why in the proof of Proposition 3.1.5, we *start* by taking $y \in f(X)$. Now what does it mean for y to be an element of $f(X)$? By the definition of $f(X)$, y must be the function f evaluated at an element of X. That is why we write $y = f(x)$ for some $x \in X$. We then have to ask ourselves what we need to do in order to show that y is in $f(Y)$. Well, by the definition of $f(Y)$, we need to show that y is the function f evaluated at an element of Y. Since $y = f(x)$, we will be done if we can show that $x \in Y$. This is where statement P comes in. The fact that we are assuming P to be a true statement means that since x is in X, x is also in Y. Finally we can conclude that $y \in f(Y)$, proving that $f(X) \subseteq f(Y)$.

In proving a theorem of the form $P \Rightarrow Q$, then, it is often advantageous to focus first on statement Q and develop a strategy for proving it. In developing this strategy you need to be mindful of the fact that statement P will be used in the proof.

The converse of Proposition 3.1.5 is false. (See Exercise 23 following this section.) The reason that it is false boils down to the fact that if $f(a) \in f(X)$ for some $a \in A$, it does not necessarily follow that $a \in X$. For example, if $f: \mathbf{R} \to \mathbf{R}$ is defined by $f(x) = x^2$ and X is the set of positive real numbers, then $f(-1) \in f(X)$ since $f(-1) = f(1)$ and $1 \in X$, but $-1 \notin X$.

The following result tells us how the union and intersection of sets are affected when we take their image under a mapping.

Proposition 3.1.6 Let A and B be sets and X and Y subsets of A. Let $f: A \to B$ be a function. Then

1. $f(X \cup Y) = f(X) \cup f(Y)$.
2. $f(X \cap Y) \subseteq f(X) \cap f(Y)$.

PROOF: We will prove 1 and leave 2 as an exercise.

Recall that if we wish to prove two sets equal, it is usually necessary to prove that each one is a subset of the other.

Let $b \in f(X \cup Y)$. Then $b = f(a)$ where $a \in X \cup Y$. So a is in X or Y. If $a \in X$, then $b \in f(X)$ and if $a \in Y$, then $b \in f(Y)$.

Thus $b \in f(X)$ or $b \in f(Y)$, which means that $b \in f(X) \cup f(Y)$. We have now proved that $f(X \cup Y) \subseteq f(X) \cup f(Y)$.

On the other hand, suppose that $b \in f(X) \cup f(Y)$. Then $b \in f(X)$ or $b \in f(Y)$.

If $b \in f(X)$, then $b = f(x)$ for some $x \in X$. Since $x \in X$, $x \in X \cup Y$ and so $b \in f(X \cup Y)$. Similarly, if $b \in f(Y)$, then $b \in f(X \cup Y)$.

In either case, we get $b \in f(X \cup Y)$ and therefore $f(X) \cup f(Y) \subseteq f(X \cup Y)$.

It now follows that $f(X \cup Y) = f(X) \cup f(Y)$. ●

Note that we do not have strict equality of sets in statement 2 of Proposition 3.1.6. One can give examples of functions where $f(X \cap Y) \neq f(X) \cap f(Y)$. See Exercise 15 following.

Inverse Image

Besides computing the image of a subset of the domain of a function, we can also talk about what is known as the *inverse* image of a subset of the codomain. It is defined as follows.

> **Definition 3.1.7** Let $f: A \to B$ be a function and let W be a subset of B. Then the **inverse image of W with respect to f** is the set $\{a \in A \mid f(a) \in W\}$. We denote this set by $f^{-1}(W)$.

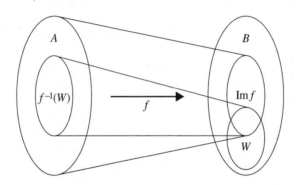

So $f^{-1}(W)$ is the set of elements of the domain of f that gets mapped to an element of W.

It is important to realize that in this context the symbol f^{-1} does not refer to the inverse of the function f. Such an inverse may not even exist! Inverse functions are discussed in the next section.

EXAMPLE 18   Let $A = \{0, 1, 2, 3, 4, 5\}$ and $B = \{7, 9, 11, 12, 13\}$. Let $f: A \to B$ be defined by $f(0) = 11$, $f(1) = 9$, $f(2) = 7$, $f(3) = 9$, $f(4) = 11$, and $f(5) = 9$.

Then, for example, $f^{-1}(\{7, 9\})$ is the set of all integers in A that get mapped to 7 or 9 by f.

Since $f(1) = 9$, $f(2) = 7$, $f(3) = 9$, and $f(5) = 9$, it follows that $f^{-1}(\{7, 9\}) = \{1, 2, 3, 5\}$.

Similarly, $f^{-1}(\{11, 12\}) = \{0, 4\}$ and $f^{-1}(\{12, 13\}) = \varnothing$.

EXAMPLE 19   Let $f: \mathbf{R} \to \mathbf{R}$ be defined by $f(x) = 4x + 5$. Let $W = \{x \in \mathbf{R} \mid x > 0\}$. Then $f^{-1}(W)$ is the set of all real numbers x such that $f(x) \in W$. By the definition of W, $f(x) \in W$ if and only if $f(x) > 0$. Thus

$$f^{-1}(W) = \{x \in \mathbf{R} \mid 4x + 5 > 0\} = (-5/4, \infty).$$

If Z is the open interval $(9, 13)$, then

$$f^{-1}(Z) = \{x \in \mathbf{R} \mid 9 < 4x + 5 < 13\} = (1, 2).$$

EXAMPLE 20 Let $f: \mathbf{Z} \to \mathbf{Z}$ be defined by

$$f(n) = \begin{cases} \dfrac{n}{2}, & \text{if } n \text{ is even} \\ 2n + 4, & \text{if } n \text{ is odd.} \end{cases}$$

It is easy to see by computation that $f^{-1}(\{6, 7\}) = \{1, 12, 14\}$, for example.

Now let's compute $f^{-1}(\mathbf{O})$ where $\mathbf{O}$ is the set of odd integers. We must find all integers n that get mapped by f to an odd integer.

Since $2n + 4$ is always even, we see that no odd integer gets mapped to an odd integer. On the other hand, some even integers get mapped to odd integers and some do not. For example, 2, 6, and 10 get mapped to odd integers and 4, 8, and 12 do not. It seems that those even integers that are not multiples of 4 get mapped to odd integers and the multiples of 4 get mapped to even integers.

An even integer that is not a multiple of 4 can be expressed in the form $4t + 2$ where t is any integer. So we will prove

$$f^{-1}(\mathbf{O}) = \{n \in \mathbf{Z} \mid n = 4t + 2 \text{ for some } t \in \mathbf{Z}\}.$$

First, let $n \in \mathbf{Z}$ such that $n = 4t + 2$ for some $t \in \mathbf{Z}$. Then $f(n) = f(4t + 2) = 2t + 1 \in \mathbf{O}$. Hence $n \in f^{-1}(\mathbf{O})$. So we have $\{n \in \mathbf{Z} \mid n = 4t + 2 \text{ for some } t \in \mathbf{Z}\} \subseteq f^{-1}(\mathbf{O})$.

Conversely, suppose that $n \in f^{-1}(\mathbf{O})$. Then $f(n)$ is odd. It follows that $f(n) \neq 2n + 4$, so n is even. Thus $f(n) = n/2$ and since $f(n)$ is odd, $f(n) = 2t + 1$ for some $t \in \mathbf{Z}$, giving us $n = 4t + 2$.

Consequently, $f^{-1}(\mathbf{O}) \subseteq \{n \in \mathbf{Z} \mid n = 4t + 2 \text{ for some } t \in \mathbf{Z}\}$ and we may now conclude that

$$f^{-1}(\mathbf{O}) = \{n \in \mathbf{Z} \mid n = 4t + 2 \text{ for some } t \in \mathbf{Z}\}.$$

Analogous to Proposition 3.1.6 is the following result on the inverse images of unions and intersections.

Proposition 3.1.8 Let $f: A \to B$ be a function. Let W and Z be subsets of B. Then

1. $f^{-1}(W \cup Z) = f^{-1}(W) \cup f^{-1}(Z)$
2. $f^{-1}(W \cap Z) = f^{-1}(W) \cap f^{-1}(Z)$

PROOF: 1. As in the proof of Proposition 3.1.6, we wish to prove that two sets are equal and therefore we will prove that each is a subset of the other. First, let $y \in f^{-1}(W \cup Z)$. Then $f(y) \in W \cup Z$, which means that $f(y) \in W$ or $f(y) \in Z$. If $f(y) \in W$, then $y \in f^{-1}(W)$ or if $f(y) \in Z$, then $y \in f^{-1}(Z)$. Thus $y \in f^{-1}(W) \cup f^{-1}(Z)$ and we have $f^{-1}(W \cup Z) \subseteq f^{-1}(W) \cup f^{-1}(Z)$.

Conversely, suppose that $y \in f^{-1}(W) \cup f^{-1}(Z)$. Then $y \in f^{-1}(W)$ or $y \in f^{-1}(Z)$. If $y \in f^{-1}(W)$, then $f(y) \in W$ or if $y \in f^{-1}(Z)$, then $f(y) \in Z$. In either case, $f(y) \in W \cup Z$, which implies that $y \in f^{-1}(W \cup Z)$. We have now proved that $f^{-1}(W) \cup f^{-1}(Z) \subseteq f^{-1}(W \cup Z)$.

We can now conclude that $f^{-1}(W \cup Z) = f^{-1}(W) \cup f^{-1}(Z)$.

2. The proof of 2 is left as an exercise. ●

We end this section with a result on functions whose domains are finite sets.

Proposition 3.1.9   Let A and B be sets and let $f: A \to B$ be a function. Assume that A is a finite set. Then Im f is a finite set and $|\text{Im } f| \le |A|$.

PROOF: Let $A = \{a_1, a_2, \ldots a_n\}$. Then the image of f consists of the distinct elements from among $f(a_1), f(a_2), \ldots f(a_n)$. Hence Im f is a finite set with at most n elements. ●

● HISTORICAL COMMENTS: THE IDEA OF A FUNCTION

The very general, abstract definition of function given in this section is probably not the view that you learned in previous mathematics courses, and it is certainly not the way that mathematicians have always approached the idea. For example, as calculus was being developed in the 17th century, geometric objects, curves, were the focus of attention. Newton and Leibniz worked by attaching other geometric objects, such as tangent lines, to curves; they did not deal with equations of curves as we do today.

It was not until the 18th century that **Leonhard Euler** (1707–1783) began the shift to the formulas that you are so familiar with from high school algebra. His "analytic expressions" encompassed more than polynomials and rational functions: he studied combinations of these and trigonometric, logarithmic, exponential, and power functions (such as roots), as well as their derivatives and integrals. Importantly, he even allowed infinite series of these functions. Euler's monumental "calculus book," the *Introductio in Analysin Infinitorum*, published in 1748, emphasizes the idea of a function as an algebraic formula by not including any art or graphs.

For a time, mathematicians hoped that all functions could be represented by formulas. In 1822, **J. B. J. Fourier** (1768–1830) even attempted to prove that any function $f: (-a, a) \to \mathbf{R}$ can be written as an infinite sum of trigonometric functions (in fact, as just a sum of sines and cosines). We understand today that Fourier's proof of this statement is not completely rigorous, but his efforts led to more intense scrutiny of the concept of a function. Soon **P. G. L. Dirichlet** (1805–1859) gave a definition of function that is essentially the same as ours, and as a counterexample to Fourier's result, defined the following remarkable function:

$$d(x) = \begin{cases} 0, & \text{if } x \text{ is rational} \\ 1, & \text{if } x \text{ is irrational.} \end{cases}$$

Obviously, this function is not given as an analytic expression, to use Euler's term. But it is also not a geometric curve that Newton or Leibniz would have understood: it's impossible to draw because it is discontinuous *everywhere*. It can only be understood in the sense of our definition, as a rule or a simple pairing of elements in the domain with elements in the codomain. (The fact that it is discontinuous everywhere is a consequence of the fact that between any two rational numbers there is an irrational number and between any two irrationals there is a rational. A proof of this fact is given in Chapter 7, Theorem 7.2.6.)

Today, mathematicians universally accept our abstract notion of a function. But there is a price: we must always suspect our intuition. The Dirichlet function led to the creation of many other "pathological" functions, perhaps culminating in 1872 when **K. Weierstrass** (1815–1897) gave an example of a function $f: \mathbf{R} \to \mathbf{R}$ that is continuous everywhere but does not have a derivative anywhere. Intuitively, its graph is unbroken but has a sharp corner at every point! In the words of **H. Poincaré** (1854–1912): "Logic sometimes makes monsters."

The article by Israel Kleiner [11] gives much more detail about the historical development of the fundamental notion of function.

Exercises 3.1

1. Write out three functions with domain $\{0, 1, 2, 3, 4, 5\}$ and codomain $\{6, 7, 8, 9, 10\}$. Make at least one of the functions have the property that its image equals its codomain.

2. Write out three functions with domain $\{2, 3, 4, 5\}$ and codomain $\{6, 7, 8, 9, 10\}$. Explain why you cannot define a function between these two sets for which its image equals its codomain.

3. For the following functions, compute first the image of f and then the image of the given subset X. (No formal proofs are necessary.)
   (a) $f: \mathbf{Z} \to \mathbf{Z}, f(n) = 2n + 1; X = \mathbf{E}$, the set of even integers
   (b) $f: \mathbf{R} \to \mathbf{R}, f(x) = 2x + 1; X = [-1, 4]$
   (c) $f: \mathbf{Z} \to \mathbf{Z}, f(n) = \begin{cases} 2n, & \text{if } n \text{ is even} \\ n, & \text{if } n \text{ is odd} \end{cases}; X = \mathbf{E}$
   (d) $f: \mathbf{R} \to \mathbf{R}, f(x) = \cos x; X = \left[0, \dfrac{\pi}{2}\right]$
   (e) $f: \mathbf{R}^* \to \mathbf{R}^*$, where $\mathbf{R}^*$ is the set of nonzero real numbers, $f(x) = \dfrac{1}{x}; X = (0, 1]$
   (f) $f: \mathbf{R} \to \mathbf{R}, f(x) = e^x; X = \{x \in \mathbf{R} \mid x > 0\}$
   (g) $f: \mathbf{Z} \times \mathbf{Z} \to \mathbf{Z}, f(n, m) = nm; X = \mathbf{E} \times \mathbf{Z}$

(h) $f: \mathbf{Z} \to \mathbf{Z} \times \mathbf{Z}$, $f(n) = (n, n)$; $X = \mathbf{E}$

(i) $f: \mathbf{R} \times \mathbf{R}^* \to \mathbf{R}$, $f(x, y) = \dfrac{x}{y}$; $X = \{(x, 1) \mid x \in \mathbf{R}\}$

4. Which of the following are *functions* from $\mathbf{R}$ to $\mathbf{R}$? For those that are functions, find their image. (Drawing a graph will help in determining the image. No formal proof is necessary.)

(a) $f(x) = \pm\sqrt{x}$

(b) $f(x) = \begin{cases} \sqrt{x}, & \text{if } x \geq 0 \\ -\sqrt{-x}, & \text{if } x < 0 \end{cases}$

(c) $f(x) = \begin{cases} 2x + 1, & \text{if } x \geq 1 \\ x^2, & \text{if } x < 1 \end{cases}$

(d) $f(x) = \begin{cases} 4x, & \text{if } x > 0 \\ x^3, & \text{if } x < 0 \end{cases}$

(e) $f(x) = \begin{cases} -x + 7, & \text{if } x \geq 2 \\ x^2, & \text{if } x < 2 \end{cases}$

5. Let $f: \mathbf{R} \to \mathbf{R}$ be defined by $f(x) = 6x + 5$.
   (a) Prove that Im $f = \mathbf{R}$.
   (b) Compute $f([1, 4])$. Prove your answer.

6. Let $f: \mathbf{Z} \to \mathbf{Z}$ be defined by $f(n) = 2n + 1$. What is wrong with the following proof that Im $f = \mathbf{Z}$?

   Let $y \in \mathbf{Z}$. Then $f\left(\dfrac{y-1}{2}\right) = 2\left(\dfrac{y-1}{2}\right) + 1 = y$. Hence $y \in$ Im f.

   Therefore, $\mathbf{Z} \subseteq$ Im f. Since Im $f \subseteq \mathbf{Z}$ by definition, it follows that Im $f = \mathbf{Z}$.

7. Let A and B be sets and let $f: A \to B$ be a function. Complete the following statement: An element b in B is *not* in Im f if . . . .

8. Let $f: \mathbf{R} \to \mathbf{R}$ be defined by $f(x) = x^2$.
   (a) Compute $f([0, 1])$. Prove your answer.
   (b) Compute $f([-1, 1])$. Prove your answer.

9. Let $f: \mathbf{R} \to \mathbf{Z}$ be the greatest integer function.
   (a) Compute $f([0, 1])$. Prove your answer.
   (b) Compute $f(\mathbf{Z})$. Prove your answer.

10. Let $f: \mathbf{Z} \to \mathbf{Z}$ be defined by $f(n) = \begin{cases} n + 2, & \text{if } n \text{ is even} \\ 2n, & \text{if } n \text{ is odd.} \end{cases}$ Let $\mathbf{E}$ denote the set of even integers and $\mathbf{O}$ the set of odd integers.
    (a) Prove that Im $f = \mathbf{E}$.
    (b) Prove that $f(\mathbf{E}) = \mathbf{E}$.
    (c) Prove that $f(\mathbf{O}) = \{n \in \mathbf{Z} \mid n = 4t + 2 \text{ for some } t \in \mathbf{Z}\}$.
    (d) Let X be the set of multiples of 4. Prove that $f(X) = f(\mathbf{O})$.

11. Let $f: \mathbf{R} \to \mathbf{R}$ be defined by $f(x) = x^3 + 4x + 1$ as in Example 15. Let $X = [0, 1]$. Prove that $f(X) = [1, 6]$. (See Example 16.)

12. Let $f: \mathbf{R} \to \mathbf{R}$ be defined by $f(x) = 2x^3 + 3x^2 - 12x + 1$. Compute the following sets.
    Hint: Draw the graph of f first.
    (a) Im f (b) $f([0, 1])$
    (c) $f([1, 2])$ (d) $f([-1, 2])$

13. Give formal proofs to your answers in Exercise 12.
    Reminder: Each proof should involve proving that two sets are equal, as in Exercise 11.

14. Let $f: \mathbf{R} \to \mathbf{R}$ be defined by $f(x) = x^4 + x^2$.
    (a) Compute the image of f.
    (b) Prove that your answer to part (a) is correct.
    (c) Compute $f([-1, 2])$.
    (d) Prove that your answer to part (c) is correct.

15. Let A and B be sets and X and Y be subsets of A. Let $f: A \to B$ be a function.
    (a) Prove that $f(X \cap Y) \subseteq f(X) \cap f(Y)$.
    (b) Give an example of sets A and B and a function $f: A \to B$ for which $f(X \cap Y) \neq f(X) \cap f(Y)$ for some subsets X and Y of A.

16. Let A and B be sets and X and Y be subsets of A. Let $f: A \to B$ be a function.
    (a) Prove that $f(X) - f(Y) \subseteq f(X - Y)$.
    (b) Give an example of sets A and B and a function $f: A \to B$ for which $f(X) - f(Y) \neq f(X - Y)$ for some subsets X and Y of A.

17. Let $A = \{1, 2, 3, 4, 5\}$, $B = \{a, b, c, d, e\}$, and define $f: A \to B$ by $f(1) = a$, $f(2) = c$, $f(3) = e$, $f(4) = a$, and $f(5) = d$. Find each of the following.
    (a) $f^{-1}(B)$ (b) $f^{-1}(\{a, c, d, e\})$
    (c) $f^{-1}(\{a, b, c, d\})$ (d) $f^{-1}(\{b\})$

18. For the following functions, compute the *inverse* image of the given subsets of the codomain. (No proofs are necessary.)
    (a) $f: \mathbf{Z} \to \mathbf{Z}$, $f(n) = 3n + 1$; $W_1 = \mathbf{E}$, the set of even integers, $W_2 = \{4\}$, $W_3 = \{1, 5, 8\}$.
    (b) $f: \mathbf{R} \to \mathbf{R}$, $f(x) = 3x + 1$; $W_1 = \{4\}$, $W_2 = \{1, 5, 8\}$, $W_3 = (4, \infty)$, $W_4 = (2, 4)$, $W_5 = \mathbf{Z}$, $W_6 = \mathbf{E}$, the set of even integers.
    (c) $f: \mathbf{R} \to \mathbf{R}$, $f(x) = \cos x$; $W_1 = [-1, 1]$, $W_2 = \{x \in \mathbf{R} \mid x \geq 0\}$, $W_3 = \mathbf{Z}$.
    (d) $f: \mathbf{R} \to \mathbf{R}$, $f(x) = e^x$; $W_1 = [-1, 0]$, $W_2 = \{x \in \mathbf{R} \mid x \geq 0\}$, $W_3 = \{1\}$.
    (e) $f: \mathbf{Z} \to \mathbf{Z}$, $f(n) = \begin{cases} n, & \text{if } n \text{ is even} \\ n - 1, & \text{if } n \text{ is odd} \end{cases}$; $W_1 = \mathbf{E}$, $W_2 = \{1\}$, $W_3 = \{6\}$, $W_4 = \mathbf{O}$, the set of odd integers.

(f) $f: \mathbf{R} \to \mathbf{Z}$, the greatest integer function; $W_1 = \{0, 1, 2\}$, $W_2 = \mathbf{Z}^+$, $W_3 = \mathbf{E}$.

(g) $f: \mathbf{Z} \to \mathbf{Z} \times \mathbf{Z}$, $f(n) = (n, n)$; $W_1 = \mathbf{Z} \times \{0\}$, $W_2 = \mathbf{Z}^+ \times \{-1\}$.

(h) $f: \mathbf{R} \times \mathbf{R}^* \to \mathbf{R}$, $f(x, y) = \dfrac{x}{y}$; $W_1 = \{1\}$, $W_2 = \mathbf{R}^*$, $W_3 = \{x \mid x > 0\}$.

19. Let $f: \mathbf{Z} \to \mathbf{Z}$ be defined by $f(n) = \begin{cases} \dfrac{n}{2}, & \text{if } n \text{ is even} \\ 2n + 4, & \text{if } n \text{ is odd.} \end{cases}$

   (a) Compute $f^{-1}(\{11, 12, 13, 14, 15\})$.
   (b) Compute $f^{-1}(\mathbf{E})$. Prove your answer.

20. Let $f: A \to B$ be a function. Let W and Z be subsets of B. Prove that $f^{-1}(W \cap Z) = f^{-1}(W) \cap f^{-1}(Z)$.

21. Let $f: A \to B$ be a function. Let X be a subset of A.
   (a) Prove that $X \subseteq f^{-1}(f(X))$.
   (b) Given an example of a function $f: A \to B$ for some A and B and a subset X of A such that $X \neq f^{-1}(f(X))$.

22. Let $f: A \to B$ be a function. Let W be a subset of B.
   (a) Prove that $f(f^{-1}(W)) \subseteq W$.
   (b) Give an example of a function $f: A \to B$ for some A and B and a subset W of B such that $f(f^{-1}(W)) \neq W$.

23. State the converse of Proposition 3.1.5 and then prove that it is false.

24. Let A and B be sets and let $f: A \to B$ be a function. Is the converse of Proposition 3.1.9 true? Specifically, if Im f is a finite set, does it follow that A is a finite set? Prove your answer.

Discussion and Discovery Exercises

D1. A real-valued function $f(x)$ is said to have a *local maximum* at a point c if there is an open interval (a, b) in the domain of f containing the point c such that $f(x) \leq f(c)$ for all x in (a, b).
   (a) Write the negation of this definition; that is, $f(x)$ does not have a local maximum at c if . . . .
   (b) A standard theorem of calculus states that if a real-valued function $f(x)$ has a local maximum at c, then either $f'(c) = 0$ or $f'(c)$ does not exist. The usual proof of this result starts out by first assuming that $f'(c)$ *does* exist and then proving that $f'(c) = 0$. Using appropriate statements P, Q, and R and rules of logic discussed in Chapter 1, explain why this is a valid method of proof.

D2. Look up the statement of the Mean Value Theorem in a calculus text.
   (a) Identify any quantifiers in the statement.
   (b) Write the negation of the Mean Value Theorem.

(c) State a theorem that is a consequence of the Mean Value Theorem. Explain why it is a consequence.

D3. (a) Write the negation of the Intermediate Value Theorem.
    (b) Give an important application of the Intermediate Value Theorem in calculus.

D4. Criticize the following "proof" that if $f: \mathbf{Z} \to \mathbf{Z}$ is defined by $f(n) = 2n$, then $f(\mathbf{O}) = \{y \in \mathbf{Z} \mid y = 4n + 2 \text{ for some } n \in \mathbf{Z}\}$.
    Let $x \in \mathbf{O}$. Then $x = 2n + 1$ for some $n \in \mathbf{Z}$. Then $f(x) = 2x = 4n + 2$. Therefore, $f(\mathbf{O}) = 4n + 2$.

D5. Prove that Dirichlet's function $d(x)$ defined at the end of this section is discontinuous at every real number c. If you are unfamiliar with the formal $\varepsilon - \delta$ definition of limit, then give an informal, intuitive proof.

D6. Give an example of a function $f: \mathbf{R} \to \mathbf{R}$ that is discontinuous at every integer but continuous at every other real number. Give at least an informal proof to justify your example. You may want to start by drawing a picture of such a function and then finding a formula for it.

3.2   SURJECTIVE AND INJECTIVE FUNCTIONS

Surjective Functions

In Section 3.1, we saw some examples of functions whose images equaled their codomains and we saw some whose images were proper subsets of their codomains. These examples suggest that we distinguish the notion of a function whose codomain and image coincide.

> **Definition 3.2.1** Let $f: A \to B$ be a function. Then f is **surjective** (or a **surjection**) if the image of f equals the codomain of f.

In the diagram following Definition 3.1.2, if $f: A \to B$ is surjective, the two ovals representing Im f and B would coincide. This leads many text authors (and many mathematicians) to use the word *onto* instead of surjective.

Proving that a function f is surjective, then, entails proving that two sets are equal, a scenario we've encountered before. But in this case, we know that the image is by definition a subset of the codomain, so it suffices to show that the codomain is a subset of the image: $\forall y \in B, \exists x \in A \ni f(x) = y$.

EXAMPLE 1 Let $A = \{1, 6, 12\}$, $B = \{0, 2, 12\}$. Define $f: A \to B$ by $f(1) = 2$, $f(6) = 0$, and $f(12) = 12$, as in Example 1 of Section 3.1. Since the image of f is $\{0, 2, 12\}$, the same as the codomain, f is surjective.

EXAMPLE 2   Let g be as defined in Example 2 of Section 3.1. Then Im $g = \{0\}$, so g is not surjective.

EXAMPLE 3   Let A be any set and let $i_A: A \rightarrow A$ be the identity function as defined in Example 3 of Section 3.1. It is easy to see that i_A is surjective because if a is in the codomain A, then $a = i_A(a)$, implying that a is in the image of i_A.

EXAMPLE 4   Let A and B be sets. Define $\pi_1: A \times B \rightarrow A$ by $\pi_1(a, b) = a$ and $\pi_2: A \times B \rightarrow B$ by $\pi_2(a, b) = b$. These functions are called **projections** and are both surjective. (See Exercise 4.) For example, if $A = B = \mathbf{R}$, then π_1 maps a point in the Cartesian plane to its x-coordinate.

EXAMPLE 5   Any nonconstant linear function $f: \mathbf{R} \rightarrow \mathbf{R}$ is surjective. Suppose $f(x) = ax + b$ where $a \neq 0$. To prove that f is surjective, we need to show that $\mathbf{R} \subseteq \text{Im } f$.

Let $y \in \mathbf{R}$. To show that y is in Im f, we need to find $x \in \mathbf{R}$ such that $f(x) = y$. To find x, we take the equation $y = f(x) = ax + b$ and solve for x in terms of y. We get $x = (y - b)/a$. Since $a \neq 0$, $(y - b)/a \in \mathbf{R}$ and $f((y - b)/a) = a((y - b)/a) + b = y - b + b = y$. Thus the x we needed to find has been found: $x = (y - b)/a$. This proves that f is surjective.

A reminder of the nature of this last proof: we were proving a statement that begins with the existential quantifier, $\exists$. Therefore, all we had to do was present the element we needed. *How* we found it (obviously we just solved a linear equation for it) might be interesting or illuminating, but was unnecessary for the proof.

EXAMPLE 6   Let $f: \mathbf{R} \rightarrow \mathbf{R}$ be defined by $f(x) = x^3$. This function is surjective. To see this, let $y \in \mathbf{R}$. Now every real number has a unique cube root. So we let $x = \sqrt[3]{y}$. Then

$$f(x) = x^3 = (\sqrt[3]{y})^3 = y.$$

Hence y is in the image of f. Since y was chosen arbitrarily, we have proven that every element of the codomain is in the image of f. It follows that f is surjective.

EXAMPLE 7   Consider the function $f: \mathbf{Z} \rightarrow \mathbf{Z}$ defined by

$$f(n) = \begin{cases} n + 2, & \text{if } n \text{ is even} \\ 2n + 1, & \text{if } n \text{ is odd.} \end{cases}$$

Note that f maps an even integer to an even integer and the odd integers get mapped to every other odd integer. So we surmise that f is *not* surjective.

To prove this formally, we need only find one element of $\mathbf{Z}$ that is not in the image of f. We claim that 5 is such an element.

Suppose, on the contrary, that $5 = f(n)$ for some $n \in \mathbf{Z}$. If n is even, then $f(n) = n + 2 = 5$. Hence $n = 3$, a contradiction to the fact that n is even. On the other hand, if n is odd, then $f(n) = 2n + 1 = 5$, which gives $n = 2$, again a contradiction since n is odd.

Thus 5 is not in the image of f and therefore f is not surjective.

EXAMPLE 8   Let $f: \mathbf{Z} \rightarrow \mathbf{Z}$ be defined by

$$f(n) = \begin{cases} n + 1, & \text{if } n \text{ is even} \\ n - 3, & \text{if } n \text{ is odd.} \end{cases}$$

We will prove that f is surjective. We need to show that every element of $\mathbf{Z}$ is in the image of f. Let $y \in \mathbf{Z}$. We want to find $n \in \mathbf{Z}$ such that $f(n) = y$. Given the definition of f, it seems reasonable to expect that n will vary depending on whether y is even or odd.

Suppose y is even. Since f maps even integers to odd integers and odd integers to even integers, the n we pick should be odd and from the definition of f, we can see that we should let $n = y + 3$. Then n is odd and $f(n) = f(y + 3) = y + 3 - 3 = y$. This proves that $y \in \text{Im } f$.

Now suppose that y is odd. Let $n = y - 1$. Then, since n is even, we get $f(n) = f(y - 1) = y - 1 + 1 = y$. Again we have $y \in \text{Im } f$. Therefore, f is surjective.

EXAMPLE 9   Let $f: \mathbf{Z} \rightarrow \mathbf{Z}$ be defined by

$$f(n) = \begin{cases} \dfrac{n}{2}, & \text{if } n \text{ is even} \\ 2n + 4, & \text{if } n \text{ is odd.} \end{cases}$$

As in the previous example, f is surjective; however (perhaps surprisingly) we do not have to divide the proof into even and odd cases. The key is to note that, whereas f maps the odd integers to some of the even integers, f maps the even integers onto all the integers. Just try some examples to convince yourself of that fact. Once you see the pattern, you will see how the proof should go.

Let $y \in \mathbf{Z}$. Then let $n = 2y$. Since n is even, we get $f(n) = f(2y) = 2y/2 = y$, proving that $y \in \text{Im } f$. We have proved that every element of $\mathbf{Z}$ is in the image of f and therefore f is surjective.

EXAMPLE 10   Consider the function $f: \mathbf{R} \rightarrow \mathbf{R}$ defined by $f(x) = x^3 + 4x + 1$. We saw in Example 15 of Section 3.1 that $\text{Im } f = \mathbf{R}$. It follows that f is surjective.

It is important to note the difference in the method of proof for this example as compared with the previous ones. In Examples 5 and 6, given $y \in \mathbf{R}$, we found an x in $\mathbf{R}$ such that $f(x) = y$ by taking the equation $y = f(x)$ and solving for x in terms of y. Essentially the same method was used in Examples 8 and 9. But for this example, getting x in terms of y requires solving a cubic equation, which is a difficult computation in general. So instead, we relied on the Intermediate Value Theorem to show that x exists without actually solving for x.

Injective Functions

Some functions have the property that two or more elements of the domain may be assigned to the same element of the codomain. In Example 2 of Section 3.1, we have $g(1) = g(6) = g(12) = 0$.

We wish to single out functions for which this is *not* the case; that is, functions for which no two different elements of the domain are assigned the same value of the codomain.

Definition 3.2.2 Let $f: A \to B$ be a function. Then f is **injective** (or an **injection**) if whenever a_1, $a_2 \in A$ and $a_1 \neq a_2$, we have $f(a_1) \neq f(a_2)$.

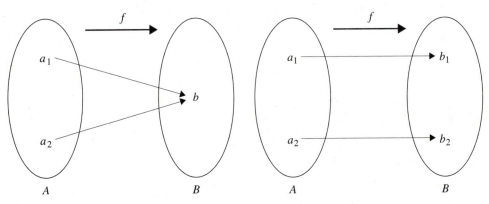

f is not an injection: f takes different elements of A to the same element of B.

f is an injection: f takes different elements of A to different elements of B. (It must do so for all pairs of distinct elements of A.)

The accompanying diagram suggests the phrase "one-to-one." As "onto" is used for surjective, this terminology is often used for injective.

EXAMPLE 11   Let $A = \{1, 6, 12\}$, $B = \{0, 2, 12\}$. Define $f: A \to B$ by $f(1) = 2$, $f(6) = 0$, and $f(12) = 12$, as defined in Example 1 of Section 3.1. Since no two elements of the domain are mapped to the same element of the codomain, f is injective.

EXAMPLE 12   Let g be as defined in Example 2 of Section 3.1. Then, since $g(1) = g(6) = g(12) = 0$, g is *not* injective.

Using the definition is not always the easiest way to prove a function f is injective. We'll try writing the definition another way.

Let $P(a_1, a_2)$ be the open sentence "$a_1 \neq a_2$" and $Q(a_1, a_2)$ be the open sentence "$f(a_1) \neq f(a_2)$." Then f is injective if

for all $a_1, a_2 \in A$, $P(a_1, a_2) \Rightarrow Q(a_1, a_2)$.

For any values of the variables a_1, a_2, $P(a_1, a_2) \Rightarrow Q(a_1, a_2)$ is equivalent to its contrapositive $\neg Q(a_1, a_2) \Rightarrow \neg P(a_1, a_2)$. $\neg Q(a_1, a_2)$ is the open sentence $f(a_1) = f(a_2)$ and $\neg P(a_1, a_2)$ is the open sentence $a_1 = a_2$. Hence we can say that f is injective if

for all $a_1, a_2 \in A$, whenever $f(a_1) = f(a_2)$, then $a_1 = a_2$.

A proof that a function f is injective usually should be done as follows. Let $a_1, a_2 \in A$, where A is the domain of the function, and assume $f(a_1) = f(a_2)$. Starting with this assumption, then *prove* that $a_1 = a_2$.

EXAMPLE 13 Let $f: \mathbf{R} \to \mathbf{R}$ be defined by $f(x) = x^3$. Suppose that $x_1, x_2 \in \mathbf{R}$ and $f(x_1) = f(x_2)$, that is, $x_1^3 = x_2^3$. Since every real number has a unique cube root, we conclude that $x_1 = \sqrt[3]{x_1^3} = \sqrt[3]{x_2^3} = x_2$, and so f is injective.

EXAMPLE 14 Any nonconstant, linear function $f: \mathbf{R} \to \mathbf{R}$ is injective. Let $f(x) = ax + b$. Suppose that $x_1, x_2 \in \mathbf{R}$ and $f(x_1) = f(x_2)$. Then $ax_1 + b = ax_2 + b$, and so $ax_1 = ax_2$. But since f is nonconstant, $a \neq 0$ and so $x_1 = x_2$, completing the proof.

EXAMPLE 15 Consider the function $f: \mathbf{Z} \to \mathbf{Z}$ defined by

$$f(n) = \begin{cases} n + 2, & \text{if } n \text{ is even} \\ 2n + 1, & \text{if } n \text{ is odd.} \end{cases}$$

To show that f is injective, suppose that $n_1, n_2 \in \mathbf{Z}$ and $f(n_1) = f(n_2)$. We consider three cases.

If n_1 and n_2 are both even, then $n_1 + 2 = n_2 + 2$, so $n_1 = n_2$.

If n_1 and n_2 are both odd, then $2n_1 + 1 = 2n_2 + 1$, and again $n_1 = n_2$.

Finally, if n_1 is even and n_2 is odd, then $f(n_1)$ is even and $f(n_2)$ is odd. Thus $f(n_1) \neq f(n_2)$, meaning that this case cannot even occur since we are assuming that $f(n_1) = f(n_2)$.

Therefore, in all cases for which $f(n_1) = f(n_2)$, we get $n_1 = n_2$, implying that f is injective.

EXAMPLE 16 Let A be any set and let $i_A: A \to A$ be the identity function. Then i_A is injective. (See Exercise 11.)

EXAMPLE 17 Let $f: \mathbf{R} \to \mathbf{R}$ be defined by $f(x) = x^3 + 4x + 1$, as in Example 10. To prove that f is injective using the method of the previous few examples is difficult because of the algebra involved. Instead we prove that if $x_1 \neq x_2$, then $f(x_1) \neq f(x_2)$.

Recall that in Example 15 of Sec. 3.1, we showed that f is an increasing function. From this fact it follows that if $x_1 < x_2$, then $f(x_1) < f(x_2)$ and if $x_1 > x_2$ then $f(x_1) > f(x_2)$. Hence f is injective.

To prove that a function is not injective, we must resort to the negation of the definition. As we noted previously, a function $f: A \to B$ is injective if for all $a_1, a_2 \in A$, $P(a_1, a_2) \Rightarrow Q(a_1, a_2)$ where $P(a_1, a_2)$ is the open sentence "$a_1 \neq a_2$" and $Q(a_1, a_2)$ is the open sentence "$f(a_1) \neq f(a_2)$." To negate this, we need a statement with an existential quantifier: $\exists a_1, a_2 \in A$ such that $P(a_1, a_2) \Rightarrow Q(a_1, a_2)$ is false. Since an implication, $P \Rightarrow Q$, is false only when P is true and Q is false, it follows that a function f is not injective if $\exists a_1, a_2 \in A$ such that $a_1 \neq a_2$ and $f(a_1) = f(a_2)$. Note that this formulation fits our understanding of what it means for a function to be "not injective"; namely, f maps two distinct elements of A to the same element of B.

EXAMPLE 18   The function $f: \mathbf{R} \to \mathbf{R}$ defined by $f(x) = x^2$ is *not* injective. We need only exhibit elements $a_1, a_2 \in \mathbf{R}$ such that $a_1 \neq a_2$ and $f(a_1) = f(a_2)$. This is easy since $f(-1) = 1 = f(1)$.

EXAMPLE 19   The function $f: \mathbf{Z} \to \mathbf{Z}$ defined by

$$f(n) = \begin{cases} n + 1, & \text{if } n \text{ is even} \\ 2n + 1, & \text{if } n \text{ is odd} \end{cases}$$

is not injective, since, for example, $f(1) = 3 = f(2)$.

Bijective Functions

As we've seen in the examples, a nonconstant linear function is both injective and surjective. Functions with this property are given a special name.

Definition 3.2.3   A function that is both injective and surjective is called **bijective** or a **bijection**.

EXAMPLE 20   Let $A = \{4, 6, 8, 10\}$ and $B = \{-1, -3, 0, 4\}$. Then the function $f: A \to B$ defined by $f(4) = 4$, $f(6) = -1$, $f(8) = 0$, $f(10) = -3$ is a bijection.

EXAMPLE 21   The function $f: \mathbf{R} \to \mathbf{R}$ defined by $f(x) = x^3$ is bijective since we have seen from Examples 6 and 13 that it is both injective and surjective.

EXAMPLE 22   Let A be any set and let $i_A: A \to A$ be the identity function. Then i_A is bijective.

EXAMPLE 23   Any nonconstant linear function is bijective by Examples 5 and 14.

EXAMPLE 24 Let $A = \{0, 1, 2\}$ and let $f: A \to A$ be defined by $f(0) = 2$, $f(1) = 0$, $f(2) = 1$. It is clear that f is a bijection. One can think of f as a rearrangement of the elements of A. Such a rearrangement is called a *permutation*.

More generally, we have the following definition.

> **Definition 3.2.4** Let A be any set. A bijection $f: A \to A$ is called a **permutation** of A.

EXAMPLE 25 Let $A = \{0, 1, 2\}$. In Example 24, we have one example of a permutation of A. It is easy to define others. In fact, there are 3! or six such permutations of A. (See the exercises.)

EXAMPLE 26 The function $f: \mathbf{R} \to \mathbf{R}$ defined by $f(x) = x^3$ is a permutation of $\mathbf{R}$.

⬤ MATHEMATICAL PERSPECTIVE: SETS OF PERMUTATIONS

As we saw at the end of this section, a permutation is the mathematical way of viewing a rearrangement. For example, if you place your CDs in a holder in some order (say alphabetically by artist and chronologically for each separate artist), you have in fact chosen a permutation of your collection. But how many different orders could you choose? That is, how many permutations are there?

In true mathematical fashion, let's ask the question in the abstract: how many permutations are there of the set $A = \{1, 2 \ldots, n\}$? We could begin by trying some easy, special cases. When $n = 2$, there are clearly just two permutations (remember that the identity function is a permutation). When $n = 3$, there are six:

$1 \to 1$	$1 \to 2$	$1 \to 3$	$1 \to 1$	$1 \to 2$	$1 \to 3$
$2 \to 2$	$2 \to 1$	$2 \to 2$	$2 \to 3$	$2 \to 3$	$2 \to 1$
$3 \to 3$	$3 \to 3$	$3 \to 1$	$3 \to 2$	$3 \to 1$	$3 \to 2$

With a little more work, you can see that there are 24 permutations of $\{1, 2, 3, 4\}$, and with a lot more work that there are 120 permutations of $\{1, 2, 3, 4, 5\}$. At this point, we can guess that, in general, there are $n! = 1 \cdot 2 \cdot 3 \ldots \cdot n$ permutations of $\{1, 2 \ldots, n\}$. (Don't forget: this is not a proof, just speculation based on a few examples. You'll be asked to rigorously prove this fact in the exercises for Section 5.2.)

One of the authors had a relative who was an inveterate player of the Massachusetts daily lottery. The winning number each day had four digits, and so was in the range 0000 ... 9999, and the player could bet that her number would come up either in exact order or in any order. Thus, an "exact" bet on 1234 had only one chance in ten thousand of winning, but an "any order" bet had 24 chances in ten thousand, because there are 24 permutations of 1234 (of course, such a win paid less). However, the chances of an exact bet on 1111 winning (1 in 10000) were the same as those for an any order bet: rearranging the digits doesn't change the number! For years, the relative mentioned bet $1 a day on her lucky number: 3331, any order. No amount of mathematical argument could convince her to improve her chances of winning by picking a number with no repeated digits. (What were her chances? In her defense, the odds are stacked against her either way.)

Both of these examples involve counting, and are simple instances of the mathematical field of *combinatorics*. In fact, an entire field can be said to have grown just from the study of permutations.

If you arrange your CD collection in a certain way, become dissatisfied with it because you can never remember where your favorite ones are, and so rearrange them, the net result is still a particular arrangement. That is, applying two permutations successively leads to another. For a more explicit example, consider doing this with the second and third permutations of $\{1, 2, 3\}$ in the previous table:

$$1 \to 2 \to 2$$
$$2 \to 1 \to 3$$
$$3 \to 3 \to 1$$

The net effect is the fifth permutation in the table. We are *composing* these two functions here; more about this in the next section.

Further, each permutation can be undone by composing it with another particular one; if you decide that your new CD arrangement is even worse, you can always return to the original one. Or

$$1 \to 2 \to 1$$
$$2 \to 3 \to 2$$
$$3 \to 1 \to 3$$

The "inverse" of the fifth permutation in the table is the sixth.

That these two properties hold (along with a third, called *associativity*) means that the collection of permutations of a set A forms what mathematicians call a *group*, and their study is the field of *group theory*. Groups arise in many contexts of mathematics and physics; they are fundamental in quantum mechanics, for example. Their properties can also be used to solve some popular puzzles, such as Rubik's Cube or the many incarnations of the 15 puzzle:

1	2	3	4
5	6	7	8
9	10	11	12
13	14	15	

(Notice that moving a tile amounts to a permutation of the 15 numbered tiles *and the blank space!*)

Exercises 3.2

---

1. Determine which of the following functions are surjective. Give a formal proof of your answer.
   (a) $f: \mathbf{R} \to \mathbf{R}$, $f(x) = 2x + 1$
   (b) $f: \mathbf{R} \to \mathbf{R}$, $f(x) = \cos x$
   (c) $f: \mathbf{R} \to [0, 1]$, $f(x) = \sin x$
   (d) $f: \mathbf{R}^* \to \mathbf{R}^*$, where $\mathbf{R}^*$ is the set of nonzero real numbers,
   $$f(x) = \frac{1}{x}$$
   (e) $f: \mathbf{R} \to \mathbf{R}$, $f(x) = x^4 + x^2$
   (f) $f: \mathbf{R} \to \mathbf{R}$, $f(x) = x^3 + x^2$
   (g) $f: \mathbf{R} \to \mathbf{R}^+$, $\mathbf{R}^+$ the set of positive real numbers, $f(x) = e^x$
   (h) $f: \mathbf{R} \to \mathbf{R}$, $f(x) = \begin{cases} 2x + 1, & \text{if } x \geq 1 \\ x^2, & \text{if } x < 1 \end{cases}$
   (i) $f: \mathbf{R} \to \mathbf{R}$, $f(x) = \begin{cases} -x - 1, & \text{if } x \geq 0 \\ x^2, & \text{if } x < 0 \end{cases}$
   (j) $f: \mathbf{Z} \to \mathbf{Z} \times \mathbf{Z}$, $f(n) = (n, n)$
   (k) $f: \mathbf{Z} \times \mathbf{Z} \to \mathbf{Z}$, $f(m, n) = m + n$
   (l) $f: \mathbf{R} \times \mathbf{R}^* \to \mathbf{R}$, $f(x, y) = \dfrac{x}{y}$

2. Determine which of the following functions are surjective. Give a formal proof of your answer. In each case, $f: \mathbf{Z} \to \mathbf{Z}$.
   (a) $f(n) = 2n + 1$
   (b) $f(n) = \begin{cases} n, & \text{if } n \text{ is even} \\ 2n - 1, & \text{if } n \text{ is odd} \end{cases}$
   (c) $f(n) = \begin{cases} n, & \text{if } n \text{ is even} \\ n - 1, & \text{if } n \text{ is odd} \end{cases}$
   (d) $f(n) = \begin{cases} n, & \text{if } n \text{ is even} \\ \dfrac{n - 1}{2}, & \text{if } n \text{ is odd} \end{cases}$
   (e) $f(n) = \begin{cases} n - 1, & \text{if } n \text{ is even} \\ 2n, & \text{if } n \text{ is odd} \end{cases}$

3. Let $f: \mathbf{Z} \to \mathbf{Z}$ be defined by
   $$f(n) = \begin{cases} n, & \text{if } n \text{ is even} \\ 2n + 3, & \text{if } n \text{ is odd.} \end{cases}$$

Is the following a correct proof that f is surjective? Explain.

Let $y \in \mathbf{Z}$. If y is even, then $f(y) = y$. If y is odd, then $f((y - 3)/2) = y$. In either case we get that y is in the image of f and therefore f is surjective.

4. Let A and B be sets. Let $\pi_1: A \times B \to A$ and $\pi_2: A \times B \to B$ be the projection functions defined in Example 4. Prove that π_1 and π_2 are surjective.

5. Let A and B be finite sets and let $f: A \to B$ be a function.
   (a) Prove that if f is surjective, then $|A| \geq |B|$.
       Hint: See Proposition 3.1.9 of Section 3.1.
   (b) State the contrapositive of part (a).
   (c) State the converse of part (a). Prove or disprove.

6. Let A and B be sets and let X be a subset of A. Let $f: A \to B$ be a surjective function.
   (a) Prove that $B - f(X) \subseteq f(A - X)$.
   (b) Give an example to show that equality does not in general hold in part (a).

7. Let $f: A \to B$ be a function. Prove that f is surjective if and only if $f^{-1}(W) \neq \emptyset$ for all nonempty subsets W of B.

8. Let $f: A \to B$ be a function. Which of the following statements are equivalent to the statement "f is surjective"?
   (a) For all $b \in B$, $\exists a \in A \ni f(a) = b$.
   (b) If $a \in A$, $f(a) = b \in B$.
   (c) $B \subseteq \operatorname{Im} f$.
   (d) $\exists b \in B \ni f(a) = b$ for some $a \in A$.

9. Write out two injective functions with domain $\{0, 1, 2, 3, 4\}$ and codomain $\{5, 6, 7, 8, 9, 10\}$.

10. Explain why you cannot define an injective function with domain $\{0, 1, 2, 3, 4\}$ and codomain $\{5, 6, 7, 8\}$.

11. Let A be any set and let $i_A: A \to A$ be the identity function. Prove that i_A is injective.

12. Determine which of the following functions are injective. Give a formal proof of your answer.
    (a) $f: \mathbf{R} \to \mathbf{R}$, $f(x) = 2x + 1$
    (b) $f: \mathbf{R} \to \mathbf{R}$, $f(x) = \cos x$
    (c) $f: \left[-\dfrac{\pi}{2}, \dfrac{\pi}{2} \right] \to \mathbf{R}$, $f(x) = \sin x$

(d) $f: \mathbf{R}^* \to \mathbf{R}^*$, where $\mathbf{R}^*$ is the set of nonzero real numbers,
$$f(x) = \frac{1}{x}$$
(e) $f: \mathbf{R} \to \mathbf{R}^+$, $\mathbf{R}^+$ the set of positive real numbers, $f(x) = e^x$
(f) $f: \mathbf{Z} \to \mathbf{Z} \times \mathbf{Z}$, $f(n) = (n, n)$
(g) $f: \mathbf{Z} \times \mathbf{Z} \to \mathbf{Z}$, $f(m, n) = m + n$
(h) $f: \mathbf{R} \times \mathbf{R}^* \to \mathbf{R}$, $f(x, y) = \dfrac{x}{y}$

13. Determine which of the following functions are injective. Give a formal proof of your answer. In each case, $f: \mathbf{Z} \to \mathbf{Z}$.
   (a) $f(n) = 2n + 1$
   (b) $f(n) = \begin{cases} n, & \text{if } n \text{ is even} \\ 2n - 1, & \text{if } n \text{ is odd} \end{cases}$
   (c) $f(n) = \begin{cases} n, & \text{if } n \text{ is even} \\ n - 1, & \text{if } n \text{ is odd} \end{cases}$
   (d) $f(n) = \begin{cases} n, & \text{if } n \text{ is even} \\ \dfrac{n - 1}{2}, & \text{if } n \text{ is odd} \end{cases}$
   (e) $f(n) = \begin{cases} n - 1, & \text{if } n \text{ is even} \\ 2n, & \text{if } n \text{ is odd} \end{cases}$

14. Determine which of the following functions are injective. Give a formal proof of your answer. In each case, $f: \mathbf{R} \to \mathbf{R}$.
   (a) $f(x) = x^4 + x^2$
   (b) $f(x) = x^3 + x^2$
   (c) $f(x) = -x^3 - x$
   (d) $f(x) = \begin{cases} 2x + 1, & \text{if } x \geq 1 \\ x^2, & \text{if } x < 1 \end{cases}$
   (e) $f(x) = \begin{cases} -x - 1, & \text{if } x \geq 0 \\ x^2, & \text{if } x < 0 \end{cases}$

15. Let A and B be sets. Let $\pi_1: A \times B \to A$ and $\pi_2: A \times B \to B$ be the projection functions defined in Example 4. Prove that π_1 is not injective unless B consists of only one element and that π_2 is not injective unless A has only one element.

16. Let A and B be sets and X and Y be subsets of A. Let $f: A \to B$ be an injective function. Prove that if $f(X) \subseteq f(Y)$, then $X \subseteq Y$. (See Proposition 3.1.5.)

17. Let A and B be sets and X and Y subsets of A. Let $f: A \to B$ be an injective function. Prove that $f(X \cap Y) = f(X) \cap f(Y)$. (See Exercise 15 of Section 3.1.)

18. Let A and B be sets and X and Y be subsets of A. Let $f: A \rightarrow B$ be an injective function. Prove that $f(X) - f(Y) = f(X - Y)$. (See Exercise 16 of Section 3.1.)

19. Let A and B be finite sets and let $f: A \rightarrow B$ be a function.
    (a) Prove that if f is injective, then $|\text{Im } f| = |A|$.
        Hint: See Proposition 3.1.9 of Section 3.1.
    (b) Is the converse of part (a) true? Prove or disprove.
    (c) Prove that if f is injective, then $|A| \leq |B|$.
    (d) State the contrapositive of part (c). Give a proof of the contrapositive that doesn't use part (c) by invoking the Pigeonhole Principle.
    (e) State the converse of part (c). Prove or disprove.

20. Let $f: A \rightarrow B$ be a function. Which of the following statements are equivalent to the statement "f is injective"?
    (a) $f(a) = f(b)$ when $a = b$.
    (b) $f(a) = f(b)$ and $a = b$ for all a, b in A.
    (c) If a and b are in A and $f(a) = f(b)$, then $a = b$.
    (d) If a and b are in A and $a = b$, then $f(a) = f(b)$.
    (e) If a and b are in A and $f(a) \neq f(b)$, then $a \neq b$.
    (f) If a and b are in A and $a \neq b$, then $f(a) \neq f(b)$.

21. Let $f: A \rightarrow B$ be an injective function. Let X be a subset of A. Prove that $f^{-1}(f(X)) = X$. (See Exercise 21 of Section 3.1.)

22. Let $f: A \rightarrow B$ be a surjective function. Let W be a subset of B. Prove that $f(f^{-1}(W)) = W$. (See Exercise 22 of Section 3.1.)

23. Write out all permutations of the set $A = \{0, 1, 2\}$.

24. (a) Write out three permutations of $\mathbf{R}$.
    (b) Write out three permutations of $\mathbf{Z}$.

25. Using only the definitions of injective and surjective, complete the following statements:
    (a) A function $f: A \rightarrow B$ is not injective if and only if . . . .
    (b) A function $f: A \rightarrow B$ is not surjective if and only if . . . .
    (c) A function $f: A \rightarrow B$ is not bijective if and only if . . . .

26. Let $f: A \rightarrow B$ be a function where A and B are finite sets such that $|A| = |B|$. Prove that f is injective $\Leftrightarrow f$ is surjective.

27. Let A and B be finite sets and let $f: A \rightarrow B$ be a function.
    (a) Prove that if f is bijective, then $|A| = |B|$.
    (b) State the contrapositive of part (a).
    (c) State the converse of part (a). Prove or disprove.

28. Let A and B be sets and let X be a subset of A. Let $f: A \rightarrow B$ be a bijection. Prove that $f(A - X) = B - f(X)$. (See Exercise 6.)

29. Let A and B be sets and let $f: A \rightarrow B$ be a function. Prove that f is bijective if and only if for every $b \in B, f^{-1}(\{b\})$ is a single element subset of A.

30. Let A and B be sets and let $f: A \rightarrow B$ be a surjective function. For each $b \in B$, let $A_b = f^{-1}(\{b\})$. Prove that the collection of sets $\{A_b \mid b \in B\}$ is a partition of A.

Discussion and Discovery Exercises

D1. Let P and Q be statements. Suppose that one wanted to prove the statement: P and Q are not both true. Explain why a valid proof of this statement would be to prove $P \Rightarrow \neg Q$.

D2. Using a first-year calculus text, look up the proof that says that if a real valued function f is differentiable at a number a, then f is continuous at a. Write out a careful, detailed explanation of this proof, listing all definitions and prior theorems that are used in the proof.

D3. Use the previous two problems to prove that a function f cannot be both differentiable at 0 and discontinuous at 0. Include in your proof the relevant statements P and Q and explain how you use the logical principle of Exercise D1.

D4. Using your knowledge of first-year calculus, explain why there cannot be an injective function $f: \mathbf{R} \rightarrow \mathbf{R}$ such that $f(x) = \ln x$ for all $x > 0$. (ln is the natural logarithm function.)

D5. A real-valued function f defined on a closed interval $[a, b]$ is said to be **decreasing** on $[a, b]$ if for all $x_1, x_2 \in [a, b]$, if $x_1 < x_2$, then $f(x_1) > f(x_2)$.
   (a) Negate this definition by completing the following sentence: f is not decreasing on $[a, b]$ if . . . .
   (b) Prove that every decreasing function on a closed interval $[a, b]$ is injective.

D6. In light of Example 17 of this section, state and prove a theorem that relates how the derivative of a function $f: \mathbf{R} \rightarrow \mathbf{R}$ can be used to determine if f is injective on an interval $[a, b]$. Be as general as possible.

D7. Let $f: \mathbf{R} \rightarrow \mathbf{R}$ be defined by $f(x) = 2x + 1$. Criticize the following "proof" that f is surjective.

   Let $y \in \mathbf{R}$. $y = 2x + 1$, $x = \dfrac{y - 1}{2}$.

   Therefore,

   $$f(x) = f\left(\frac{y-1}{2}\right) = 2\frac{y-1}{2} = y.$$

   Hence $y \subseteq \operatorname{Im} f$.

3.3   COMPOSITION AND INVERTIBLE FUNCTIONS

The concept of function is supremely important and is fundamental to virtually all of mathematics. We often step back from viewing functions as "rules" relating objects from different sets and study functions as objects themselves. In this way, we enrich other concepts; we can, for example, consider sets of functions, and devise ways to combine functions and create others.

> **Definition 3.3.1** Let A and B be nonempty sets. We define $\mathbf{F}(A, B) = \{f \mid f: A \to B\}$, the set of all functions from A to B. If $A = B$, we simply write $\mathbf{F}(A)$.

As we go along, we will study these sets in many different contexts, and find them interesting sources of unusual examples. For now, notice that we can abbreviate the statement that $f: A \to B$ is a function to $f \in \mathbf{F}(A, B)$.

Composition of Functions

One of the most important ways of creating new functions in calculus is by taking the composition of two functions. We can define composition of functions in general as long as the domains and codomains match up appropriately.

> **Definition 3.3.2** Let A, B, and C be nonempty sets, and let $f \in \mathbf{F}(A, B), g \in \mathbf{F}(B, C)$. We define a function $gf \in \mathbf{F}(A, C)$, called the **composition** of f and g, by $gf(a) = g(f(a))$, $\forall a \in A$.

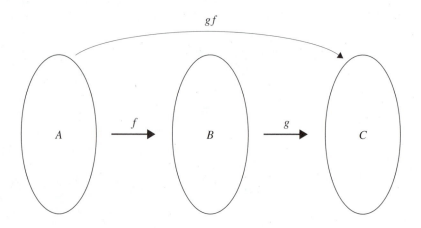

Note: We will use the simple juxtaposition *gf* to denote composition of functions instead of the more standard *g* ∘ *f*.

Notice that if $f, g \in \mathbf{F}(A)$, then both *gf* and *fg* are defined.

EXAMPLE 1 Let $A = \{0, 1\}$, $B = \{\text{red, green, blue}\}$, and $C = \{\pi, e\}$. Define a function $f: A \to B$ by $f(0) = $ red and $f(1) = $ blue; also define a function $g: B \to C$ by $g(\text{red}) = \pi$, $g(\text{green}) = e$, and $g(\text{blue}) = e$.

Then $gf(0) = \pi$ and $gf(1) = e$.

Notice that the function *fg* (composing in the reverse order) is not defined; this is because the codomain of *g* is not the domain of *f*. In simpler terms, *f*, for example, does not assign a value to $\pi = g(\text{red})$.

EXAMPLE 2 Let $f, g \in \mathbf{F}(\mathbf{R})$ be defined by $f(x) = x^2$ and $g(x) = x + 1$. Then $gf(x) = g(x^2) = x^2 + 1$.

Here we can consider the function *fg* as well, and we see that $fg(x) = f(x + 1) = (x + 1)^2$. Thus *fg* and *gf* do exist, but $fg \neq gf$!

In this example, you should clearly understand why we concluded that $fg \neq gf$. Recall that two functions in $\mathbf{F}(\mathbf{R})$ are equal if and only if they take the same values at *all* $x \in \mathbf{R}$; the statement has a universal quantifier. Thus they are *not* equal if we can find *at least one* x for which they do not agree. But $gf(1) = 1^2 + 1 = 2$ and $fg(1) = (1 + 1)^2 = 4$.

EXAMPLE 3 Let $f, g \in \mathbf{F}(\mathbf{Z})$ be defined by

$$f(n) = \begin{cases} n + 2, & \text{if } n \text{ is even} \\ 2n + 1, & \text{if } n \text{ is odd} \end{cases} \quad \text{and} \quad g(n) = \begin{cases} 2n, & \text{if } n \text{ is even} \\ \dfrac{n + 1}{2}, & \text{if } n \text{ is odd.} \end{cases}$$

If *n* is even, $gf(n) = g(n + 2) = 2(n + 2)$ since $n + 2$ is even. If *n* is odd, $gf(n) = g(2n + 1) = (2n + 1 + 1)/2 = n + 1$. Thus we have

$$gf(n) = \begin{cases} 2n + 4, & \text{if } n \text{ even} \\ n + 1, & \text{if } n \text{ is odd.} \end{cases}$$

Now we compute the composition *fg*. If *n* is even, $fg(n) = f(2n) = 2n + 2$. If *n* is odd, $fg(n) = f((n + 1)/2)$. The value of $f((n + 1)/2)$ depends on whether $(n + 1)/2$ is even or odd.

If $(n + 1)/2$ is even, then $(n + 1)/2 = 2t$ for some integer *t* or equivalently $n = 4t - 1$. In this case,

$$f\left(\frac{n + 1}{2}\right) = \frac{n + 1}{2} + 2 = \frac{n + 5}{2}.$$

If $(n + 1)/2$ is odd, then $(n + 1)/2 = 2t + 1$ for some integer t or equivalently $n = 4t + 1$. We get

$$f\left(\frac{n+1}{2}\right) = 2\left(\frac{n+1}{2}\right) + 1 = n + 2.$$

Thus

$$fg(n) = \begin{cases} 2n + 2, & \text{if } n \text{ is even} \\ \dfrac{n+5}{2}, & \text{if } n = 4t - 1 \text{ for some } t \in \mathbf{Z} \\ n + 2, & \text{if } n = 4t + 1 \text{ for some } t \in \mathbf{Z}. \end{cases}$$

In computing composition of functions or computing any function for that matter, take care to differentiate between the symbol for the function and the value of the function at an element of its domain. In the last example, we computed the composition gf by finding $gf(n)$ for any n in $\mathbf{Z}$. In doing the computation of gf it is incorrect to write $gf = 2n + 4$ when n is even since gf and $2n + 4$ do not belong to the same set. gf is a function and $2n + 4$ is an integer. The expression $gf(n)$, however, is the value of the function gf at n and therefore is an integer. So the correct expression is $gf(n) = 2n + 4$.

Let A be a set. Then the identity function i_A has a special property with respect to composition.

Proposition 3.3.3   Let $f: A \to B$ be a function. Then $f\, i_A = f$ and $i_B f = f$.

PROOF: The proof is left as an exercise.

A question one might ask is "Which properties of f and g carry over to the composition gf?"

Proposition 3.3.4   Let $f \in \mathbf{F}(A, B)$ and $g \in \mathbf{F}(B, C)$.

1. If f and g are surjections, then gf is also a surjection.
2. If f and g are injections, then gf is also an injection.
3. If f and g are bijections, then gf is also a bijection.

PROOF: 1. Suppose that f and g are surjective functions. Since gf is a function from A to C, to show that gf is surjective, it is necessary to take an arbitrary element c in C and show that it is in the image of gf; that is, given c in C, we must show that there is an element a in A such that $gf(a) = c$.

So let $c \in C$. Then since g is surjective, there exists $b \in B$ such that $g(b) = c$. By the same token, since f is surjective and $b \in B$, there exists $a \in A$ such that $f(a) = b$. But then $gf(a) = g(b) = c$, proving that $c \in \text{Im } gf$. Therefore, gf is surjective.

2. Suppose that f and g are injective functions. To prove that gf is injective, we must take arbitrary elements a_1, $a_2 \in A$, and assuming that $gf(a_1) = gf(a_2)$, we must then show that $a_1 = a_2$.

So let a_1, $a_2 \in A$ and suppose that $gf(a_1) = gf(a_2)$. Now $gf(a_1) = gf(a_2)$ means that $g(f(a_1)) = g(f(a_2))$ and since g is injective it follows that $f(a_1) = f(a_2)$. But then we can conclude that $a_1 = a_2$, since f is also injective. Hence gf is injective.

3. This follows immediately from 1 and 2. ●

Corollary 3.3.5 Let A be a nonempty set. Let f and g be two permutations of A. Then gf is a permutation of A.

PROOF: The proof is left as an exercise. ●

You may wonder whether the converses of the statements in Proposition 3.3.4 are also true. It seems reasonable, for example, that if gf is bijective, then f and g must be as well; however, it's false, as you will see in the exercises.

Composition of functions shares a property with the operations of addition and multiplication of real numbers: associativity. The next proposition gives a proof of this result.

Proposition 3.3.6 Let A, B, C, and D be nonempty sets. Let $f \in \mathbf{F}(A, B)$, $g \in \mathbf{F}(B, C)$, $h \in \mathbf{F}(C, D)$. Then $(hg)f = h(gf)$.

PROOF: Recall that two functions are equal if they have the same domain and codomain and if they assign each element of the domain to the same element of the codomain. In this case, both $(hg)f$ and $h(gf)$ have domain A and codomain D.

Now let $a \in A$. Then

$$[(hg)f](a) = (hg)(f(a)) = h(g(f(a)))$$

and

$$[h(gf)](a) = h(gf(a)) = h(g(f(a))).$$

Thus $[(hg)f](a) = [h(gf)](a)$ for all $a \in A$. It now follows that $(hg)f = h(gf)$. ●

Because of the previous proposition, we can delete the parentheses and simply write hgf for the composition of three functions without ambiguity.

The above proposition can be extended to four or more functions so that if we have n functions $f_1, f_2, \ldots f_n$ for any positive integer n we can write $f_1 f_2 \ldots f_n$ for their composition without need of parentheses, provided of course that these compositions are all defined.

Inverse Functions

You are probably familiar with the notion of an inverse function from calculus. What the inverse of a function f does, if it can be defined, is map an element $f(a)$ of the image of f back to a. For example, if $f(x) = x^2$, then the inverse of f, if it exists, should map $4 = f(2)$ to 2. In general, it should send $f(a) = a^2$ back to a and if $a \geq 0$, the function that will do this is the square root function. (See Example 7 for a more detailed discussion of this example.)

Therefore, the inverse of a function f, when composed with f, should be the same as the identity function, which maps every element to itself.

So let's try to define the inverse of a function $f: A \to B$ where A and B are any sets. Since f sends an element of A to an element of B, the inverse function must map elements of B to A. More specifically, if f maps a in A to b in B, then the inverse of f, which we will denote f^{-1}, must map b to a.

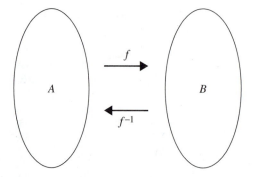

Thus the composition of f and its inverse must be the identity mapping of the appropriate set.

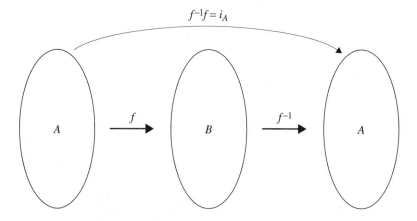

This leads to the following definition.

> **Definition 3.3.7** Let A and B be sets, and let $f \in \mathbf{F}(A, B)$. Then f is **invertible** if there is a function $f^{-1} \in \mathbf{F}(B, A)$ such that $ff^{-1} = i_B$ and $f^{-1}f = i_A$. If f^{-1} exists, it is called the **inverse** of f.

Note that the symmetry of this definition implies that if f is invertible, then its inverse function f^{-1} is also invertible and that the inverse of f^{-1} is f.

Note also that we have to justify calling f^{-1} *the* inverse of f. To see that there is only one function that can be the inverse of a function f, suppose there exists another function $g: B \to A$ such that $gf = i_A$ and $fg = i_B$. Then, using Propositions 3.3.3 and 3.3.6 as well as the definition of invertible function, we get $g = gi_B = g(ff^{-1}) = (gf)f^{-1} = i_A f^{-1} = f^{-1}$.

Thus f^{-1} is unique.

EXAMPLE 4　Let $f \in \mathbf{F}(\mathbf{R})$ be defined by $f(x) = x^3$. Then f is invertible and $f^{-1}(x) = \sqrt[3]{x}$.

EXAMPLE 5　Define $f \in \mathbf{F}(\{1, 2\}, \{1\})$ by $f(1) = f(2) = 1$. (It's the only such function, really.) Then f is not invertible because there are only two functions in $\mathbf{F}(\{1\}, \{1, 2\})$, and neither one satisfies the definition of inverse.

EXAMPLE 6　Let A be any set and let $i_A: A \to A$ be the identity function. Then i_A is invertible and $i_A^{-1} = i_A$.

EXAMPLE 7　Let $f \in \mathbf{F}(\mathbf{R})$ be defined by $f(x) = x^2$. As previously noted, the inverse of this function, if it exists, must be the function $f^{-1}(x) = \sqrt{x}$. But we run into a couple of problems with this example.

First of all, because $\mathbf{R}$ is the domain and codomain of f, it should also be the domain and codomain of f^{-1}. But $\sqrt{x}$ is not defined for all real numbers, only for nonnegative ones. This problem can be fixed up easily by just redefining the codomain of f to coincide with its image, namely, the set $[0, \infty)$ of nonnegative real numbers. In effect, we have made f a surjective function.

The second problem is that since $f(2) = f(-2) = 4$, f^{-1} must map 4 to both 2 and -2. But no function can send an element of its domain to more than one element of its codomain by the very definition of function. The problem is that f is not injective. So our function f, even with the new codomain, is not invertible because it is not injective.

We can, however, make f injective by restricting its domain to the set $[0, \infty)$. We now have a new function $f \in \mathbf{F}([0, \infty))$ defined by $f(x) = x^2$, which is injective and surjective and its inverse function f^{-1} is in $\mathbf{F}([0, \infty))$ and is defined by $f^{-1}(x) = \sqrt{x}$.

In light of the last example, we should expect that an invertible function must be injective and surjective. The converse is also true. We prove this in the following theorem.

Theorem 3.3.8   Let A and B be sets, and let $f \in \mathbf{F}(A, B)$. Then f is invertible if and only if f is bijective.

PROOF: Suppose that f is invertible, so that f^{-1} exists. We need to show that f is both injective and surjective.

To show that f is injective, suppose that $a_1, a_2 \in A$ and $f(a_1) = f(a_2)$. Applying f^{-1} to both sides of this equation gives $f^{-1}f(a_1) = f^{-1}f(a_2)$. Since $f^{-1}f = i_A$, we get $i_A(a_1) = i_A(a_2)$ and this implies that $a_1 = a_2$. Hence f is injective.

To see that f is also surjective, we need to show that B, the codomain of f, equals the image of f. Since the image is by definition a subset of the codomain, we only have to show that every element of B is in the image of f.

Let $b \in B$. If we let $a = f^{-1}(b)$, we get $f(a) = f(f^{-1}(b)) = ff^{-1}(b) = i_B(b) = b$. This proves that b is in the image of f and therefore f is surjective.

Conversely, suppose that f is bijective. To prove that f is invertible, we must define a function f^{-1} such that $ff^{-1} = i_B$ and $f^{-1}f = i_A$. We start by letting $b \in B$. Since f is surjective, we can find $a \in A$ such that $f(a) = b$; since f is injective, only one such a exists. Define $f^{-1}(b) = a$.

Then $ff^{-1}(b) = f(a) = b$, so $ff^{-1} = i_B$.

Now let $a \in A$ and let $b = f(a)$. Then by the definition of f^{-1}, $f^{-1}(b) = a$. Consequently, $f^{-1}f(a) = f^{-1}(b) = a$. Therefore, $f^{-1}f = i_A$.

Thus f is invertible. ●

EXAMPLE 8   We saw in Example 23 of Section 3.2 that any nonconstant linear function is bijective. It follows then from Theorem 3.3.8 that any nonconstant linear function is invertible. But how do we find its inverse? Let $f(x) = ax + b$, $a \neq 0$. To find f^{-1}, we merely let $y = ax + b$ and solve for x in terms of y. We get $f^{-1}(y) = (y - b)/a$. The reader should verify that this is indeed the inverse of f.

EXAMPLE 9   The function $f: [-\pi/2, \pi/2] \to [-1, 1]$ defined by $f(x) = \sin x$ is bijective and therefore invertible. Its inverse is the arcsine function. Note that we restricted the domain of f to the interval $[-\pi/2, \pi/2]$ to make it injective and used the image of the sine function as the codomain to make it surjective.

Sometimes, when a function f is invertible, it may be very difficult or impossible to write out a formula for f^{-1}. Nevertheless, thanks to Theorem 3.3.8, it is possible to determine that f^{-1} *exists* even if we cannot find a specific formula for it.

EXAMPLE 10   Let $f: \mathbf{R} \to \mathbf{R}$ be defined by $f(x) = x^3 + 4x + 1$. In Section 3.2, we showed that f is bijective. (See Examples 10 and 17 of Section 3.2.) Then by Theorem 3.3.8, f is invertible.

MATHEMATICAL PERSPECTIVE: MULTI-VALUED FUNCTIONS

You might remember from your calculus course that many of the functions studied there do not have inverses, because they fail the "horizontal line test." For example, the sine function has no inverse because, in particular, the horizontal line $y = 1$ hits the graph at every peak.

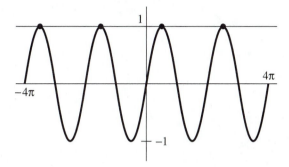

This means that if we try to construct the inverse, we cannot decide how to define $\sin^{-1}(1)$. One choice is to restrict the domain of the sine to $[-\pi/2, \pi/2]$, as we did in Example 9. However, we can try to draw the graph of the inverse by simply reflecting the above graph about the line $y = x$.

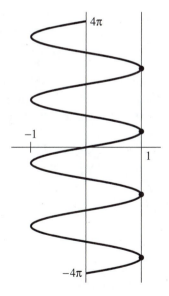

The resulting graph does not represent a function because, of course, it fails the vertical line test: there are now many y-values corresponding to $x = 1$. Such graphs are said to represent *multi-valued functions*. They are studied not by restricting the domain, but instead by extending it. That is, we would "paste" together copies of the interval $[-1, 1]$ so that each y-value corresponds to a single x-value in only one of the copies.

This ingenious (and hard to visualize) idea was formulated by **Bernhard Riemann** (1826–1866), and grew from his study of functions of a complex variable (see Section 7.3). The extended domain is called the *Riemann Surface* of the multi-valued function. For example, when the natural logarithm function is extended to the complex numbers in the most appropriate way, it leads to a multi-valued function whose Riemann surface can be visualized as a spiral:

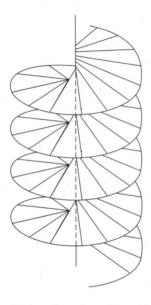

Remember that this picture represents the domain of the function; a graph of the function itself would require four dimensions.

Exercises 3.3

1. Let $f(x) = x^2 + 1$, $g(x) = \sin x$, and $h(x) = e^{2x}$ in $\mathbf{F}(\mathbf{R})$.
   (a) Compute fg and gf.
   (b) Compute fh and hf.
   (c) Compute gh and hg.

2. For the following functions $f, g \in \mathbf{F}(\mathbf{Z})$, compute fg and gf.

   (a) $f(n) = 2n$ and $g(n) = \begin{cases} n, & \text{if } n \text{ is even} \\ 2n - 1, & \text{if } n \text{ is odd.} \end{cases}$

   (b) $f(n) = \begin{cases} \dfrac{n}{2}, & \text{if } n \text{ is even} \\ n+1, & \text{if } n \text{ is odd} \end{cases}$ and $g(n) = \begin{cases} n-1, & \text{if } n \text{ is even} \\ 2n, & \text{if } n \text{ is odd.} \end{cases}$

3. Express each of the following functions as a composition $f = gh$. Be sure to give appropriate sets A, B, and C so that $h \in \mathbf{F}(A, B)$ and $g \in \mathbf{F}(B, C)$. (Neither g nor h should be an identity function, but there may be many possible answers.)

   (a) $f \in \mathbf{F}(\mathbf{R})$ defined by $f(x) = (3x^2 - 11)^8$

   (b) $f \in \mathbf{F}(\mathbf{R})$ defined by $f(x) = \cos\left(\dfrac{x+1}{4}\right)$

   (c) $f \in \mathbf{F}(\mathbf{Z}, \mathbf{R})$ defined by $f(x) = \cos\left(\dfrac{x+1}{4}\right)$

   (d) $f \in \mathbf{F}(\{0, 1\}, \mathbf{R})$ defined by $f(0) = 1$ and $f(1) = 0$

   (e) $f \in \mathbf{F}(\mathbf{R}, \mathbf{Z}^+)$ defined by $f(x) = \|[x]\|$

4. Prove Proposition 3.3.3.

5. Prove Corollary 3.3.5.

6. State the converse of each part of Proposition 3.3.4. Prove or disprove each one.

7. Let A, B, and C be nonempty sets, and let $f \in \mathbf{F}(A, B)$ and $g \in \mathbf{F}(B, C)$.

   (a) Prove that if gf is injective, then f is injective.

   (b) Prove that if gf is surjective, then g is surjective.

8. Let A and B be nonempty sets. Let $f \in \mathbf{F}(A, B)$ and $g \in \mathbf{F}(B, A)$. Prove that if gf and fg are bijective, then so are f and g.

   Hint: Use the previous exercise.

9. Let $f: A \to B$ and suppose that $S \subseteq A$, $S \neq \varnothing$. We can define a function $f_{|S}: S \to B$, called **f restricted to S**, by simply taking $f_{|S}(a) = f(a)$, $\forall a \in S$.

   (a) If f is injective, prove that $f_{|S}$ is injective.

   (b) Prove or disprove the converse of part (a).

   (c) Give an example to show that if f is surjective, then $f_{|S}$ need not be surjective.

   (d) Give an example in which f is surjective, S is a proper subset of A, and $f_{|S}$ is also surjective.

   (e) If $f \in \mathbf{F}(A)$ and $S \subseteq A$, what can you say about $ff_{|S}$ and $f_{|S}f$?

10. Determine which of the following functions are invertible. If a function f is invertible, find f^{-1}.

    (a) $f: \{-1, 0, 1\} \to \{\pi, e, \sqrt{2}\}$ defined by $f(-1) = e$, $f(0) = e$, and $f(1) = \sqrt{2}$

    (b) $f: \{-1, 0, 1\} \to \{\pi, e, \sqrt{2}\}$ defined by $f(-1) = e$, $f(0) = \pi$, and $f(1) = \sqrt{2}$

    (c) $f: \mathbf{Z} \to \mathbf{Z}$ defined by $f(x) = 7x + 3$

    (d) $f: \mathbf{R} \to \mathbf{R}$ defined by $f(x) = 7x + 3$

    (e) $f: \mathbf{Z} \to \mathbf{Z}$ defined by $f(n) = 2n$

    (f) $f: \mathbf{Z} \to \mathbf{E}$, $\mathbf{E}$ the set of even integers, defined by $f(n) = 2n$

    (g) $f: \mathbf{R} \to \mathbf{R}$ defined by $f(x) = \cos x$

(h) $f: [0, \pi] \to [-1, 1]$ defined by $f(x) = \cos x$

(i) $f: \mathbf{R} \to \mathbf{R}$ defined by $f(x) = e^x$

(j) $f: \mathbf{R} \to \{x \in \mathbf{R} \mid x > 0\}$ defined by $f(x) = e^x$

(k) $f: \mathbf{Z} \to \mathbf{Z} \times \{0, 1\}$ defined by $f(n) = \begin{cases} \left(\dfrac{n}{2}, 0\right), & \text{if } n \in \mathbf{E} \\[2mm] \left(\dfrac{n-1}{2}, 1\right), & \text{if } n \in \mathbf{O} \end{cases}$

(l) $f: \mathbf{Z} \to \mathbf{Z}^+$ defined by $f(n) = \begin{cases} 4n, & \text{if } n > 0 \quad \text{and} \quad n \in \mathbf{E} \\ 4|n| + 1, & \text{if } n \le 0 \quad \text{and} \quad n \in \mathbf{E} \\ 4n + 2, & \text{if } n > 0 \quad \text{and} \quad n \in \mathbf{O} \\ 4|n| + 3, & \text{if } n < 0 \quad \text{and} \quad n \in \mathbf{O} \end{cases}$

11. Define $f: \mathbf{R} - \{1\} \to \mathbf{R} - \{1\}$ by $f(x) = \dfrac{x + 1}{x - 1}$.

    (a) Prove that f is injective.

    (b) Prove that f is surjective.

    (c) Find $f^{-1}(x)$.

12. Let $a, b, c, d \in \mathbf{R}$ such that $ad - bc \ne 0$ and $c \ne 0$. Define $f: \mathbf{R} - \{-d/c\} \to \mathbf{R} - \{a/c\}$ by $f(x) = \dfrac{ax + b}{cx + d}$.

    (a) Prove that f is injective.

    (b) Prove that f is surjective.

    (c) Find $f^{-1}(x)$.

13. Let $A = \{a, b, c\}$, a set with three elements. Let $S(A)$ be the set of all permutations of A.

    (a) Write out the six elements of $S(A)$.

    (b) Compute the inverses of each element of $S(A)$.

    Note: Each inverse is also an element of $S(A)$.

14. Let $f \in \mathbf{F}(A, B)$. Prove that if f is bijective, so is f^{-1}.

15. Let A be a set and let $f \in \mathbf{F}(A)$. Prove that f is invertible and $f = f^{-1}$ if and only if $ff = i_A$.

16. Let $f \in \mathbf{F}(A, B)$ and $g \in \mathbf{F}(B, C)$. Prove that if f and g are invertible, so is gf and that $(gf)^{-1} = f^{-1}g^{-1}$.

17. Let A be a nonempty set. Let $f, g, h \in \mathbf{F}(A)$. Suppose that fg and gh are bijective. Prove that $f, g,$ and h are all bijective.

18. Let $A, B,$ and C be nonempty sets, and let $f \in \mathbf{F}(A, B)$ and $g, h \in \mathbf{F}(B, C)$.

    (a) Prove that if f is surjective and $gf = hf$, then $g = h$.

    (b) Give an example in which $gf = hf$, but $g \ne h$.

19. Let A, B, and C be nonempty sets, and let $f, g \in \mathbf{F}(A, B)$ and $h \in \mathbf{F}(B, C)$.
    (a) Prove that if h is injective and $hf = hg$, then $f = g$.
    (b) Give an example in which $hf = hg$, but $f \neq g$.

20. Let $f(x) = x^3 + x$. Prove that f is invertible without actually solving for f^{-1}.

21. Let A and B be sets and let $f: A \to B$ be a function. Suppose that f has a *left inverse* g; that is, there exists a function $g: B \to A$ such that $gf = i_A$. Prove that f is injective.

22. Let A and B be sets and let $f: A \to B$ be a function. Suppose that f has a *right inverse* h; that is, there exists a function $h: B \to A$ such that $fh = i_B$. Prove that f is surjective.

Discussion and Discovery Exercises

D1. One of the most important theorems concerning real-valued functions is the Fundamental Theorem of Calculus. It is often stated in two parts. Part 1 says that if $f(x)$ is continuous on the closed interval $[a, b]$ and a function g is defined by $g(x) = \int_a^x f(t)dt$, then g is differentiable on $[a, b]$ and $g'(x) = f(x)$ for all $x \in [a, b]$; that is, g is an antiderivative for f.
    (a) Is the following statement true? If f is continuous on $[a, b]$ and g is an antiderivative for f, then $g(x) = \int_a^x f(t)dt$.
    (b) State part 2 of the Fundamental Theorem of Calculus and give a proof.

D2. Give an example of a bijective function $f: \mathbf{R} \to \mathbf{R}$ that is neither increasing nor decreasing.

D3. Let $f: \mathbf{R} \to \mathbf{R}$ be defined by $f(x) = x^3 + 4x + 1$. In Example 10 we showed that f is invertible without finding a specific formula for f^{-1}. The function f is differentiable everywhere. What about the function f^{-1}? At what numbers is it differentiable?

BINARY OPERATIONS AND RELATIONS

4

4.1 BINARY OPERATIONS

We all learn early in school (if not before) about the usual arithmetic operations: addition and multiplication, as well as subtraction and division. We will be studying many other operations that have some of the same properties as these, and so we need to introduce a general term for them. Now addition is an operation that takes two integers and produces a third; 2 and 3 produce 5, for example. We can think of addition then as a function from $\mathbf{Z} \times \mathbf{Z}$ to $\mathbf{Z}$. This function would map $(2, 3)$ to 5, $(7, -6)$ to 1, and so on. With this example in mind, we define the notion of a binary operation on a set A.

> **Definition 4.1.1** A **binary operation** on a nonempty set A is a function from $A \times A$ to A.

A binary operation on a set A, then, can be thought of as a way of combining two elements a and b of A to produce a third element. Strictly speaking, we should use functional notation, say $f(a, b)$, to denote that element, but that's a bit cumbersome. It's much more common to introduce an operation symbol analogous to $+$ or $\times$, such as $*$, to denote the binary operation and write $a * b$ instead of using the functional notation $*(a, b)$.

EXAMPLE 1 The usual operations of addition, multiplication, and subtraction are, of course, binary operations on $\mathbf{Z}$ (and for that matter, on $\mathbf{Q}$ and $\mathbf{R}$).

123

It is important to note that when combining two elements of a set by means of a binary operation, the third element must be in the set. So, for example, addition is a binary operation on **Q** because the sum of two fractions is a fraction.

EXAMPLE 2   Division is not a binary operation on **Z**, since, for example, 1/2 is not an element of **Z**. Worse, 1/0 is undefined, so division is not a binary operation on **Q** or **R** either. It's interesting (and instructive!), though, to notice that division *is* a binary operation on **Q** − {0} and on **R** − {0}.

These examples are the most important binary operations on **Z**, **Q**, and **R**. We can of course define many others. Here are two examples.

EXAMPLE 3   We will define a binary operation on **Z** as follows. If $a, b \in$ **Z**, let $a * b = 2a + b$. Here the + sign is ordinary addition of integers. So for example, $5 * 8 = 18$ and $(-2) * 6 = 2$.

EXAMPLE 4   Let $*$ be the binary operation on **Z** defined by $a * b = a + b + 1$. For example, $6 * 10 = 17$.

These last two examples are contrived, of course, and thus not "important" examples of binary operations. But examples such as these serve to illustrate properties that a binary operation may or may not have. For instance, notice that, using the binary operation of Example 3, $8 * 5 = 21$. Thus $5 * 8 \neq 8 * 5$. Hence a commutative law that we take for granted for addition and multiplication of integers does not hold for all binary operations. We will discuss such questions in this section.

EXAMPLE 5   Let A be a nonempty set. Recall that the set **F**(A) is the set of all functions from A to A. If f and g are two elements of **F**(A), then their composition fg is also an element of **F**(A). Thus composition is a binary operation on the set **F**(A). This example is no doubt less familiar to you than the others, and has many "unusual" properties, as you will see.

EXAMPLE 6   Let A be a set. The set **P**(A) is the set of all subsets of A. We can define a binary operation on **P**(A) as follows. If X and Y are in **P**(A), let $X * Y = X \cap Y$, the intersection of the sets X and Y. Since $X \cap Y$ is a subset of A and therefore in **P**(A), we have defined a binary operation on **P**(A). (Note that we could also define a binary operation on **P**(A) by taking the union of X and Y. This example will be considered in the exercises.)

EXAMPLE 7 Let $f, g \in \mathbf{F}(\mathbf{R})$, the set of all functions from $\mathbf{R}$ to $\mathbf{R}$. Let $f * g = f + g$, the usual addition of functions: $(f + g)(x) = f(x) + g(x)$ for all x in $\mathbf{R}$. Since $f + g$ is a function from $\mathbf{R}$ to $\mathbf{R}$, we get a binary operation on $\mathbf{F}(\mathbf{R})$.

Another binary operation on $\mathbf{F}(\mathbf{R})$ is ordinary multiplication of functions: $(f * g)(x) = (f \cdot g)(x) = f(x)g(x)$.

We will call these two operations on $\mathbf{F}(\mathbf{R})$ simply addition and multiplication on $\mathbf{F}(\mathbf{R})$. We will use the usual $+$ sign to denote the sum of two functions and $\cdot$ for multiplication; that is, the product of the functions f and g will be written $f \cdot g$ so as not to be confused with composition.

EXAMPLE 8 Let $M_2(\mathbf{R})$ denote the set of all symbols of the form

$$\begin{pmatrix} a & b \\ c & d \end{pmatrix},$$

where $a, b, c, d \in \mathbf{R}$. Such a symbol is called a 2×2 *matrix*. Matrices play an important role in the study of systems of linear equations. We define a binary operation called addition on $M_2(\mathbf{R})$ as follows:

$$\begin{pmatrix} a & b \\ c & d \end{pmatrix} + \begin{pmatrix} e & f \\ g & h \end{pmatrix} = \begin{pmatrix} a + e & b + f \\ c + g & d + h \end{pmatrix}.$$

Another binary operation, called multiplication, can also be defined on $M_2(\mathbf{R})$. Its definition is a bit more complicated than that of addition. It is not just multiplication of the corresponding terms. We will not go into the justification of this definition here except to note that it has important applications in linear algebra. We define

$$\begin{pmatrix} a & b \\ c & d \end{pmatrix}\begin{pmatrix} e & f \\ g & h \end{pmatrix} = \begin{pmatrix} ae + bg & af + bh \\ ce + dg & cf + dh \end{pmatrix}.$$

Associative and Commutative Laws

Using as our guide the operations that you're very familiar with, we now begin to classify binary operations by some of their properties.

> **Definition 4.1.2** A binary operation $*$ on A is **associative** if $(a * b) * c = a * (b * c)$, $\forall a, b, c \in A$.
>
> A binary operation $*$ on A is **commutative** if $a * b = b * a$, $\forall a, b \in A$.

Note that the definitions of associative and commutative binary operations contain universal quantifiers. Proving a binary operation associative or commutative requires proving the appropriate property for *all* elements of *A*. On the other hand, to prove that a binary operation is not associative or not commutative requires only that the definition break down for *some* elements of *A*. Thus a single counterexample will suffice.

EXAMPLE 9   Addition and multiplication are associative and commutative on **Z**. These properties are actually axioms of the integers and will be discussed in Section 5.1. Addition and multiplication are also associative and commutative on **Q** and **R**. Subtraction, however, is neither associative nor commutative. For example, $(3 - 2) - 1 = 1 - 1 = 0$, but $3 - (2 - 1) = 3 - 1 = 2$; and $8 - 4 \neq 4 - 8$.

Similarly, division fails to be associative or commutative on **Q** − {0} or **R** − {0}. (See Exercise 1.)

EXAMPLE 10   Consider the binary operation on **Z** defined in Example 3: $a * b = 2a + b$ for all $a, b \in$ **Z**. We saw that this operation is not commutative. It is not associative either since, for example, $(2 * 3) * 4 = 7 * 4 = 18$ and $2 * (3 * 4) = 2 * 10 = 14$.

EXAMPLE 11   The binary operation on **Z** defined in Example 4, $a * b = a + b + 1$, is both commutative and associative. The proofs go as follows.

Let $a, b \in$ **Z**. Then $a * b = a + b + 1 = b + a + 1 = b * a$. Thus $*$ is commutative. Note that we used the fact that ordinary addition in **Z** is commutative.

Let $a, b, c \in$ **Z**. Then

$$(a * b) * c = (a + b + 1) * c = (a + b + 1) + c + 1 = a + b + c + 2$$

and

$$a * (b * c) = a * (b + c + 1) = a + (b + c + 1) + 1 = a + b + c + 2.$$

Hence $(a * b) * c = a * (b * c)$ for all $a, b, c \in$ **Z**, proving that $*$ is associative. In this proof, we used both the associativity and commutativity of ordinary addition in **Z**.

EXAMPLE 12   Composition is associative on **F**(*A*), for any set *A*. This fact was proved in Proposition 3.3.6. Interestingly, composition is not usually commutative on **F**(*A*). For example, if $A = $ **R**, consider the functions $f(x) = x + 1$ and $g(x) = x^2$:

$$f(g(x)) = x^2 + 1 \quad \text{but} \quad g(f(x)) = (x + 1)^2.$$

(In Exercise 3, you will see exactly what we mean here by "usually.")

EXAMPLE 13   Let *A* be a set. The binary operation on **P**(*A*) defined in Example 6 is both associative and commutative. This fact follows from parts 2 and 4 of Proposition 2.2.2.

EXAMPLE 14   Addition on $\mathbf{F}(\mathbf{R})$ is both associative and commutative. To see that addition is associative, let $f, g, h \in \mathbf{F}(\mathbf{R})$. We must show that $(f + g) + h = f + (g + h)$. It is clear that the domain and codomain of both these functions is $\mathbf{R}$. We must show that they take on the same values. Let $x \in \mathbf{R}$. Then

$$
\begin{aligned}
[(f + g) + h](x) &= (f + g)(x) + h(x) &&\text{(def. of addition on } \mathbf{F}(\mathbf{R})\text{)} \\
&= [f(x) + g(x)] + h(x) &&\text{(def. of addition on } \mathbf{F}(\mathbf{R})\text{)} \\
&= f(x) + [g(x) + h(x)] &&\text{(assoc. of addition on } \mathbf{R}\text{)} \\
&= f(x) + (g + h)(x) &&\text{(def. of addition on } \mathbf{F}(\mathbf{R})\text{)} \\
&= [f + (g + h)](x) &&\text{(def. of addition on } \mathbf{F}(\mathbf{R})\text{).}
\end{aligned}
$$

This proves that $(f + g) + h = f + (g + h)$.

Note that we use the fact that addition on $\mathbf{R}$ is associative. It is also important for proofs involving functions to note that the symbol f and the expression $f(x)$ mean two different things. The symbol f refers to the function, an element of $\mathbf{F}(\mathbf{R})$, and $f(x)$ is the value of the function at the real number x and hence is a real number itself.

We leave the proof that addition is commutative as an exercise. (See Exercise 4(a).)

EXAMPLE 15   Multiplication on $\mathbf{F}(\mathbf{R})$ is both commutative and associative. The proofs are left as exercises. (See Exercises 4(b) and 4(c).)

EXAMPLE 16   Addition on $M_2(\mathbf{R})$ is associative and commutative. Here is a proof that addition is commutative: let $a, b, c, d, e, f, g, h \in \mathbf{R}$.

$$
\begin{pmatrix} a & b \\ c & d \end{pmatrix} + \begin{pmatrix} e & f \\ g & h \end{pmatrix} = \begin{pmatrix} a + e & b + f \\ c + g & d + h \end{pmatrix}
$$

$$
= \begin{pmatrix} e + a & f + b \\ g + c & h + d \end{pmatrix}
$$

$$
= \begin{pmatrix} e & f \\ g & h \end{pmatrix} + \begin{pmatrix} a & b \\ c & d \end{pmatrix}.
$$

We leave the proof that addition is associative as an exercise. (See Exercise 5(a).)

EXAMPLE 17   Multiplication on $M_2(\mathbf{R})$ is associative but not commutative. We prove that multiplication is associative and leave the proof that it is not commutative as an exercise. (See Exercise 5(b).)

Let $a, b, c, d, e, f, g, h, x, y, z, w \in \mathbf{R}$.

$$\left(\begin{pmatrix} a & b \\ c & d \end{pmatrix}\begin{pmatrix} e & f \\ g & h \end{pmatrix}\right)\begin{pmatrix} x & y \\ z & w \end{pmatrix} = \begin{pmatrix} ae+bg & af+bh \\ ce+dg & cf+dh \end{pmatrix}\begin{pmatrix} x & y \\ z & w \end{pmatrix}$$

$$= \begin{pmatrix} (ae+bg)x+(af+bh)z & (ae+bg)y+(af+bh)w \\ (ce+dg)x+(cf+dh)z & (ce+dg)y+(cf+dh)w \end{pmatrix}$$

$$= \begin{pmatrix} aex+bgx+afz+bhz & aey+bgy+afw+bhw \\ cex+dgx+cfz+dhz & cey+dgy+cfw+dhw \end{pmatrix}$$

$$= \begin{pmatrix} a(ex+fz)+b(gx+hz) & a(ey+fw)+b(gy+hw) \\ c(ex+fz)+d(gx+hz) & c(ey+fw)+d(gy+hw) \end{pmatrix}$$

$$= \begin{pmatrix} a & b \\ c & d \end{pmatrix}\begin{pmatrix} ex+fz & ey+fw \\ gx+hz & gy+hw \end{pmatrix}$$

$$= \begin{pmatrix} a & b \\ c & d \end{pmatrix}\left(\begin{pmatrix} e & f \\ g & h \end{pmatrix}\begin{pmatrix} x & y \\ z & w \end{pmatrix}\right).$$

Identities

We are all familiar with the fact that $x + 0 = 0 + x = x$ and $x \cdot 1 = 1 \cdot x = x$ for all real numbers x. Thus any number added to 0 is itself and any number multiplied by 1 is itself. Because of these special properties of 0 and 1, they are called *identity* elements with respect to addition and multiplication, respectively.

Not all sets with binary operations have such special elements but many of them do. We make the following definition.

Definition 4.1.3 If $*$ is a binary operation of A, an element e of A is an **identity element** of A with respect to $*$ if $a * e = e * a = a$, $\forall a \in A$.

If a set A has an identity element with respect to a binary operation $*$, we will sometimes just say that $*$ has an identity element if it is clear from the context what the underlying set A is. Strictly speaking, of course, it is the set A together with its binary operation that has the identity element. In addition, we will often just say identity instead of identity element.

EXAMPLE 18   The identity element for **Z**, **Q**, and **R** with respect to addition is 0, as we just noted, and with respect to multiplication is 1.

There is no identity for subtraction, though. The proof will be by contradiction. Suppose then that e is an identity for subtraction on **Z**. Then $e = e - 0 = 0 - e = -e$, so $2e = 0$, implying that e must be 0. But 0 doesn't work: $0 - 1 \neq 1 - 0$, for example. (Question: What about division on **Q** $- \{0\}$ or **R** $- \{0\}$? See Exercise 6.)

We often use contradiction to show that a set has no identity with respect to a given binary operation; that is, we suppose that it has an identity and derive a contradiction. Here is another example of that technique.

EXAMPLE 19 Consider the binary operation $a * b = 2a + b$ on $\mathbf{Z}$ defined in Example 3. Suppose that $\mathbf{Z}$ has an identity e with respect to $*$. Then $e * 1 = 2e + 1 = 1$, which implies $2e = 0$ or $e = 0$. But $1 * 0 = 2 + 0 = 2 \neq 1$, so 0 cannot be the identity. Therefore $\mathbf{Z}$ has no identity with respect to $*$.

EXAMPLE 20 Let $*$ be the binary operation on $\mathbf{Z}$ defined in Example 4, $a * b = a + b + 1$. We will show that $\mathbf{Z}$ has an identity with respect to $*$. First we have to discover what the identity is. If e is the identity, then for all a in $\mathbf{Z}$, we have $a * e = a + e + 1 = a$, from which it follows that e must be -1. It is easy to verify that $a * (-1) = (-1) * a = a$ for all $a \in \mathbf{Z}$.

EXAMPLE 21 Fortunately for our terminology, the identity function i_A is an identity element for composition on $\mathbf{F}(A)$. This fact is a consequence of Proposition 3.3.3.

EXAMPLE 22 Let A be a nonempty set. In Example 6, we defined the binary operation $*$ on $\mathbf{P}(A)$ by $X * Y = X \cap Y$ if X and Y are subsets of A. This binary operation has an identity, namely A itself, since for any $X \in \mathbf{P}(A)$, $X \cap A = A \cap X = X$.

EXAMPLE 23 In Example 7 we defined two binary operations on $\mathbf{F}(\mathbf{R})$, addition and multiplication of functions. Each of these binary operations has an identity. It is easy to see that the identity for addition is the constant function that maps every real number to 0 and the identity for multiplication is the constant function that maps every real number to 1.

EXAMPLE 24 Addition on $M_2(\mathbf{R})$ has an identity, the matrix $\begin{pmatrix} 0 & 0 \\ 0 & 0 \end{pmatrix}$. The identity on $M_2(\mathbf{R})$ with respect to multiplication is not so clear. It is the matrix $\begin{pmatrix} 1 & 0 \\ 0 & 1 \end{pmatrix}$. We leave the justification of this as an exercise. (See Exercise 9.)

You may have noticed that we occasionally used the definite article here: we talked about "the" identity element. This really isn't justified until we prove the next proposition.

Proposition 4.1.4 If $*$ is a binary operation on A with identity element e, then e is unique.

Again, a preliminary word about the method of proof. We are assuming the existence of one object, an identity, and must show that there are no others. The conclusion is not existentially quantified; we will *not* exhibit an identity. Rather, a clever and common way to prove uniqueness is to assume that there *is* another such object, another identity, and then prove that in fact the two must coincide; thus there is really only one.

PROOF: Suppose that e' is also an identity element. Then $e * e' = e$, since $e \in A$. But on the other hand, e is an identity element, so we also have $e * e' = e'$. Hence $e = e'$. ●

Inverses

Another important concept associated with a binary operation is the notion of the inverse of an element.

> **Definition 4.1.5** Suppose that $*$ is a binary operation on A with identity e, and let $a \in A$. We say that a is **invertible** with respect to $*$ if there exists $b \in A$ such that $a * b = b * a = e$. If b exists, we say that b is an **inverse** of a with respect to $*$.

EXAMPLE 25   Inverses may or may not exist. Every element x of $\mathbf{Z}$ has an inverse with respect to addition, namely its negative $-x$. However, very few integers have multiplicative inverses; in fact, only 1 and -1 do. (This fact is proved in Section 5.1.) For example, the reason that 2 does not have a multiplicative inverse is that 1/2 is not in $\mathbf{Z}$.

But each element of $\mathbf{Q}$ or $\mathbf{R}$, except for 0, has a multiplicative inverse— its reciprocal.

EXAMPLE 26   Consider again the binary operation $a * b = 2a + b$ on $\mathbf{Z}$ defined in Example 3. We've seen that this binary operation does not have an identity. Since the definition of the inverse of an element presupposes the existence of an identity, it follows that no element of $\mathbf{Z}$ has an inverse with respect to $*$.

EXAMPLE 27   Let $*$ be the binary operation on $\mathbf{Z}$ defined by $a * b = a + b + 1$. We have seen that -1 is the identity of $\mathbf{Z}$ with respect to this binary operation.

Let $a \in \mathbf{Z}$. If a has an inverse x with respect to $*$, then $a * x = a + x + 1 = -1$. Solving for x gives

$$x = -a - 2.$$

One easily checks that

$$a * (-a - 2) = (-a - 2) * a = -1.$$

Therefore every element of $\mathbf{Z}$ has an inverse with respect to $*$.

EXAMPLE 28   Let A be a nonempty set. As we have seen, composition is a binary operation on $\mathbf{F}(A)$ and i_A is the identity element. It is easy to see that the elements of

F(*A*) that have inverses with respect to composition are precisely the invertible functions in **F**(*A*). (See Definition 3.3.7.) And these functions are exactly those functions in **F**(*A*) that are bijective. (See Theorem 3.3.8.)

EXAMPLE 29 Let *A* be a nonempty set. Consider again the binary operation defined on **P**(*A*) by $X * Y = X \cap Y$. We have seen that *A* is the identity element of **P**(*A*) with respect to $*$. Since $A * A = A \cap A = A$, it follows that *A* is its own inverse. (In fact, an identity element is always its own inverse.)

However, if $X \in$ **P**(*A*) and $X \neq A$, then $X \cap Y \neq A$ for all $Y \in$ **P**(*A*) since $X \cap Y \subseteq X$. Therefore if $X \in$ **P**(*A*) and $X \neq A$, *X* does not have an inverse with respect to $*$.

EXAMPLE 30 Every element of **F**(**R**) has an inverse with respect to addition of functions: namely, its negative $-f$. For example, the inverse of the function $f: \mathbf{R} \to \mathbf{R}$ defined by $f(x) = 2x^3 - 3x + \cos x$ is the function $-f: \mathbf{R} \to \mathbf{R}$ defined by $(-f)(x) = -2x^3 + 3x - \cos x$.

For multiplication in **F**(**R**), the situation is a bit more complicated. If a function $f \in$ **F**(**R**) takes on the value 0 at even one point, say, $f(x_0) = 0$, then no function multiplied by $f(x)$ can equal 1 at x_0. Hence the functions in **F**(**R**) that have inverses are those that are never 0. The inverse of such a function is its reciprocal. For example, the multiplicative inverse of the function $f: \mathbf{R} \to \mathbf{R}$ defined by $f(x) = x^2 + 1$ is the function $1/f: \mathbf{R} \to \mathbf{R}$, defined by

$$\left(\frac{1}{f}\right)(x) = \frac{1}{x^2 + 1}.$$

(*Note:* We use the symbol $1/f$ for the multiplicative inverse of a function and reserve the symbol f^{-1} for the inverse with respect to composition.)

EXAMPLE 31 Not every element of $M_2(\mathbf{R})$ has an inverse. In fact, the invertible elements of $M_2(\mathbf{R})$ are precisely those matrices $\begin{pmatrix} a & b \\ c & d \end{pmatrix}$ for which $ad - bc \neq 0$. (See Exercise 11.)

In some cases, inverses have a uniqueness property like identities do. The result is stated next and the proof is left as an exercise (Exercise 12).

Proposition 4.1.6 Let $*$ be an associative binary operation on a set *A* with identity element *e*. If $a \in A$ has an inverse with respect to $*$, then that inverse is unique.

If the inverse of an element $a \in A$ is unique, we denote it by a^{-1} and call it the inverse of *a* with respect to $*$.

Closure

A subset of a set with a binary operation "inherits" a way to combine pairs of elements: they are, after all, elements of the larger set, and the operation can be applied. But where is the result? There is no guarantee that it is in the subset. We single out the case when in fact we have an operation on the subset.

> **Definition 4.1.7** Let $*$ be a binary operation on the set A, and suppose that $X \subseteq A$. If $*$ is also a binary operation on X (that is, if $x * y \in X$, $\forall x, y \in X$), then X is said to be **closed in A under $*$**.

Note that the definition of closure is a statement with a universal quantifier.

EXAMPLE 32 $\mathbf{Z}^+$ is closed in $\mathbf{Z}$ under addition and multiplication. This fact is one of the axioms of the integers and will be discussed in Section 5.1.

EXAMPLE 33   The set $N = \{n \in \mathbf{Z} \mid n < 0\}$, the set of negative integers, is closed in $\mathbf{Z}$ under addition, but not under multiplication. (The formal proof of these facts will be an exercise in Section 5.1.)

EXAMPLE 34   Let $\mathbf{E} = \{n \in \mathbf{Z} \mid n \text{ is even}\}$. $\mathbf{E}$ is closed in $\mathbf{Z}$ under both addition and multiplication. Here is a proof for the closure of addition.
Let $a, b \in \mathbf{E}$. Then $a = 2a_1$ and $b = 2b_1$ where $a_1, b_1 \in \mathbf{Z}$. Hence

$$a + b = 2a_1 + 2b_1 = 2(a_1 + b_1).$$

Since $a_1 + b_1 \in \mathbf{Z}$, $a + b = 2(a_1 + b_1) \in \mathbf{E}$.
The proof for closure of multiplication is left as an exercise.

Let A be a set with a binary operation $*$. Let X be a subset of A. To prove that X is *not* closed under $*$, we need to express the negation of the definition of closure. Let $P(x, y)$ be the open sentence "$x, y \in X$" and $Q(x, y)$ be the open sentence "$x * y \in X$." Then X is closed under $*$ if $\forall x, y \in A$, $P(x, y) \Rightarrow Q(x, y)$. Thus X is not closed under $*$ if there exist $x, y \in A$ such that $P(x, y)$ is true and $Q(x, y)$ is false. In other words, $\exists x, y \in A$ such that $x, y \in X$ but $x * y \notin X$.

EXAMPLE 35   The set $\mathbf{O} = \{n \in \mathbf{Z} \mid n \text{ is odd}\}$ is closed in $\mathbf{Z}$ under multiplication but not under addition. We leave the proof that $\mathbf{O}$ is closed under multiplication as an exercise. (See Exercise 16.)
To see that $\mathbf{O}$ is not closed under addition, we need only find integers x and y such that x and y are odd but $x + y$ is even. One example is $x = 3$ and $y = 5$.

EXAMPLE 36 Let $C(R) = \{f \in F(R) \mid f \text{ is continuous on } R\}$. Then $C(R)$ is closed in $F(R)$ under both addition and multiplication. This follows from a result of calculus that says that the sum and product of continuous functions are continuous.

EXAMPLE 37 Let $X = \{f \in F(R) \mid f(x) \in Z, \forall x \in R\}$. This is the set of *integer-valued* functions in $F(R)$. X is closed in $F(R)$ under composition.

To see this, let $f, g \in X$; we choose $x \in R$, and we must show that $fg(x) \in Z$. But $fg(x) = f(g(x)) \in Z$, since $g(x) \in Z \subseteq R$.

EXAMPLE 38 Let $S = \left\{ \begin{pmatrix} a & b \\ c & d \end{pmatrix} \in M_2(R) \mid a = 0 \right\}$. We will show that S is closed under addition but not closed under multiplication.

Let $\begin{pmatrix} a & b \\ c & d \end{pmatrix}, \begin{pmatrix} x & y \\ z & w \end{pmatrix} \in S$. Then $a = x = 0$. Hence

$$\begin{pmatrix} a & b \\ c & d \end{pmatrix} + \begin{pmatrix} x & y \\ z & w \end{pmatrix} = \begin{pmatrix} 0+0 & b+y \\ c+z & d+w \end{pmatrix} = \begin{pmatrix} 0 & b+y \\ c+z & d+w \end{pmatrix} \in S.$$

Therefore, S is closed under addition.

To prove that S is not closed under multiplication, we merely have to show that there exist two elements of S whose product is not in S. Two such elements are $\begin{pmatrix} 0 & 1 \\ 1 & 0 \end{pmatrix}$ and $\begin{pmatrix} 0 & 0 \\ -1 & 0 \end{pmatrix}$.

MATHEMATICAL PERSPECTIVE: GROUPS

In a number of the examples of binary operations $*$ on a set A done in this section, the operation was associative, the set A had an identity with respect to $*$, and every element of A had an inverse. Mathematicians have given a name to any set A with this property. It is called a *group*. The notion of group was mentioned informally in the comments at the end of Section 3.2.

The set Z of integers is a group with respect to the binary operation of ordinary addition. Z is also a group with respect to the binary operation defined in Example 4. It is not a group with respect to the binary operation in Example 3, however, because that binary operation is not associative. Z is also not a group with respect to the binary operation of multiplication because, although multiplication is associative and there is an identity, only 1 and -1 have inverses with respect to multiplication.

Now consider the set $F(A)$ of all functions from a nonempty set A to itself. We saw in Example 5 that composition is a binary operation on $F(A)$ and composition is associative. Moreover, the identity function i_A is the identity element of $F(A)$ with respect to composition. However, not every function in $F(A)$ has an inverse with respect to composition. In fact the ones that do are the bijective functions from A to A.

To get a group from this set of functions, we focus on a subset of $\mathbf{F}(A)$. Let $\mathbf{S}(A)$ be the set of all bijective functions from A to A. Then $\mathbf{S}(A)$ is closed in $\mathbf{F}(A)$ with respect to composition. (See Exercise 22 following.) This means that composition is a binary operation on $\mathbf{S}(A)$. It is an associative binary operation on $\mathbf{S}(A)$, since the associative property is inherited from $\mathbf{F}(A)$; that is, $(fg)h = f(gh)$ for all f, g, h in $\mathbf{S}(A)$ because $(fg)h = f(gh)$ for all f, g, h in $\mathbf{F}(A)$. The identity function i_A is bijective and is therefore in $\mathbf{S}(A)$. Thus $\mathbf{S}(A)$ has an identity with respect to composition. Finally, all of the elements of $\mathbf{S}(A)$ have inverses with respect to composition because they are invertible functions. Moreover, these inverses are themselves invertible and therefore in $\mathbf{S}(A)$.

Thus $\mathbf{S}(A)$ is a group with respect to composition. This type of group is called a *permutation* group, because the elements of $\mathbf{S}(A)$ are permutations of the set A. We discussed some examples of permutation groups at the end of Section 3.2. Groups appear in many areas of mathematics including number theory, geometry, and topology. The study of groups, called *group theory*, has been a thriving area of mathematical research for many years.

Exercises 4.1

1. (a) Prove that division is a binary operation on $\mathbf{R} - \{0\}$.
   (b) Prove that division on $\mathbf{R} - \{0\}$ is neither associative nor commutative.

2. Determine whether each of the following binary operations is associative and/or commutative. Prove your answers.
   (a) On $\mathbf{Z}$, $n * m = n + m + 2$
   (b) On $\mathbf{Z}$, $n * m = 6 + nm$
   (c) On $\mathbf{Z}$, $n * m = n^2 m^2$
   (d) On $\mathbf{Z}^+$, $n * m = \max(n, m)$, the larger of n and m
   (e) On $\mathbf{Z}^+$, $n * m = \min(n, m)$, the smaller of n and m
   (f) On $\mathbf{P}(A)$, for any set A, $X * Y = X \cup Y$
   (g) On $\mathbf{R}$, $x * y = 2^{xy}$
   (h) On $\mathbf{Z}^+$, $n * m = n^m$

3. (a) Let A be a nonempty set. Prove that if A has more than one element, then composition on $\mathbf{F}(A)$ is not commutative.
   (b) Prove that if $|A| = 1$, then composition on $\mathbf{F}(A)$ is commutative.

4. (a) Prove that addition on $\mathbf{F}(\mathbf{R})$ is commutative.
   (b) Prove that multiplication on $\mathbf{F}(\mathbf{R})$ is associative.
   (c) Prove that multiplication on $\mathbf{F}(\mathbf{R})$ is commutative.

5. (a) Prove that addition on $M_2(\mathbf{R})$ is associative.
   (b) Prove that multiplication on $M_2(\mathbf{R})$ is not commutative.

6. Does division on $\mathbf{R} - \{0\}$ have an identity element? If so, what is it?

7. For each of the binary operations in Exercise 2, determine which ones have identities. Prove your answers.

8. Compute:

(a) $\begin{pmatrix} 6 & 3 \\ 1 & 4 \end{pmatrix} + \begin{pmatrix} 8 & 9 \\ -7 & 27 \end{pmatrix}$ (b) $\begin{pmatrix} -3 & 7 \\ 2 & 0 \end{pmatrix} \begin{pmatrix} 16 & -8 \\ -9 & -3 \end{pmatrix}$.

9. Prove that the matrix $\begin{pmatrix} 1 & 0 \\ 0 & 1 \end{pmatrix}$ is the identity of $M_2(\mathbf{R})$ with respect to multiplication.

10. For each of the binary operations in Exercise 2 that has an identity, determine which elements of the underlying set have inverses.

11. Prove that $\begin{pmatrix} a & b \\ c & d \end{pmatrix}$ is an invertible element of $M_2(\mathbf{R})$ if and only if $ad - bc \neq 0$.

12. Prove Proposition 4.1.6.

13. Let $*$ be an associative binary operation on a set A with identity element e.
    (a) Prove that e is invertible.
    (b) Prove that if $a \in A$ is invertible, then a^{-1} is invertible and $(a^{-1})^{-1} = a$.

14. Let $*$ be an associative binary operation on a set A with identity element e, and let $a, b \in A$.
    (a) Prove that if a and b are invertible, then $a * b$ is invertible.
    (b) Prove that if A is the set of real numbers $\mathbf{R}$ and $*$ is ordinary multiplication, then the converse of part (a) is true.
    (c) Give an example of a set A with a binary operation $*$ for which the converse of part (a) is false.

15. Consider the statement P: "A commutative binary operation is associative."
    (a) What is $\neg P$?
    (b) Precisely one of the statements P and $\neg P$ is true. Prove it.

16. (a) Prove that the set $\mathbf{E}$ of even integers is closed in $\mathbf{Z}$ under multiplication.
    (b) Prove that the set $\mathbf{O}$ of odd integers is closed in $\mathbf{Z}$ under multiplication.

17. Let S be the set of nonzero real numbers.
    (a) Is S closed under addition? Prove your answer.
    (b) Is S closed under multiplication? Prove your answer.

18. Let $n \in \mathbf{Z}$ and let $n\mathbf{Z}$ be the set of multiples of n: $n\mathbf{Z} = \{a \in \mathbf{Z} \mid a = nt$ for some $t \in \mathbf{Z}\}$. Prove that $n\mathbf{Z}$ is closed in $\mathbf{Z}$ under ordinary addition and multiplication of integers.

19. Let $*$ be the binary operation on $\mathbf{Z}$ defined by $a * b = a + 2b$. Prove or disprove that each of the following subsets is closed in $\mathbf{Z}$ under $*$.
    (a) The set $\mathbf{E}$ of even integers

     (b) The set $\mathbf{O}$ of odd integers

     (c) The set $n\mathbf{Z}$ of multiples of a fixed integer n

20. Let $*$ be the binary operation on $\mathbf{Z}$ defined by $a * b = a + b + 1$. Prove or disprove that each of the following subsets is closed in $\mathbf{Z}$ under $*$.

     (a) The set $\mathbf{E}$ of even integers

     (b) The set $\mathbf{O}$ of odd integers

     (c) The set $n\mathbf{Z}$ of multiples of a fixed integer n

21. Let $\mathbf{D(R)} = \{f \in \mathbf{F(R)} \mid f \text{ is differentiable on } \mathbf{R}\}$. Prove that $\mathbf{D(R)}$ is closed in $\mathbf{F(R)}$ under addition and multiplication. (You may use without proof any result from calculus you need, but state that result carefully.)

22. Let A be a nonempty set. Determine whether or not the following sets are closed in $\mathbf{F}(A)$ under composition. Prove your answers.

     (a) $\{f \in \mathbf{F}(A) \mid f \text{ is injective}\}$.

     (b) $\{f \in \mathbf{F}(A) \mid f \text{ is surjective}\}$.

     (c) $\{f \in \mathbf{F}(A) \mid f \text{ is bijective}\}$.

23. Determine whether or not the following sets are closed in $\mathbf{F(R)}$ under addition. Prove your answers.

     (a) $\{f \in \mathbf{F(R)} \mid f(0) = 0\}$.

     (b) $\{f \in \mathbf{F(R)} \mid f(0) = 1\}$.

     (c) $\{f \in \mathbf{F(R)} \mid f'(0) = 0\}$.

     (d) $\{f \in \mathbf{F(R)} \mid f(-x) = f(x) \text{ for all } x \in \mathbf{R}\}$.

     (e) $\{f \in \mathbf{F(R)} \mid f(-x) = -f(x) \text{ for all } x \in \mathbf{R}\}$.

24. Determine whether or not the sets in Exercise 23 are closed in $\mathbf{F(R)}$ under multiplication. Prove your answers.

25. Determine whether or not the sets in Exercise 23 are closed in $\mathbf{F(R)}$ under composition. Prove your answers.

26. Determine whether or not the following sets S are closed in $M_2(\mathbf{R})$ under addition. Prove your answers.

     (a) $S = \left\{ \begin{pmatrix} a & b \\ c & d \end{pmatrix} \in M_2(\mathbf{R}) \mid b = 0 \right\}$.

     (b) $S = \left\{ \begin{pmatrix} a & b \\ c & d \end{pmatrix} \in M_2(\mathbf{R}) \mid c = 1 \right\}$.

     (c) $S = \left\{ \begin{pmatrix} a & b \\ c & d \end{pmatrix} \in M_2(\mathbf{R}) \mid a = d \right\}$.

     (d) $S = \left\{ \begin{pmatrix} a & b \\ c & d \end{pmatrix} \in M_2(\mathbf{R}) \mid 3a + 4b - 7c + d = 0 \right\}$.

27. Determine whether or not the following sets S are closed in $M_2(\mathbf{R})$ under multiplication. Prove your answers.

     (a) $S = \left\{ \begin{pmatrix} a & b \\ c & d \end{pmatrix} \in M_2(\mathbf{R}) \mid b = 0 \right\}$.

(b) $S = \left\{ \begin{pmatrix} a & b \\ c & d \end{pmatrix} \in M_2(\mathbf{R}) \mid d = 0 \right\}.$

(c) $S = \left\{ \begin{pmatrix} a & b \\ c & d \end{pmatrix} \in M_2(\mathbf{R}) \mid a = d \right\}.$

(d) $S = \left\{ \begin{pmatrix} a & b \\ c & d \end{pmatrix} \in M_2(\mathbf{R}) \mid a, b, c, d \in \mathbf{Z} \right\}.$

28. A matrix A in $M_2(\mathbf{R})$ is called a **nonzero** matrix if A is not the matrix $\begin{pmatrix} 0 & 0 \\ 0 & 0 \end{pmatrix}$. Let S be the statement "The product of two nonzero matrices in $M_2(\mathbf{R})$ is a nonzero matrix."
    (a) Write S in the form: "For all ... if ... then ... ."
    (b) Write the negation of S.
    (c) Prove or disprove S.

29. Let $*$ be an associative binary operation on a set A with identity element e. Let $B = \{a \in A \mid a \text{ is invertible with respect to } *\}$. Prove that B is closed in A under $*$.

30. Let $*$ be an associative binary operation on a set A with identity element e. Suppose $a, b, c \in A$ and that a is invertible.
    (a) Prove that if $a * b = a * c$, then $b = c$.
    (b) Give an example to show that this "cancellation law" is not true if a is not invertible.

31. Let $*$ be a binary operation on a set A. Complete the following statements:
    (a) $*$ is not associative if ... .
    (b) $*$ is not commutative if ... .
    (c) $x \in A$ is not an identity element of A if ... .
    (d) $a \in A$ does not have an inverse if ... .

32. Let $*$ be a binary operation on a set A. Let B and C be subsets of A that are closed under $*$.
    (a) Prove that $B \cap C$ is closed under $*$.
    (b) Give an example to show that $B \cup C$ need not in general be closed under $*$.

33. Let A be a nonempty set. In Example 6, we defined the binary operation $*$ on $\mathbf{P}(A)$ by $X * Y = X \cap Y$ if X and Y are subsets of A. Let $X \in \mathbf{P}(A)$. Since $X * X = X \cap X = X$, does this mean that X is the identity for this binary operation? Explain.

34. Let $*$ be an associative binary operation on a set A. Suppose that A has an identity e with respect to $*$. Let $a \in A$. We say that a has a *left inverse* with respect to $*$ if there exists $b \in A$ such that $b * a = e$. Similarly, we say that a has a *right inverse* with respect to $*$ if there exists $c \in A$ such that $a * c = e$.

(a) Prove that if an element $a \in A$ has a left inverse b and a right inverse c, then a is invertible and $a^{-1} = b = c$.

(b) Suppose that $a \in A$ has a left inverse b and that b is *not* a right inverse for a. Prove that a has no right inverse and that a is not invertible.

(c) Give an example of a set with an associative binary operation having an identity for which there is an element with a left inverse but no right inverse. *Hint: See Exercise 2 of Section 3.3.*

35. Prove that the set $M_2(\mathbf{R})$ is a group with respect to matrix addition but is not a group with respect to multiplication.

36. Prove that the set $\left\{ \begin{pmatrix} a & b \\ c & d \end{pmatrix} \in M_2(\mathbf{R}) \,\middle|\, ad - bc \neq 0 \right\}$ is a group with respect to matrix multiplication.

37. Let $S = \left\{ \begin{pmatrix} a & a \\ 0 & 0 \end{pmatrix} \,\middle|\, a \in \mathbf{R} \right\}$.

(a) Prove that S is closed in $M_2(\mathbf{R})$ with respect to matrix multiplication.

(b) Prove that S has an identity with respect to matrix multiplication.

(c) What elements of S have inverses with respect to matrix multiplication and the identity you found in part (b)? Prove your answer.

(d) Prove that the set T of nonzero matrices in S forms a group with respect to matrix multiplication.

Discussion and Discovery Exercises

D1. Let $*$ be an associative operation on a set A. Explain how the definition of invertible element implies that if an element a in A is invertible, then a^{-1} is invertible and $(a^{-1})^{-1} = a$.

D2. Let $*$ be a binary operation on a set A. Suppose that A has an identity e with respect to $*$. Let B be subset of A and suppose that B is closed in A with respect to $*$. Does it follow that B must have an identity with respect to $*$? Must that identity, if it exists, be e? Explain with proofs and examples.

D3. Consider the set $M_2(\mathbf{R})$ with respect to the binary operation of matrix multiplication. We say that a matrix A has finite order if some power of A is the identity matrix $\begin{pmatrix} 1 & 0 \\ 0 & 1 \end{pmatrix}$. Specifically, we say that A has order n if $A^n = \begin{pmatrix} 1 & 0 \\ 0 & 1 \end{pmatrix}$ where n is a positive integer and n is the smallest positive integer with this property. For example, the matrix $\begin{pmatrix} -1 & 0 \\ 0 & -1 \end{pmatrix}$ has order 2, since $\begin{pmatrix} -1 & 0 \\ 0 & -1 \end{pmatrix}^2 = \begin{pmatrix} 1 & 0 \\ 0 & 1 \end{pmatrix}$.

(a) Find a matrix of order 4.

    Hint: The entries of the matrix will just involve the integers 0, 1, and −1.

(b) Find a matrix of order 3.

    Hint: The entries of the matrix will not be integers.

(c) Find a matrix that does not have finite order.

D4. Consider the set $M_2(\mathbf{R})$ with respect to the binary operation of matrix multiplication. Let I denote the identity matrix $\begin{pmatrix} 1 & 0 \\ 0 & 1 \end{pmatrix}$. It is shown in linear algebra that if a 2×2 matrix A has a left inverse B then A is invertible and $B = A^{-1}$. That is, if there exists a 2×2 matrix B such that $BA = I$, then A is invertible and $B = A^{-1}$. (See Exercise 34.)

(a) Look up the proof of this result in a linear algebra text and write a clear explanation of the proof.

    Hint: The result is usually stated for $n \times n$ matrices and involves the connection between an $n \times n$ matrix being invertible and the solution of $n \times n$ linear systems.

(b) Prove that if a 2×2 matrix A has a right inverse C then A is invertible and $C = A^{-1}$.

    Hint: This problem can be done just using part (a) and Exercise 13(b) of this section.

D5. In Proposition 4.1.6, we proved that if ∗ is an associative binary operation on a set A with identity element e and if $a \in A$ has an inverse with respect to ∗, then that inverse is unique. The proof uses the fact that ∗ is associative. The following example is cooked up to show that for some binary operations, an element may have more than one inverse.

    We define a binary operation on $\mathbf{R}$ as follows. If $a, b \in \mathbf{R}$, let $a * b = a + b - (ab)^2$.

(a) Prove that ∗ is commutative but not associative.

(b) Find the identity element of $\mathbf{R}$ with respect to ∗.

(c) Prove that 1 has two inverses with respect to ∗.

(d) Prove that any real number a has at most 2 inverses with respect to ∗.

(e) Which real numbers have no inverse with respect to ∗? Which ones have exactly one inverse? Which ones have two inverses?

4.2   # EQUIVALENCE RELATIONS

One of the most basic concepts of mathematics, one that we are all familiar with from our earliest days in a math class, is equality. But this very strict idea of two objects being equal is often not what we mean when we use the word in everyday life. For example, baseball fans throughout America can be classified by their team allegiance: there are Red Sox fans, Cubs fans, and so on. From the point of view of a team trying to sell tickets, any two of its fans (with money!) can be considered equivalent, although they are certainly not the same person. This "equality" is a generalization of the strict notion, considering two objects equivalent because they share some property, or are in some relation to each other. In this section we try to capture this generalization mathematically.

Relations

> **Definition 4.2.1** A **relation** R on a set A is a subset of $A \times A$. If $(a, b) \in R$, we write aRb.

EXAMPLE 1   The set $R = \{(0, 0), (0, 1), (2, 2), (7, 18)\}$ is a relation on the set $\mathbf{Z}^+$. We would thus write $0R0$, $0R1$, $2R2$, and $7R18$.

EXAMPLE 2   Let $R = \{(a, b) \in \mathbf{Z} \times \mathbf{Z} \mid b = na$ for some $n \in \mathbf{Z}\}$, a relation on the set $\mathbf{Z}$ of integers. Then, for example, $2R4$, (-3), $R(-42)$, and $7R0$. In general, aRb if b is a multiple of a.

EXAMPLE 3   On the set of baseball fans, define aRb if a and b are fans of the same team.

EXAMPLE 4   We can define a relation R on the set of all people by aRb if persons a and b have at least one parent in common.

EXAMPLE 5   A different relation S on the set of all people can be defined by aSb if persons a and b are siblings; that is, they have both parents in common. Notice that S is a subset of the relation R in Example 4. In fact, it's a proper subset, since for example Elizabeth I and Edward VI of England were both children of Henry VIII, but had different mothers (with different fortunes).

EXAMPLE 6   Define a relation R on the set $\mathbf{R}$ of real numbers by aRb if $a < b$. So for example, $5R5.5$ and $\pi R4$.

EXAMPLE 7   Let S be the set of all finite subsets of $\mathbf{Z}$. Define R on S by ARB if $|A| = |B|$.

Properties of Relations

The following definition lists four important properties that a relation may have.

> **Definition 4.2.2** Let R be a relation on a set A. We say:
>
> 1. R is **reflexive** if aRa, $\forall a \in A$.
> 2. R is **symmetric** if for all $a, b \in A$, if aRb, then bRa.
> 3. R is **transitive** if for all $a, b, c \in A$, if aRb and bRc, then aRc.
> 4. R is **antisymmetric** if for all $a, b \in A$, if aRb and bRa, then $a = b$.

EXAMPLE 8   Let R be the relation of Example 1. Then R is not reflexive because 1 is not related to 1.

R is not symmetric because $0R1$ but 1 is not related to 0.

R is transitive since the only instance in which aRb and bRc is when $a = 0$, $b = 0$, and $c = 1$, and in this case aRc.

R is antisymmetric because there are no integers a and b such that aRb and bRa and $a \neq b$.

EXAMPLE 9   Let R be the relation on $\mathbf{Z}$ defined by aRb if $b = na$ for some integer n, as in Example 2.

Let $a \in A$. Since $a = 1a$, aRa. Therefore R is reflexive.

R is not symmetric since, for example, $2R4$ but 4 is not related to 2.

R is transitive. Let a, b, $c \in \mathbf{Z}$ and suppose that aRb and bRc. Then $b = na$ for some $n \in \mathbf{Z}$ and $c = mb$ for some $m \in \mathbf{Z}$. Hence $c = mna$ and because mn is an integer, it follows that aRc.

R is not antisymmetric since $2R(-2)$ and $(-2)R2$ but $2 \neq -2$.

EXAMPLE 10   It follows from the order properties of the real numbers (see Section 5.1) that the relation R on the set $\mathbf{R}$ of real numbers defined by aRb if $a < b$ is transitive but not reflexive or symmetric.

This relation may not seem to be antisymmetric since for any real numbers a and b, we cannot have $a < b$ and $b < a$. But it is for precisely this reason that it *is* antisymmetric. The definition of antisymmetric requires that for all a, $b \in \mathbf{R}$, the implication $a < b$ and $b < a$ imply $a = b$. But remember that an implication $P \Rightarrow Q$ is true whenever the premise P is false. In this case the premise $a < b$ and $b < a$ is always false. Therefore for every choice of real numbers for the variables a and b, the implication $a < b$ and $b < a$ implies that $a = b$ is true. Hence R is antisymmetric.

Note that to prove a relation on a set A symmetric requires proving a statement with a universal quantifier and an implication. For *every a, b* in A, it is necessary to prove that *if aRb, then bRa*. Similarly, the transitive property is a statement with a universal quantifier and an implication. For *every a, b, c* in A, *if aRb and bRc, then aRc*. The same is true for the antisymmetric property. The reflexive property however has a universal quantifier but does not contain an implication. It is merely necessary to show that aRa for all $a \in A$.

Relations that have the properties of being reflexive, symmetric, and transitive are particularly important.

Equivalence Relations

Definition 4.2.3   A relation R on a set A is called an **equivalence relation** if it is reflexive, symmetric, and transitive.

EXAMPLE 11 On any set A, the relation of equality, aRb if $a = b$, is clearly an equivalence relation. We would certainly hope this were true, since we are attempting to generalize this very notion! But there are many other relations that can be seen to satisfy the three properties of the definition.

EXAMPLE 12 Consider again the set of baseball fans, and the relation defined in Example 3. Let's assume, quite reasonably, that no one can really be a fan of two different teams. Then it's easy to see that the three properties are satisfied.

EXAMPLE 13 On $\mathbf{Z}$, define aRb if $|a| = |b|$, so for example, $2R(-2)$ and $4R4$.
Since $|a| = |a|$, $\forall a \in \mathbf{Z}$, R is immediately seen to be reflexive.
To see that R is symmetric, let $a, b \in \mathbf{Z}$ and suppose that aRb. Then $|a| = |b|$, which implies that $|b| = |a|$ and thus bRa.
Now suppose that $a, b, c \in \mathbf{Z}$ and that aRb and bRc. Then $|a| = |b|$ and $|b| = |c|$ and hence $|a| = |c|$. Therefore aRc, proving that R is transitive.

EXAMPLE 14 For a less trivial example, consider the relation R on $\mathbf{Z}$ defined by aRb if a and b are either both even or both odd. Because every integer is either even or odd, aRa, $\forall a \in A$, and R is reflexive. Symmetry is obvious. Finally, suppose that $a, b, c \in \mathbf{Z}$ and aRb and bRc. Then if b is even, so are a and c, and we have aRc. Similarly, if b is odd, we get aRc. So R is transitive.

EXAMPLE 15 The relation R in the previous example could be defined in a different way by recognizing that aRb if $a - b$ is even. An integer is even if and only if it is divisible by 2. This insight leads us to try the same thing with an integer other than 2, say 3.
Define aRb if $a - b$ is divisible by 3; that is, $a - b = 3k$, for some $k \in \mathbf{Z}$. We must check the three properties to prove that R is an equivalence relation.
Reflexive: If $a \in \mathbf{Z}$, $a - a = 0 = 3 \cdot 0$, so aRa.
Symmetric: If aRb, then $a - b = 3k$, for some $k \in \mathbf{Z}$. But then $b - a = 3(-k)$, and so bRa.
Transitive: If aRb and bRc, then $a - b = 3k_1$, $b - c = 3k_2$, for some k_1, $k_2 \in \mathbf{Z}$. Then $a - c = (a - b) + (b - c) = 3k_1 + 3k_2 = 3(k_1 + k_2)$, so aRc.

EXAMPLE 16 You may have noticed that the number 3 played no important role in the previous argument, and could just as easily be replaced by any positive integer. Therefore, for $n \in \mathbf{Z}^+$, we define a relation R on $\mathbf{Z}$, called **congruence mod n**, by aRb if $a - b = nk$ for some $k \in \mathbf{Z}$. This is usually written $a \equiv b(\mathbf{mod}\ n)$. You are asked in the exercises to check that congruence mod n is in fact an equivalence relation. We will see a good deal more of congruence later.

When R is an equivalence relation, it is common to write $a \sim b$ instead of aRb, read as "a is equivalent to b." As we mentioned before, we want to generalize equality and view two elements related by an equivalence relation as "equivalent." If we group related elements together, we get what is known as an *equivalence class*.

Equivalence Classes

Definition 4.2.4 If R is an equivalence relation on A, and $a \in A$, the set $[a] = \{x \in A \mid x \sim a\}$ is called the **equivalence class** of a. Elements of the same class are said to be **equivalent**.

The following examples refer to the six examples of equivalence relations (11–16) previously given.

EXAMPLE 17 Since in any set A the only element equal to an element a is itself, $[a] = \{a\}$.

EXAMPLE 18 The class of a single baseball fan is simply the set of all fans who root for the same team as that person. Thus, for example, [John Updike] is the set of all Red Sox fans.

EXAMPLE 19 Every nonzero element of $\mathbf{Z}$ can be associated with exactly one other element, its opposite, with the same absolute value. Thus, $[a] = \{a, -a\}$. Notice that every class has two elements except $[0]$.

EXAMPLE 20 Even numbers are equivalent, as are odd numbers. Hence there are two classes: the set of even integers $\mathbf{E}$ and the set of odd integers $\mathbf{O}$.

EXAMPLE 21 Integers with the same remainder are equivalent, so there are three classes:

$[0] = \{\ldots -6, -3, 0, 3, 6, \ldots\}$, those integers exactly divisible by 3;
$[1] = \{\ldots -5, -2, 1, 4, \ldots\}$, the integers with a remainder of 1; and
$[2] = \{\ldots -4, -1, 2, 5, \ldots\}$, the integers with a remainder of 2.

Notice that we could also have called these classes $[6]$, $[-5]$, and $[20]$.

EXAMPLE 22 This example is similar to the previous one, but there are n possible remainders; namely, $0, 1, 2, \ldots n - 1$. Thus there are n equivalence classes, and we usually write them as $[0], [1], \ldots [n - 1]$.

You may have sensed that an equivalence relation divides the set in question into classes; we have distinct collections of Cubs fans, of Dodgers fans, and so on. In fact, the set of baseball fans is *partitioned* into sets with allegiance to different teams, as the following theorem shows.

Theorem 4.2.5   If R is an equivalence relation on a nonempty set A, then the set of equivalence classes of R forms a partition of A.

PROOF: Recall that a partition of the set A is a collection of nonempty subsets of A such that the union of these sets equals A and no two of the sets intersect. Since each equivalence class is a subset of A, so is the union of these equivalence classes. To prove that the union of the equivalence classes equals A, we must show that every element of A is in some equivalence class. But if $a \in A$, then $a \sim a$ since R is reflexive, so $a \in [a]$. Note that this also shows that every equivalence class is a nonempty set.

Now we must also show that different equivalence classes do not intersect. Suppose, on the contrary, that there exist distinct equivalence classes $[a]$ and $[b]$ such that $[a] \cap [b] \neq \emptyset$. Let $x \in [a] \cap [b]$. Then, in particular $x \sim a$, and since R is symmetric, $a \sim x$. But also $x \sim b$ since $x \in [b]$. Since R is transitive, it follows that $a \sim b$. Therefore $a \in [b]$. Now take an arbitrary $y \in [a]$, and notice that since $y \sim a$ and $a \sim b$, we have $y \sim b$ and hence $y \in [b]$. We have shown that in fact $[a] \subseteq [b]$. However, we could start this same argument with $y \in [b]$ and conclude that $[b] \subseteq [a]$, and so $[a] = [b]$. This is a contradiction to our assumption that $[a] \neq [b]$, so we can conclude that $[a] \cap [b] = \emptyset$. ●

The previous theorem shows that any equivalence relation on a set A leads to a partition of A. The next theorem shows that any partition of a set A gives rise to an equivalence relation on A.

Theorem 4.2.6   Let $\mathscr{P}$ be a partition of a nonempty set A. Define a relation R_1 on A by aR_1b if a and b are in the same element of the partition. Then R_1 is an equivalence relation on A.

PROOF: If $a \in A$, then $a \in X$ for some $X \in \mathscr{P}$, so clearly aR_1a. It's equally obvious that if aR_1b, so that $a, b \in X$, for some $X \in \mathscr{P}$, then we can just as well say $b, a \in X$, and bR_1a. Finally, if aR_1b and bR_1c, we have $a, b \in X$ and $b, c \in Y$, for some $X, Y \in \mathscr{P}$. But since $\mathscr{P}$ is a partition of A and $b \in X \cap Y \neq \emptyset$, we must have $X = Y$. Therefore $a, c \in X$, and aR_1c. Thus R_1 is reflexive, symmetric, and transitive, and hence an equivalence relation. ●

Theorems 4.2.5 and 4.2.6 are related in the following way. Let R be an equivalence relation on the nonempty set A. Theorem 4.2.5 says that the set $\mathscr{P}$ of equivalence classes of R is a partition of A. Theorem 4.2.6 says that this partition gives rise to an equivalence relation R_1 of A. From the definition of this equivalence relation given in the proof of Theorem 4.2.6, it follows that $R_1 = R$.

Similarly, if we start with a partition $\mathscr{P}$ of A, then there is a corresponding equivalence relation R_1 by Theorem 4.2.6. Theorem 4.2.5 says that there is a partition determined by this equivalence relation, namely, the set of equivalence classes of R_1. But this set is just the partition $\mathscr{P}$ that we started with.

These remarks tell us that there is a bijection between the set of all equivalence relations of A and the set of all partitions of A.

Partial and Linear Orderings

The relation R on the real numbers **R** defined by aRb if $a \leq b$ is not an equivalence relation because it is not symmetric. However, R is reflexive, transitive, and antisymmetric. It is an important relation because it gives an *ordering* of the real numbers: it defines what it means for one number to be "less than" another. The notion of *partial ordering* is a generalization of this example.

> **Definition 4.2.7** A relation R on a set A is called a **partial ordering** on A if R is reflexive, transitive, and antisymmetric.
>
> If A is a set and there exists a partial ordering on A, then we say that A is a **partially ordered set**.

EXAMPLE 23 As just noted, the relation $\leq$ is a partial ordering on **R**. It is also a partial ordering on **Z**. Note, however, that the relation $<$ is not a partial ordering on **R**.

EXAMPLE 24 Let A be a set and $\mathbf{P}(A)$ be the power set of A. The relation R on $\mathbf{P}(A)$ defined by XRY if $X \subseteq Y$ is a partial ordering on $\mathbf{P}(A)$. We leave the verification as an exercise.

The partial ordering $\leq$ on **R** has the property that for all $a, b \in \mathbf{R}$, either $a \leq b$ or $b \leq a$. Any partial ordering with this property is called a *linear ordering*.

> **Definition 4.2.8** Let A be a set and R be a partial ordering on A. We say that R is a **linear ordering** on A if for all $a, b \in A$, either aRb or bRa.
>
> If A is a set and there exists a linear ordering on A, then we say that A is a **linearly ordered set**.

EXAMPLE 25 As we just saw, the relation $\leq$ is a linear ordering on **R**.

EXAMPLE 26 Let A be a set and $\mathbf{P}(A)$ be the power set of A. The relation R on $\mathbf{P}(A)$ defined by XRY if $X \subseteq Y$ is not a linear ordering on $\mathbf{P}(A)$ except in the special cases where A is the empty set or A contains only one element.

● HISTORICAL COMMENTS: THE RATIONAL NUMBERS

After the theory of irrational numbers was firmly established by Weierstrass, Dedekind, and Cantor (see Section 2.2) among others, a formal definition of the rationals was undertaken, with the properties of the integers assumed as known. This definition gives us a nice application of equivalence relations.

Assuming the existence of the set of integers $\mathbf{Z}$, we look at the set $\mathbf{Z} \times \mathbf{Z}$ of ordered pairs of integers. We can think of a rational number as an ordered pair (a, b) of integers and intuitively associate the pair (a, b) with the fraction a/b. So, for instance, the fraction $1/2$ is the ordered pair $(1, 2)$, the fraction $2/3$ is the ordered pair $(2, 3)$, and so on. The denominator of a fraction cannot be zero, of course, so we consider just the set $S = \{(a, b) \in \mathbf{Z} \times \mathbf{Z} \mid b \neq 0\}$. Thus we define the rational numbers using only a subset of $\mathbf{Z} \times \mathbf{Z}$.

We know, however, from our introduction to fractions in grade school that a rational number has more than one representation. For example, $1/2 = 2/4 = 3/6 = 12/24$, $2/3 = 4/6 = 12/18$, and so forth. Hence the ordered pairs $(1, 2)$ and $(3, 6)$ represent the same rational number. For the purposes of a formal definition of the rational numbers, then, we need to express certain ordered pairs such as $(1, 2)$ and $(3, 6)$ as one element. We do this by defining a relation on the set S of ordered pairs of integers (a, b) with $b \neq 0$. How should such a relation be defined? Well, we need to look at the reason why the fractions $1/2$ and $3/6$ are equal and generally why any two fractions a/b and c/d are equal. An easy algebraic calculation tells us that $a/b = c/d$ if and only if $ad = bc$. Therefore, we define a relation R on S by $(a, b)R(c, d)$ if $ad = bc$. The next step is to verify that R is an equivalence relation on S. We leave the proof as an exercise. (See Exercise 18.)

Once we know that R is an equivalence relation, we can look at the set of equivalence classes. If $(a, b) \in S$, for simplicity of notation, we write $[a, b]$ for the equivalence class of (a, b), instead of $[(a, b)]$. Thus $[a, b]$ is the set of all ordered pairs (x, y) where $y \neq 0$ and $ay = bx$. For example, $[1, 2] = \{(1, 2), (2, 4), (3, 6), (4, 8), (-5, -10), \ldots \}$. So all of the different representations of the fraction $1/2$ are united in one element, the class $[1, 2]$. Finally, we can define the set of rational numbers $\mathbf{Q}$ as the set of all equivalence classes $[a, b]$.

We learn early in grade school the rules for adding and multiplying fractions. Using these rules as a model, we can give a formal definition of addition and multiplication on $\mathbf{Q}$.

If $[a, b], [c, d] \in \mathbf{Q}$, we define

$$[a, b] + [c, d] = [ad + bc, bd] \quad \text{and} \quad [a, b] \cdot [c, d] = [ac, bd].$$

Note that since $b \neq 0$ and $d \neq 0$, then $bd \neq 0$, so that $[ad + bc, bd]$ and $[ac, bd]$ are actually in $\mathbf{Q}$.

Our definitions of addition and multiplication on $\mathbf{Q}$ are not yet complete, however. They seem to depend on the choices of a, b, c, and d to represent the equivalence classes. For example, using the definition of addition, $[1, 2] + [2, 3] = [(1)(3) + (2)(2), (2)(3)] = [7, 6]$. But we can also write $[1, 2]$ as $[4, 8]$ and $[2, 3]$ as $[12, 18]$. Do we get the same answer if we use these new representations? Let's do the computation and see:

$$[4, 8] + [12, 18] = [(4)(18) + (8)(12), (8)(18)] = [168, 144].$$

But $[168, 144] = [7, 6]$ since $(168)(6) = (144)(7) = 1008$. So in this instance at least we can use different representations of $[1, 2]$ and $[2, 3]$ and arrive at the same sum.

Now we must verify that in *all* cases addition in **Q** does not depend on the representation of the equivalence classes. So suppose that we wish to add $[a, b]$ and $[c, d]$. Let $[x, y]$ be another representation of $[a, b]$ and $[z, w]$ be another representation of $[c, d]$. Then $(a, b)R(x, y)$ and $(c, d)R(z, w)$. This means that $ay = bx$ and $cw = dz$.

$[a, b] + [c, d] = [ad + bc, bd]$ and $[x, y] + [z, w] = [xw + yz, yw]$. To show that these two answers are the same we must show that $(ad + bc)(yw) = (bd)(xw + yz)$:

$$\begin{aligned}
(ad + bc)(yw) &= adyw + bcyw \\
&= aydw + cwby \\
&= bxdw + dzby \\
&= bd(xw + yz).
\end{aligned}$$

This completes the proof that addition on **Q** does not depend on the choices of integers to represent the equivalence classes. We say that addition on **Q** is *well defined*.

We leave the proof that multiplication on **Q** is well defined as an exercise.

As you might expect, addition and multiplication on **Q** are both associative and commutative and **Q** has an identity for addition and an identity for multiplication. Also, every element of **Q** has an inverse with respect to addition and every element of **Q** except the additive identity has an inverse with respect to multiplication. All of these properties are left to the exercises.

Exercises 4.2

1. For each of the following relations on **Z**, determine whether it is reflexive, symmetric, transitive, or antisymmetric.
   (a) aRb if $a > b$.
   (b) aRb if $a = -b$.
   (c) aRb if $b = 5a$.
   (d) aRb if $a \le b + 1$.

2. Let $A = \{1, 2, 3\}$. Let R be the relation on A defined by aRb if $a + b \ne 3$. Determine whether R is reflexive, symmetric, transitive, or antisymmetric.

3. Let A be a nonempty finite set and let $\mathbf{P}(A)$ be the power set of A. Determine whether each of the following relations on $\mathbf{P}(A)$ is reflexive, symmetric, transitive, or antisymmetric.
   (a) XRY if $X \subset Y$. (Recall: The symbol $\subset$ means that X is a *proper* subset of Y.)
   (b) XRY if $X \cap Y = \emptyset$.
   (c) XRY if $X \cap Y \neq \emptyset$.
   (d) XRY if $|X| = |Y|$.
   (e) XRY if $|X| \leq |Y|$.
   (f) XRY if $|X| < |Y|$.

4. For each of the following relations, determine whether it is reflexive, symmetric, or transitive. For those that are equivalence relations, give the equivalence classes.
   (a) On $\mathbf{F}(\mathbf{R})$, fRg if $f(0) = g(0)$.
   (b) On $\mathbf{Z} \times \mathbf{Z}$, $(a, b)R(c, d)$ if $ad = bc$.
   (c) On $A \times B$, for any sets A and B, $(a_1, b_1)R(a_2, b_2)$ if $a_1 = a_2$.
   (d) Let A be a nonempty set and let $f \in \mathbf{F}(A)$. Define a relation on A by aRb if $f(a) = f(b)$.
   (e) On $\mathbf{Q}$, $rRs \Leftrightarrow r - s \in \mathbf{Z}$.

5. Let R be the relation of congruence mod 4 on $\mathbf{Z}$: aRb if $a - b = 4k$, for some $k \in \mathbf{Z}$.
   (a) What integers are in the equivalence class of 18?
   (b) What integers are in the equivalence class of 31?
   (c) How many distinct equivalence classes are there? What are they?

6. Repeat Exercise 4 for congruence mod 5.

7. Let $n \in \mathbf{Z}^+$. Prove that congruence mod n is an equivalence relation on $\mathbf{Z}$.

8. Let R be a relation on a set A. Complete the following sentences.
   (a) R is not reflexive if . . . .
   (b) R is not symmetric if . . . .
   (c) R is not transitive if . . . .
   (d) R is not antisymmetric if . . . .

9. Let A be the set of all people in the United States. Let R be the relation on A defined by: a is related to b if a and b were born on the same day of the week.
   (a) Prove that R is an equivalence relation on A.
   (b) Describe the equivalence classes for R. How many are there?

10. Let R be the relation on $\mathbf{Z}$ defined by aRb if $a^2 - b^2$ is divisible by 4.
    (a) Prove that R is an equivalence relation on $\mathbf{Z}$.
    (b) What are the equivalence classes for this equivalance relation?

11. Let A be a nonempty set and $\mathbf{P}(A)$ be the power set of A.
    (a) Prove that the relation R on $\mathbf{P}(A)$ defined by XRY if $X \subseteq Y$ is a partial ordering on $\mathbf{P}(A)$.

(b) Prove that R is a linear ordering if A has only one element and is not a linear ordering if A has more than one element.

12. Let R be the relation on $\mathbf{Z}^+$ defined by aRb if b is a multiple of a; that is, if there exists $n \in \mathbf{Z}^+$ such that $b = na$.
    (a) Prove that R is a partial ordering on $\mathbf{Z}^+$.
    (b) Determine whether or not R is a linear ordering on $\mathbf{Z}^+$.

13. Let R be a relation on a set A. Complete the following sentence: R is not a partial ordering on A if $\ldots$.

14. Let R be a partial ordering on a set A. Complete the following sentence: R is not a linear ordering on A if $\ldots$.

15. Let $*$ be a binary operation on a set A. Assume that $*$ is associative and that A has an identity e with respect to $*$. Let R be the relation on A defined as follows: if $a, b \in A$, then aRb if there exists an invertible element $c \in A$ such that $b = c^{-1} * a * c$. Prove that R is an equivalence relation on A.

16. Let $\mathbf{F}(\mathbf{R})$ be the set of all functions $f : \mathbf{R} \to \mathbf{R}$. Define a relation R on $\mathbf{F}(\mathbf{R})$ by fRg if $f(x) \le g(x)$ for all $x \in \mathbf{R}$.
    (a) Prove that R is a partial ordering on $\mathbf{F}(\mathbf{R})$.
    (b) Determine whether or not R is a linear ordering on $\mathbf{F}(\mathbf{R})$.

17. Let R be a relation on a set A.
    (a) Prove that R is both symmetric and antisymmetric if and only if $R \subseteq \{(a, a) \mid a \in A\}$. Equivalently, R is both symmetric and antisymmetric if and only if for all $a, b \in A$, either a is not related to b or $a = b$.
    (b) Prove that if R is both symmetric and antisymmetric, then R is transitive.
    (c) If R is both symmetric and antisymmetric, does it follow that R is reflexive? Explain.

In the following exercises, we use the definition of the rational numbers $\mathbf{Q}$ as given in the Historical Comments at the end of this section.

18. Let $S = \{(a, b) \in \mathbf{Z} \times \mathbf{Z} \mid b \ne 0\}$. Let R be the relation on S defined by $(a, b)R(c, d)$ if $ad = bc$. Prove that R is an equivalence relation on S.

19. Prove that multiplication on $\mathbf{Q}$ is well defined in the same sense that we showed that addition on $\mathbf{Q}$ is well defined.

20. (a) Prove that addition on $\mathbf{Q}$ is associative.
    (b) Prove that addition on $\mathbf{Q}$ is commutative.

21. (a) Prove that multiplication on $\mathbf{Q}$ is associative.
    (b) Prove that multiplication on $\mathbf{Q}$ is commutative.

22. (a) Prove that $[0, x] = [0, y]$ for all $x, y \in \mathbf{Z}, x \ne 0, y \ne 0$.
    (b) Prove that $[ax, ay] = [x, y]$ for all $a, x, y \in \mathbf{Z}, a \ne 0, y \ne 0$.
    (c) Prove that $[x, x] = [1, 1]$ for all $x \in \mathbf{Z}, x \ne 0$.

23. (a) Prove that $[0, 1]$ is the additive identity of $\mathbf{Q}$.
    (b) Prove that every element of $\mathbf{Q}$ has an inverse with respect to addition.

24. (a) Prove that $[1, 1]$ is the multiplicative identity of $\mathbf{Q}$.
    (b) Prove that every element of $\mathbf{Q}$ except $[0, 1]$ has an inverse with respect to multiplication.

25. Prove that the following distributive law holds in $\mathbf{Q}$: $[a, b]([x, y] + [z, w]) = [a, b][x, y] + [a, b][z, w]$.

Discussion and Discovery Exercises

D1. In Theorem 4.2.5, it is proved that different equivalence classes have empty intersection. Explain the logic of this proof using symbols P and Q to stand for statements in the proof.

D2. Let R be the relation on $\mathbf{Z}$ defined by aRb if $a^2 + b^2$ is even. Criticize the following "proof" that R is an equivalence relation on $\mathbf{Z}$.

$aRa = a^2 + a^2 = 2a^2 =$ even. Therefore, $\mathbf{Z}$ is reflexive.
$aRb = a^2 + b^2 =$ even $= b^2 + a^2$. Therefore, aRb and bRa. $\mathbf{Z}$ is symmetric.
aRb and $bRc = a^2 + b^2$ and $b^2 + c^2$ are even.
$a^2 + c^2 = (a^2 + b^2) - (b^2 - c^2) =$ even $-$ even $=$ even. Therefore, transitive.

The following two exercises are for those who have completed an introductory linear algebra course.

D3. What does it mean for two $m \times n$ matrices to be *row equivalent*? Prove that row equivalence defines an equivalence relation on the set of all $m \times n$ matrices.

D4. What does it mean for two $n \times n$ matrices to be *similar*? Prove that similarity defines an equivalence relation on the set of all $n \times n$ matrices.

D5. In the remarks following the proof of Theorem 4.2.6, it is asserted that if A is a set, then there is a bijection between the set of all equivalance relations of A and the set of all partitions of A. Using these remarks as a guide, give a formal proof of this assertion by explicitly defining a function between these two sets and proving that it is a bijection.

D6. The definition of a relation R on a set A does not rule out the possibility that R is the empty set. For example, if A is the set $\mathbf{R}$ of real numbers and we say that aRb if $a < b$ and $b < a$, then $R = \varnothing$. We say that R is the empty relation. Determine whether the empty relation is reflexive, symmetric, transitive, or antisymmetric.

THE INTEGERS

5

AXIOMS AND BASIC PROPERTIES

In this chapter we study some of the most important properties of the set **Z** of integers. The study of the integers, called **number theory**, has held a fascination throughout history for both the professional and the amateur mathematician. Some of the most intricate and elegant mathematical ideas have developed as a result of attempts to solve difficult problems in number theory. We will discuss some of those ideas and problems and some of their history in this chapter.

The Axioms of the Integers

We will not construct the integers or give a formal definition. Our starting point will be the familiar properties of addition and multiplication in **Z**.

> The set **Z** of integers has two binary operations: **addition**, denoted by $+$, and **multiplication**, denoted by $\cdot$, with the following properties:
>
> **A1.** Addition is **associative**: $(x + y) + z = x + (y + z)$ for all x, y, $z \in \mathbf{Z}$.
> **A2.** Addition is **commutative**: $x + y = y + x$ for all $x, y \in \mathbf{Z}$.
> **A3.** **Z** has an **identity with respect to addition**, namely, the integer 0.
> **A4.** Every integer x in **Z** has an **inverse with respect to addition**, namely its negative $-x$.

> **A5.** Multiplication is **associative**: $(x \cdot y) \cdot z = x \cdot (y \cdot z)$ for all x, y, $z \in \mathbf{Z}$.
>
> **A6.** Multiplication is **commutative**: $x \cdot y = y \cdot x$ for all x, $y \in \mathbf{Z}$.
>
> **A7.** $\mathbf{Z}$ has an **identity with respect to multiplication**, namely, the integer 1 and $1 \neq 0$.
>
> **A8.** For all integers x, y, z, $x \cdot (y + z) = x \cdot y + x \cdot z$. These are called the **distributive laws**.

We do not prove these properties of $\mathbf{Z}$; rather we take them as **axioms**: statements we *assume* to be true about the integers. We will list three more axioms for the integers later in this section. The last axiom, the **Well-Ordering Principle**, leads to the **Principle of Induction**, an important tool for proving theorems in all areas of mathematics. We will discuss induction in Section 5.2.

Note that it follows from Proposition 4.1.4 that 0 is the only additive identity of $\mathbf{Z}$ and that 1 is the only multiplicative identity.

Note also that no mention is made in these properties of inverses with respect to multiplication. The reason as we have noted, is that only 1 and -1 have multiplicative inverses in the set of integers. For a formal proof of this fact, see Corollary 5.1.8 later in the section.

Notation: From now on, we will use the juxtaposition xy instead of $x \cdot y$ to denote multiplication of integers.

We will also use the familiar notation $x - y$ to mean $x + (-y)$.

Many of the familiar algebraic properties of the integers can be proven from the eight axioms just listed. We state and prove some of these properties in the next proposition.

Proposition 5.1.1   Let a, b, $c \in \mathbf{Z}$.

> **P1.** If $a + b = a + c$, then $b = c$.
> **P2.** $a0 = 0a = 0$.
> **P3.** $(-a)b = a(-b) = -(ab)$.
> **P4.** $-(-a) = a$.

PROOF: P1. Suppose that $a + b = a + c$. Then, by first adding $-a$ to both sides of this equation, we get the following sequence of steps:

$$(-a) + (a + b) = (-a) + (a + c).$$
$$(-a + a) + b = (-a + a) + c \quad \text{by the associative law } \mathbf{A1}.$$
$$0 + b = 0 + c \quad \text{by } \mathbf{A4}.$$
$$b = c \quad \text{by } \mathbf{A3}.$$

P2. Applying **A8** and the fact that $0 + 0 = 0$, we get:

$$a0 = a(0 + 0) = a0 + a0.$$

Therefore, $a0 + 0 = a0 + a0$.

Now we can apply **P1** to conclude that $a0 = 0$.

Since multiplication in **Z** is commutative, we also get $0a = 0$.

P3. We will prove that $(-a)b = -(ab)$. (The proof that $a(-b) = -(ab)$ is similar.)

Notice that the left- and right-hand sides of this equation are conceptually different. The left-hand side is the additive inverse of a multiplied on the right by b. The right-hand side is the additive inverse of ab. So there is something to prove!

Now $ab + (-a)b = (a + (-a))b = 0b = 0$, by **P2**.

We conclude that $(-a)b$ is the additive inverse of ab. Therefore, $(-a)b = -(ab)$.

The proof of **P4** is left as an exercise. ●

The following additional properties are corollaries of Proposition 5.1.1. The proofs are left as exercises.

Proposition 5.1.2   Let $a, b, c \in \mathbf{Z}$.

P5. $(-a)(-b) = ab$.
P6. $a(b - c) = ab - ac$.
P7. $(-1)a = -a$.
P8. $(-1)(-1) = 1$.

As usual, we let $\mathbf{Z}^+$ be the set of positive integers: $\mathbf{Z}^+ = \{1, 2, 3, \ldots\}$ where $2 = 1 + 1$, $3 = 2 + 1$, $\ldots$, and so on. In addition to the eight axioms of **Z** previously listed, we have three additional axioms that are really properties of $\mathbf{Z}^+$. The first two are:

> **A9. Closure Property** $\mathbf{Z}^+$ is closed with respect to addition and multiplication: if $x, y \in \mathbf{Z}^+$, then $x + y \in \mathbf{Z}^+$ and $xy \in \mathbf{Z}^+$.
>
> **A10. Trichotomy Law** For every integer x, exactly one of the following statements is true: $x \in \mathbf{Z}^+$, $-x \in \mathbf{Z}^+$, or $x = 0$.

As we did with the first eight axioms, we can prove familiar properties of **Z** using Axioms 9 and 10.

Proposition 5.1.3   If $x \in \mathbf{Z}$, $x \neq 0$, then $x^2 \in \mathbf{Z}^+$.

PROOF: By Axiom 10, either $x \in \mathbf{Z}^+$ or $-x \in \mathbf{Z}^+$. If $x \in \mathbf{Z}^+$, then $x^2 = xx \in \mathbf{Z}^+$ by closure of multiplication. On the other hand, if $-x \in \mathbf{Z}^+$, then $x^2 = xx = (-x)(-x) \in \mathbf{Z}^+$ by closure of multiplication and **P5**. ●

Inequalities

In the set $\mathbf{Z}$ of integers, there is a notion of one element being "smaller" than another. A formal definition of this notion is the following.

> **Definition 5.1.4** Let $x, y \in \mathbf{Z}$. We say $x < y$ (read: x is less than y) if $y - x \in \mathbf{Z}^+$.

Note: If $x < y$, we can also write $y > x$ (read: y is greater than x). We also write $x \leq y$ if $x < y$ or $x = y$. Similarly, $y \geq x$ if $y > x$ or $y = x$.

Note also that the symbols $<$ and $\leq$ define relations on $\mathbf{Z}$ and that the symbol $\leq$ defines a linear ordering on $\mathbf{Z}$.

It follows from Definition 5.1.4 that for every integer n, the statement $n > 0$ is equivalent to the statement $n \in \mathbf{Z}^+$ and the statement $n < 0$ is equivalent to the statement $-n \in \mathbf{Z}^+$. Therefore, we can say: $\mathbf{Z}^+ = \{n \in \mathbf{Z} \mid n > 0\}$.

The following well-known properties of inequalities are consequences of this definition as well as Axioms 9 and 10.

Proposition 5.1.5 Let $a, b, c \in \mathbf{Z}$.

Q1. Exactly one of the following holds: $a < b$, $b < a$, or $a = b$.
Q2. If $a > 0$, then $-a < 0$ and if $a < 0$, then $-a > 0$.
Q3. If $a > 0$ and $b > 0$, then $a + b > 0$ and $ab > 0$.
Q4. If $a > 0$ and $b < 0$, then $ab < 0$.
Q5. If $a < 0$ and $b < 0$, then $ab > 0$.
Q6. If $a < b$ and $b < c$, then $a < c$.
Q7. If $a < b$, then $a + c < b + c$.
Q8. If $a < b$ and $c > 0$, then $ac < bc$.
Q9. If $a < b$ and $c < 0$, then $ac > bc$.

PROOF: Q1. It follows immediately from Axiom 10 that exactly one of the following statements is true:

$$b - a \in \mathbf{Z}^+, \qquad -(b - a) = a - b \in \mathbf{Z}^+, \quad \text{or} \quad b - a = 0.$$

Therefore, exactly one of the following holds:

$$a < b, \qquad b < a, \quad \text{or} \quad a = b.$$

Q2. If $a > 0$, then $0 - (-a) = a \in \mathbf{Z}^+$, implying that $-a < 0$. A similar proof shows that if $a < 0$, then $-a > 0$.

Q3. This is just a restatement of Axiom 9.

Q4. Suppose that $a > 0$ and $b < 0$. By **Q2**, $-b > 0$ and so by **Q3**, $a(-b) > 0$. But $a(-b) = -(ab)$ by **P3** so $-(ab) > 0$.

Applying **Q2** again gives $-(-(ab)) < 0$. But $-(-(ab)) = ab$, so we can conclude that $ab < 0$.

The proofs of **Q5** through **Q9** are left as exercises.

Note: Statements **Q2** through **Q9** of Proposition 5.1.5 remain true if $<$ is replaced by $\leq$ and $>$ is replaced by $\geq$.

The Well-Ordering Principle

Our final axiom of **Z** is known as the Well-Ordering Principle and is another property of $\mathbf{Z}^+$.

> **A11. The Well-Ordering Principle** Every nonempty subset of $\mathbf{Z}^+$ has a smallest element; that is, if S is a nonempty subset of $\mathbf{Z}^+$, then there exists $a \in S$ such that $a \leq x$ for all $x \in S$.

Despite its apparent simplicity, the Well-Ordering Principle is an important tool for proving properties of the integers. In the next section, we will see how the Well-Ordering Principle is used to establish the Principle of Mathematical Induction.

We now prove what seems an obvious result but whose formal proof requires the Well-Ordering Principle.

Proposition 5.1.6   There is no integer x such that $0 < x < 1$.

PROOF: We will assume that the proposition is false and derive a contradiction. So suppose there is an integer x such that $0 < x < 1$.

Let $S = \{n \in \mathbf{Z}^+ \mid 0 < n < 1\}$. Now $x \in S$ by our assumption, so $S \neq \varnothing$. Thus, by the Well-Ordering Principle, S has a smallest element. Call it x_0. Since $x_0 < 1$, $x_0 - 1 < 0$.

Now by **Q4**, $x_0(x_0 - 1) < 0$, implying that $x_0^2 - x_0 < 0$ or $x_0^2 < x_0$. Since $x_0 < 1$, it follows from **Q6** that $x_0^2 < 1$. But $x_0^2 \in \mathbf{Z}^+$ (Why?), which means that $x_0^2 \in S$.

Thus x_0^2 is an element of S, which is smaller than x_0—a contradiction to the choice of x_0 as the smallest element of S. Therefore our assumption that there is an integer x such that $0 < x < 1$ must be false and the proposition follows.

Corollary 5.1.7   1 is the smallest element of $\mathbf{Z}^+$.

PROOF: We leave it as an exercise.

We can now prove that only 1 and -1 have multiplicative inverses in **Z**.

Corollary 5.1.8   The only integers having multiplicative inverses in **Z** are ± 1.

PROOF: First note that the statement of Corollary 5.1.8 is a statement containing a universal quantifier and an equivalence. It can be restated as:

For all $a \in \mathbf{Z}$, a has a multiplicative inverse in $\mathbf{Z} \Leftrightarrow a = \pm 1$.

To prove the first implication, suppose $a \in \mathbf{Z}$ has a multiplicative inverse in $\mathbf{Z}$. Then there exists $x \in \mathbf{Z}$ such that $ax = 1$. Clearly $a \neq 0$, so $a \in \mathbf{Z}^+$ or $-a \in \mathbf{Z}^+$. Suppose $a \in \mathbf{Z}^+$ and $a \neq 1$. Then $a > 1$ by Corollary 5.1.7, and since $ax = 1 \in \mathbf{Z}^+$, $x \in \mathbf{Z}^+$ also. (See Exercise 11(a).) Now since $x \neq 0$, $x \geq 1$. Multiplying both sides of the inequality $a > 1$ by x, we get $1 = ax > 1x \geq 1$, a contradiction. Therefore $a = 1$.

If $-a \in \mathbf{Z}^+$, a similar proof shows that $a = -1$. It now follows that if a has a multiplicative inverse in $\mathbf{Z}$, then $a = \pm 1$.

Conversely, if $a = \pm 1$, then a has a multiplicative inverse in $\mathbf{Z}$ since $(1)(1) = 1$ and $(-1)(-1) = 1$. ●

● HISTORICAL COMMENTS: PEANO'S AXIOMS

The axiomatic construction of the integers given in this section came fairly late in the history of mathematics, in the late 19th century in fact. Not all mathematicians thought that a logical development of the natural numbers (that is, positive integers) was possible. "God made the integers, all else is the work of man," is the famous quote of **Leopold Kronecker** (1823–1891). Nonetheless, formal definitions of the integers were devised, most notably by **Giuseppe Peano** (1858–1932) in 1889. Peano listed five axioms, or postulates as they were called, of the natural numbers:

$P1$: 1 is a natural number.
$P2$: Every natural number has a successor.
$P3$: 1 is not the successor of any natural number.
$P4$: If the successors of the natural numbers n and m are equal, then so are n and m.
$P5$: If S is a subset of the natural numbers with the two properties that (1) $1 \in S$ and (2) whenever $n \in S$, the successor of n is also in S, then S is the set of all natural numbers.

The notion of successor is left undefined but intuitively we think of the successor of n as being the number $n + 1$.

From these five postulates, Peano was able to *prove*, as *theorems*, all of the axioms of $\mathbf{Z}$ given in this section. Postulate 5, for example, is the Principle of Induction that we will discuss in the next section and is equivalent to the Well-Ordering Principle. Remarkably, with the set of natural numbers and Peano's postulates as a starting point, one can construct the set of integers $\mathbf{Z}$, the set of rational numbers $\mathbf{Q}$ (see the Historical Comments in Section 4.2), and the set of real numbers $\mathbf{R}$, without any additional axioms. We have chosen not to adopt Peano's approach here so that we may more quickly get to important algebraic properties of the integers.

Exercises 5.1

The following exercises, many of them having to do with very familiar properties of the integers, should be done using only the 11 axioms of the integers or the results proved in this section.

1. Complete the proofs of **P4** through **P8** by proving the following:
   (a) $-(-a) = a$ for all $a \in \mathbf{Z}$.
   (b) $(-a)(-b) = ab$ for all $a, b \in \mathbf{Z}$.
   (c) $a(b - c) = ab - ac$ for all $a, b, c \in \mathbf{Z}$.
   (d) $(-1)a = -a$ for all $a \in \mathbf{Z}$.
   (e) $(-1)(-1) = 1$.

2. Let $a, b \in \mathbf{Z}$. Prove that $-(a + b) = -a - b$.

3. Let $a, b \in \mathbf{Z}$. Suppose that $a < b$. Prove that $-a > -b$.

4. (a) State the negation of Proposition 5.1.3.
   (b) State the converse of Proposition 5.1.3. Prove or disprove.

5. Complete the proof of Proposition 5.1.5 by proving the following:
   (a) If $a < 0$ and $b < 0$, then $ab > 0$.
   (b) If $a < b$ and $b < c$, then $a < c$.
   (c) If $a < b$, then $a + c < b + c$.
   (d) If $a < b$ and $c > 0$ then $ac < bc$.
   (e) If $a < b$ and $c < 0$, then $ac > bc$.

6. Let $a, b, c, d \in \mathbf{Z}$. Prove that if $a < b$ and $c < d$, then $a + c < b + d$.

7. Let $a, b \in \mathbf{Z}$. Suppose that $ab > 0$. Prove that either $a > 0$ and $b > 0$ or $a < 0$ and $b < 0$.

8. Let $a, b \in \mathbf{Z}$. Prove the following converse of statement **Q3** in Proposition 5.1.5: if $a + b > 0$ and $ab > 0$, then $a > 0$ and $b > 0$.

9. Let $a, b, c \in \mathbf{Z}$.
   (a) Suppose that $ac < bc$ and $c > 0$. Prove that $a < b$.
   (b) Suppose that $ac < bc$ and $c < 0$. Prove that $a > b$.

10. Consider the statement S: "If the sum of two integers is less than or equal to 100, then one of the numbers is less than or equal to 50."
    (a) Rewrite S using variables and mathematical symbols.
    (b) State the contrapositive of S.
    (c) Prove or disprove S.
    (d) State the converse of S. Prove or disprove.

11. Let $a, b \in \mathbf{Z}$.
    (a) Suppose that $a \in \mathbf{Z}^+$ and $ab \in \mathbf{Z}^+$. Prove that $b \in \mathbf{Z}^+$.
    (b) Suppose that $a < b$ and $a \in \mathbf{Z}^+$. Prove that $b \in \mathbf{Z}^+$.

12. Write the negation of Corollary 5.1.8.

13. Prove that $x^2 \geq x$ for all $x \in \mathbf{Z}$.

14. Prove that if $x \in \mathbf{Z}$ and $x < 0$, then $x^3 < 0$.

15. Let P be the statement "For all $x, y \in \mathbf{Z}$, if $xy = 0$, then $x = 0$ or $y = 0$."
    (a) Write the negation of P.
    (b) Write the contrapositive of P.
    (c) Prove or disprove P.
    (d) Write the converse of P. Prove or disprove.

16. Let $x, y, z \in \mathbf{Z}$. Prove that if $xy = xz$ and if $x \neq 0$, then $y = z$.

17. Let $n \in \mathbf{Z}$. Prove that there is no integer x such that $n < x < n + 1$.

18. Prove that there is no smallest element of $\mathbf{Z}$ and no largest element of $\mathbf{Z}$.

19. Let $\mathbf{Z}^-$ be the set of negative integers.
    (a) Prove that -1 is the largest element of $\mathbf{Z}^-$.
    (b) Prove that every nonempty subset of $\mathbf{Z}^-$ has a largest element.
    (c) Prove that $\mathbf{Z}^-$ is closed in $\mathbf{Z}$ under addition but not closed under multiplication.

20. Let $n_0 \in \mathbf{Z}$ and let $T = \{n \in \mathbf{Z} \mid n \geq n_0\}$. Prove that every nonempty subset of T has a smallest element.

21. (a) Write the negation of Axiom 9.
    (b) Write the negation of Axiom 10.
    (c) Write the negation of Axiom 11.

22. Prove Corollary 5.1.7.

23. Let S be a nonempty subset of $\mathbf{Z}^+$. Complete the following sentence: An element a is not the smallest element of S if . . . .

24. (a) Prove that the symbol $<$ defines a relation on $\mathbf{Z}$ that is transitive but not reflexive and not symmetric.
    (b) Is $<$ an antisymmetric relation? Prove your answer.

25. (a) Prove that the symbol $\leq$ defines a partial ordering on $\mathbf{Z}$.
    (b) Prove that the symbol $\leq$ defines a linear ordering on $\mathbf{Z}$.

Discussion and Discovery Exercises

D1. Criticize the following "proof" of Exercise 5(c).

   If $a < b$, then $b - a > 0$. Hence $b - a + c > c$. Therefore, $b + c > a + c$ or in other words, $a + c < b + c$.

D2. What is wrong with the following "proof" of Corollary 5.1.8?

   Suppose $a \in \mathbf{Z}$ has a multiplicative inverse in $\mathbf{Z}$. Then there exists $x \in \mathbf{Z}$ such that $ax = 1$. So $a = 1/x$. But if $x > 1$, or if $x < -1$, then

$1/x$ is not an integer, contradicting the fact that a is an integer. Therefore, $x = \pm 1$. Since $ax = 1$, it follows that $a = \pm 1$.

D3. Let a and b be positive integers. State and prove a theorem that gives necessary and sufficient conditions on a and b for the existence of two real numbers whose sum is a and whose product is b. For example, there are real numbers whose sum is 13 and whose product is 36, but there are no real numbers whose sum is 12 and whose product is 40. Your answer should be in the form of a biconditional.

D4. Let **N** be the set of natural numbers as defined by Peano with the axioms $P1$, $P2$, $P3$, $P4$, $P5$ that were given at the end of this section. These axioms allow us to define a function f from **N** to **N** by sending each $n \in \mathbf{N}$ to its successor; that is, if we let n_s denote the successor of n, then $f: \mathbf{N} \to \mathbf{N}$ is defined by $f(n) = n_s$. Note that axiom $P2$ guarantees that every $n \in \mathbf{N}$ has a successor. In the following, you may use only Peano's axioms.
   (a) Prove that 1 is not in the image of f.
   (b) Prove that f is injective.
   (c) Prove that Im $f = \mathbf{N} - \{1\}$.
   (d) Prove that **N** is not a finite set.
   (e) Prove that if $n \in \mathbf{N}$, then $f(n) \neq n$.

5.2 INDUCTION

Induction: A Method of Proof

Induction is a method for proving statements about the positive integers. Let $P(n)$ be such a statement. $P(n)$ may be a formula, such as "the sum of the first n positive integers is $n(n + 1)/2$" or "the sum of the first n odd positive integers is a perfect square." Or $P(n)$ may be a statement such as "every polynomial of degree n with real coefficients has at most n zeros."

The purpose of an induction proof is to show that a statement $P(n)$ is true for *every* positive integer n. The first step is to verify that $P(1)$ is true; that is, that the statement is true when $n = 1$. Once $P(1)$ is established to be true, we want to verify $P(2)$, $P(3)$, $P(4)$, and so on. Since there are an infinite number of these statements, they cannot all be verified separately. In an induction proof, we show that, given a positive integer k for which $P(k)$ is true, it follows that the statement $P(k + 1)$ is true. In other words, if the statement is true for some integer, then it is true for the following integer. This establishes that in the sequence of statements $P(1)$, $P(2)$, $P(3), \ldots$, whenever one statement in this sequence is true, then the next statement in the sequence must also be true.

Since we have already verified that $P(1)$ is true, it now follows that the next statement $P(2)$ is true. Once $P(2)$ is known to be true then $P(3)$ must also be true, then $P(4)$, then $P(5)$, and so on.

Every $P(n)$, for any given n, has now been verified in a finite number of steps. For example, $P(500)$ is true because $P(1)$ is true, which implies $P(2)$ is true, which implies $P(3)$ is true, . . . , which implies $P(499)$ is true, which finally implies $P(500)$ is true.

Thus in an induction proof, there are two steps required: first, prove that $P(1)$ is true; and second, assuming $P(k)$ is true for a positive integer k, prove that $P(k + 1)$ is true.

What follows is a formal proof of the induction principle, called the First Principle of Mathematical Induction.

Theorem 5.2.1   **First Principle of Mathematical Induction** Let $P(n)$ be a statement about the positive integer n. Suppose that

1. $P(1)$ is true.
2. Whenever k is a positive integer for which $P(k)$ is true, then $P(k + 1)$ is also true.

Then $P(n)$ is true for every positive integer n.

PROOF: Let S be the set of all positive integers for which $P(n)$ is false. If we can show that $S = \varnothing$, it will follow that $P(n)$ is true for all positive integers n. We will assume that $S \neq \varnothing$ and derive a contradiction.

Since $S \neq \varnothing$, by the Well-Ordering Principle, S has a smallest element. Call it k_0.

Now 1 is not in S since $P(1)$ is true, so $k_0 > 1$ by Corollary 5.1.7. Therefore $k_0 - 1$ is a positive integer and is not in S by the choice of k_0 as the smallest element of S. Therefore $P(k_0 - 1)$ is true. But then by condition 2, $P(k_0 - 1 + 1) = P(k_0)$ is true. But $P(k_0)$ must be false since k_0 is in S.

This gives us our contradiction and so $S = \varnothing$. ◖

The assumption that $P(k)$ is true in condition 2 is called the **induction hypothesis**.

EXAMPLE 1   The formula for the sum of the first n positive integers

$$1 + 2 + 3 + \ldots + n = \frac{n(n + 1)}{2}$$

can be proven by induction.

Given a positive integer n, let $P(n)$ be the statement

$$1 + 2 + 3 + \ldots + n = \frac{n(n + 1)}{2}.$$

Since the formula is obviously true for $n = 1$, $P(1)$ is true. Therefore condition 1 of Theorem 5.2.1 holds.

Suppose now that k is a positive integer for which $P(k)$ is true. Then

$$1 + 2 + 3 + \ldots + k = \frac{k(k+1)}{2}.$$

(This is the induction hypothesis.) We now want to establish that $P(k + 1)$ is true, which is equivalent to showing that

$$1 + 2 + 3 + \ldots + k + (k+1) = \frac{(k+1)(k+2)}{2}.$$

Adding $k + 1$ to both sides of the induction hypothesis, we get

$$1 + 2 + 3 + \ldots + k + (k+1) = \frac{k(k+1)}{2} + (k+1)$$

$$= \frac{k^2 + k}{2} + \frac{2k + 2}{2}$$

$$= \frac{k^2 + 3k + 2}{2}$$

$$= \frac{(k+1)(k+2)}{2}.$$

Hence $P(k + 1)$ is true. It now follows by induction that $P(n)$ is true for every positive integer n.

EXAMPLE 2 We will prove by induction that for every positive integer n,

$$1 + 2 + 2^2 + 2^3 + \ldots + 2^n = 2^{n+1} - 1.$$

Call this statement $P(n)$.
Since $1 + 2 = 2^{1+1} - 1$, we see that $P(1)$ is true.
Suppose now that k is a positive integer for which $P(k)$ is true. Then

$$1 + 2 + 2^2 + 2^3 + \ldots + 2^k = 2^{k+1} - 1.$$

So

$$1 + 2 + 2^2 + 2^3 + \ldots + 2^k + 2^{k+1} = 2^{k+1} - 1 + 2^{k+1}$$
$$= 2 \cdot 2^{k+1} - 1$$
$$= 2^{k+2} - 1,$$

which proves that $P(k + 1)$ is true. Therefore, by induction, the formula is true for all $n \in \mathbf{Z}^+$.

Some Other Forms of Induction

Sometimes a statement $P(n)$ may be true not for *all* positive integers n but rather for all integers beyond a certain point, say for all integers $n > 5$ or for all $n \geq 0$. A modified form of the Principle of Induction can be used to prove such statements. It goes as follows.

Theorem 5.2.2 **First Principle of Mathematical Induction, Modified Form** Let $P(n)$ be a statement about the integer n. Suppose that there is an integer n_0 such that

1. $P(n_0)$ is true.
2. Whenever $k \geq n_0$ is an integer for which $P(k)$ is true, then $P(k + 1)$ is also true.

Then $P(n)$ is true for every integer $n \geq n_0$.

PROOF: We leave it as an exercise.

EXAMPLE 3 We will prove by induction that for every integer $n \geq 3$, $n^2 > 2n + 1$.

Given a positive integer n, let $P(n)$ be the statement $n^2 > 2n + 1$. It is clear that $P(3)$ is true.

Now suppose $k \geq 3$ is an integer such that $P(k)$ is true; that is, $k^2 > 2k + 1$. Then we want to prove that $P(k + 1)$ is true; that is, we want to prove that $(k + 1)^2 > 2(k + 1) + 1 = 2k + 3$.

But

$$
\begin{aligned}
(k + 1)^2 &= k^2 + 2k + 1 \\
&> 2k + 1 + 2k + 1 \\
&= 4k + 2 \\
&> 2k + 3.
\end{aligned}
$$

The first inequality comes from the induction hypothesis and the second, $4k + 2 > 2k + 3$, is clear since $2k > 1$. Therefore, $P(k + 1)$ is true and so by Theorem 5.2.2, $P(n)$ is true for all $n \geq 3$.

Sometimes, in order to do a proof by induction, it is necessary to modify the induction hypothesis and assume more than just that $P(k)$ is true for a given k but rather that $P(i)$ is true for all positive integers $i \leq k$ and then prove that $P(k + 1)$ is true. This stronger induction hypothesis is still sufficient to prove that $P(n)$ is true for all positive integers n. But of course it is necessary to prove another theorem to establish that fact. We state it next and leave the proof as an exercise as it is very similar to the proof of Theorem 5.2.1.

Theorem 5.2.3 **Second Principle of Mathematical Induction** Let $P(n)$ be a statement about the positive integer n. Suppose that

1. $P(1)$ is true.
2. If $k \in \mathbf{Z}^+$ and $P(i)$ is true for every positive integer $i \leq k$, then $P(k + 1)$ is true.

Then $P(n)$ is true for every positive integer n.

EXAMPLE 4   Suppose that f is function from $\mathbf{Z}^+$ to $\mathbf{Z}^+$ such that $f(1) = 1$, $f(2) = 5$, and $f(n + 1) = f(n) + 2f(n - 1)$ for all $n \geq 2$. This is an example of a **recursively defined** function. All values of f are uniquely determined by these equations but f is not defined by an explicit formula. Rather the value of f at a positive integer k is determined by its values at positive integers preceding k.

Many functions that are defined recursively do have explicit formulas and induction is often the way to verify such a formula. In this example we will use the Second Principle of Induction to verify an explicit formula for $f(n)$.

Induction, however, does not tell us what this formula is. One can make an educated guess for f by plugging values of n into f and then trying to detect a pattern.

Since $f(3) = 7$, $f(4) = 17$, $f(5) = 31$, and $f(6) = 65$, we see that $f(n)$ is one more or one less than a power of 2 and a little trial and error will lead us to the formula

$$f(n) = 2^n + (-1)^n.$$

Let $P(n)$ be the statement $f(n) = 2^n + (-1)^n$. Since $f(1) = 1 = 2^1 + (-1)^1$ and $f(2) = 5 = 2^2 + (-1)^2$, $P(1)$ and $P(2)$ are true.

Next suppose that k is a positive integer greater than 2 such that $P(i)$ is true for all $i = 1, 2, \ldots k$. Then, in particular, $P(k)$ and $P(k - 1)$ are true. So we have

$$f(k) = 2^k + (-1)^k \quad \text{and} \quad f(k - 1) = 2^{k-1} + (-1)^{k-1}.$$

Hence

$$
\begin{aligned}
f(k + 1) &= f(k) + 2f(k - 1) \\
&= 2^k + (-1)^k + 2(2^{k-1} + (-1)^{k-1}) \\
&= 2^k + (-1)^k + 2^k + 2(-1)^{k-1} \\
&= 2 \cdot 2^k + (-1)^k(1 - 2) \\
&= 2^{k+1} + (-1)^{k+1}.
\end{aligned}
$$

This shows that $P(k + 1)$ is true and therefore, by the Second Principle of Induction, $P(n)$ is true for all positive integers n.

One can state a modified form of the Second Principle of Induction similar to the modified form of the First Principle. We leave the statement as an exercise.

The Principle of Induction is often stated in the language of sets. The theorem that follows is a restatement of the First Principle of Mathematical Induction in set theory language. The other forms of induction may be restated in a similar way.

Theorem 5.2.4   Let S be a subset of $\mathbf{Z}^+$. Suppose that

   1. $1 \in S$.
   2. If k is a positive integer for which $k \in S$, then $k + 1 \in S$.

Then $S = \mathbf{Z}^+$.

PROOF: We leave it as an exercise. ●

   As an example of Theorem 5.2.4, let's revisit Example 1.

EXAMPLE 5   Let

$$S = \left\{ n \in \mathbf{Z}^+ \;\middle|\; 1 + 2 + 3 + \ldots + n = \frac{n(n + 1)}{2} \right\}.$$

Then clearly $1 \in S$. Suppose now that $k \in S$. Then

$$1 + 2 + 3 + \ldots + k = \frac{k(k + 1)}{2}.$$

It follows that

$$1 + 2 + 3 + \ldots + k + (k + 1) = \frac{k(k + 1)}{2} + (k + 1)$$

$$= \frac{k^2 + k}{2} + \frac{2k + 2}{2}$$

$$= \frac{k^2 + 3k + 2}{2}$$

$$= \frac{(k + 1)(k + 2)}{2}$$

proving that $k + 1 \in S$. By Theorem 5.2.4, $S = \mathbf{Z}^+$. We can now conclude that

$$1 + 2 + 3 + \ldots + n = \frac{n(n + 1)}{2}$$

for all positive integers n. Note that the algebraic content of this proof is the same as Example 1, so that there is no substantive difference in the two approaches.

   Often induction can be used to generalize the results of a theorem. Recall that in Theorem 2.2.3, we proved that if A, B, and C are sets, then $A \cap (B \cup C) = (A \cap B) \cup (A \cap C)$. The following theorem is a generalization of that fact.

Theorem 5.2.5 Let $A, B_1, B_2, \ldots, B_n$ be sets. Then $A \cap (B_1 \cup B_2 \cup \ldots \cup B_n) = (A \cap B_1) \cup (A \cap B_2) \cup \ldots \cup (A \cap B_n)$.

PROOF: The induction proof will be done with respect to the variable n. Let $P(n)$ be the statement: "Whenever $A, B_1, B_2, \ldots, B_n$ are sets, then $A \cap (B_1 \cup B_2 \cup \ldots \cup B_n) = (A \cap B_1) \cup (A \cap B_2) \cup \ldots \cup (A \cap B_n)$."

Clearly $P(1)$ is true since it is merely an identity.

Suppose now that k is a positive integer such that $P(k)$ is true. Then $A \cap (B_1 \cup B_2 \cup \ldots \cup B_k) = (A \cap B_1) \cup (A \cap B_2) \cup \ldots \cup (A \cap B_k)$ whenever we have sets $A, B_1, B_2, \ldots, B_k$. This is the induction hypothesis.

Now we wish to prove that $P(k + 1)$ is true. So we let $A, B_1, B_2, \ldots, B_k, B_{k+1}$ be sets. Then

$$
\begin{aligned}
A \cap (B_1 \cup B_2 \cup &\ldots \cup B_k \cup B_{k+1}) \\
&= A \cap ((B_1 \cup B_2 \cup \ldots \cup B_k) \cup B_{k+1}) \\
&= (A \cap (B_1 \cup B_2 \cup \ldots \cup B_k)) \cup (A \cap B_{k+1}) \\
&= ((A \cap B_1) \cup (A \cap B_2) \cup \ldots \cup (A \cap B_k)) \cup (A \cap B_{k+1}) \\
&= (A \cap B_1) \cup (A \cap B_2) \cup \ldots \cup (A \cap B_k) \cup (A \cap B_{k+1}).
\end{aligned}
$$

The second equality follows from Theorem 2.2.3, which is the case $k = 2$, and the third one from the induction hypothesis. Since $P(k + 1)$ is now proved, the proof of the theorem is complete. ●

The Binomial Theorem

An important application of mathematical induction is the **Binomial Theorem**.

Definition 5.2.6 Let $n \in \mathbf{Z}^+$ and $r \in \mathbf{Z}$ such that $0 \le r \le n$. The **binomial coefficient** $\binom{n}{r}$ is defined as $\binom{n}{r} = \dfrac{n!}{r!(n-r)!}$.

As usual, $n! = (n)(n - 1)(n - 2) \ldots 2 \cdot 1$ if $n \ge 1$ and $0! = 1$.

The binomial coefficient $\binom{n}{r}$ is the number of ways of choosing r objects from a collection of n objects. An equivalent interpretation is that it is the number of subsets of r elements contained in a set with n elements.

Theorem 5.2.7 Let $a, b \in \mathbf{Z}$ and let $n \in \mathbf{Z}^+$. Then

$$
(a + b)^n = \sum_{k=0}^{n} \binom{n}{k} a^{n-k} b^k.
$$

PROOF: We'll assume a and b are given and that, for each $n \in \mathbf{Z}^+$, $P(n)$ is the statement: $(a + b)^n = \sum\limits_{k=0}^{n} \binom{n}{k} a^{n-k} b^k$.

It is easy to see that $P(1)$ is true. Now suppose that n is a given positive integer and that $P(n)$ is true. We will prove that $P(n + 1)$ is true. So we are assuming that $(a + b)^n = \sum\limits_{k=0}^{n} \binom{n}{k} a^{n-k} b^k$ (the induction hypothesis) and we want to prove that $(a + b)^{n+1} = \sum\limits_{k=0}^{n+1} \binom{n + 1}{k} a^{n+1-k} b^k$.

$$
\begin{aligned}
(a + b)^{n+1} &= (a + b)(a + b)^n \\
&= a(a + b)^n + b(a + b)^n \\
&= a \sum_{k=0}^{n} \binom{n}{k} a^{n-k} b^k + b \sum_{k=0}^{n} \binom{n}{k} a^{n-k} b^k \\
&= \sum_{k=0}^{n} \binom{n}{k} a^{n+1-k} b^k + \sum_{k=0}^{n} \binom{n}{k} a^{n-k} b^{k+1} \\
&= a^{n+1} + \sum_{k=1}^{n} \binom{n}{k} a^{n+1-k} b^k + \sum_{k=0}^{n-1} \binom{n}{k} a^{n-k} b^{k+1} + b^{n+1}.
\end{aligned}
$$

Now by a change of index, we can write

$$
\sum_{k=0}^{n-1} \binom{n}{k} a^{n-k} b^{k+1} = \sum_{k=1}^{n} \binom{n}{k - 1} a^{n+1-k} b^k.
$$

Hence we have

$$
\begin{aligned}
(a + b)^{n+1} &= a^{n+1} + \sum_{k=1}^{n} \left[\binom{n}{k} + \binom{n}{k - 1} \right] a^{n+1-k} b^k + b^{n+1} \\
&= a^{n+1} + \sum_{k=1}^{n} \binom{n + 1}{k} a^{n+1-k} b^k + b^{n+1} \\
&= \sum_{k=0}^{n+1} \binom{n + 1}{k} a^{n+1-k} b^k.
\end{aligned}
$$

Note that we have used the identity

$$
\binom{n}{k} + \binom{n}{k - 1} = \binom{n + 1}{k} \qquad \text{for } 1 \le k \le n.
$$

This is a straightforward algebraic calculation. The details are left as an exercise.

This proves $P(n + 1)$ and by induction it follows that $P(n)$ is true for all positive integers n. ●

Corollary 5.2.8

$$\sum_{k=0}^{n} \binom{n}{k} = 2^n.$$

PROOF: Apply the Binomial Theorem with $a = b = 1$. ●

Note: The Binomial Theorem is also true when a and b are real numbers. The reason is that the axioms of addition and multiplication that hold in **Z** (Axioms 1–8) also hold in **R**. These matters are discussed in detail in Chapter 7.

● MATHEMATICAL PERSPECTIVE: BERNOULLI NUMBERS

We conclude this section with a discussion of the Bernoulli numbers, which were alluded to in the Introduction. The Bernoulli numbers are a sequence of rational numbers defined by **Jakob Bernoulli** (1654–1705). They have had some very interesting applications in mathematics over the last several hundred years. As background, consider the following formulas:

$$1 + 2 + 3 + \ldots + (n - 1) = \frac{n(n - 1)}{2}$$

$$1^2 + 2^2 + 3^2 + \ldots + (n - 1)^2 = \frac{n(n - 1)(2n - 1)}{6}$$

$$1^3 + 2^3 + 3^3 + \ldots + (n - 1)^3 = \frac{n^2(n - 1)^2}{4}.$$

These formulas for sums of powers can be proven by induction. The first is an easy consequence of the formula in Example 1 and the other two are exercises.

Can you see any pattern to these formulas? Nothing is immediately obvious, but note that the right-hand side of each equation is a polynomial expression in the variable n. The first is $\frac{1}{2}n^2 - \frac{1}{2}n$, the second $\frac{1}{3}n^3 - \frac{1}{2}n^2 + \frac{1}{6}n$, and the third $\frac{1}{4}n^4 - \frac{1}{2}n^3 + \frac{1}{4}n^2$. Note that each of these polynomials has 0 as its constant term and that the highest power of n that appears is one more than the exponent on the left-hand side of the equation. In looking at other formulas of the form $1^k + 2^k + 3^k + \ldots + (n - 1)^k$, Bernoulli noted that the sum was a polynomial in n of degree $k + 1$ and 0 constant term. The coefficient of n in these polynomials takes on the values $-\frac{1}{2}, \frac{1}{6}, 0, -\frac{1}{30}, 0$ for $k = 1, 2, 3, 4, 5$. These are the first five Bernoulli numbers. Bernoulli was led to the following recursive definition.

Definition 5.2.9 The sequence of numbers $B_0, B_1, B_2, \ldots$, called **Bernoulli numbers**, are defined by $B_0 = 1$ and if $B_0, B_1, B_2, \ldots, B_{t-1}$ are defined then

$$B_t = -\frac{1}{t+1} \sum_{j=0}^{t-1} \binom{t+1}{j} B_j.$$

Using this formula for $t = 1, 2, 3, 4$, we obtain the following equations:

$$B_1 = -\frac{1}{2} \binom{2}{0} B_0 = -\frac{1}{2}$$

$$B_2 = -\frac{1}{3} \left[\binom{3}{0} B_0 + \binom{3}{1} B_1 \right] = \frac{1}{6}$$

$$B_3 = -\frac{1}{4} \left[\binom{4}{0} B_0 + \binom{4}{1} B_1 + \binom{4}{2} B_2 \right] = 0$$

$$B_4 = -\frac{1}{5} \left[\binom{5}{0} B_0 + \binom{5}{1} B_1 + \binom{5}{2} B_2 + \binom{5}{3} B_3 \right] = -\frac{1}{30}.$$

Other Bernoulli numbers can be computed in similar fashion. It should be noted that if t is an odd positive integer >1, then $B_t = 0$. This fact is not easy to prove from our definition.

Using these numbers Bernoulli was able to give a formula for the sum of the first n kth powers. We state the result without proof. You are encouraged to learn the proof on your own. One source is *A Classical Introduction to Modern Number Theory* by K. Ireland and M. Rosen [10].

Theorem 5.2.10   **Bernoulli** If k is a positive integer, then

$$1^k + 2^k + 3^k + \ldots + (n-1)^k = \frac{1}{k+1} \sum_{j=0}^{k} \binom{k+1}{j} B_j n^{k+1-j}.$$

The reader should verify the formula for $k = 1, 2, 3$ to see that it agrees with the formulas given.

Another application of the Bernoulli numbers is to the Riemann zeta function, which was discussed in the Introduction. Recall that for a real number $s > 1$, $\varsigma(s) = \sum_{n=1}^{\infty} 1/n^s$. Euler proved that $\varsigma(2) = \pi^2/6$ and in the Introduction we gave his informal proof of that fact. Euler generalized that result to even integers and again Bernoulli numbers are involved. More specifically, he proved that if k is a positive integer, then

$$s(2k) = (-1)^{k+1} \frac{(2\pi)^{2k}}{2(2k)!} B_{2k}.$$

As an example, we see that the series $\sum_{n=1}^{\infty} 1/n^4$ converges to $\pi^4/90$.

Exercises 5.2

---

1. Prove the following formulas using mathematical induction.
   (a) $1 + 3 + 5 + \ldots + (2n - 1) = n^2$.
   (b) $1^2 + 2^2 + 3^2 + \ldots + n^2 = \dfrac{n(n + 1)(2n + 1)}{6}$.
   (c) $1^3 + 2^3 + 3^3 + \ldots + n^3 = \dfrac{n^2(n + 1)^2}{4}$.

2. Prove the following:
   (a) $1^2 + 3^2 + 5^2 + \ldots + (2n - 1)^2 = \dfrac{(2n - 1)(2n)(2n +1)}{6}$.
   (b) $2^2 + 4^2 + 6^2 + \ldots + (2n)^2 = \dfrac{(2n)(2n + 1)(2n + 2)}{6}$.

3. Prove that if a is any real number except 1, then
$$1 + a + a^2 + a^3 + \ldots + a^n = \frac{(a^{n+1} - 1)}{a - 1}.$$

4. (a) Prove that $2^n > n^2$ for all integers $n \geq 5$.
   (b) Prove that $2^n < n!$ for all $n \geq 4$.

5. Let $a, b_1, b_2, \ldots, b_n \in \mathbf{Z}$. Prove that $a(b_1 + b_2 + \ldots + b_n) = ab_1 + ab_2 + \ldots + ab_n$.

6. Let $f: \mathbf{Z}^+ \rightarrow \mathbf{Z}^+$ be defined recursively by $f(1) = 1$ and $f(n + 1) = f(n) + 2^n$ for all $n \in \mathbf{Z}^+$. Prove that $f(n) = 2^n - 1$.

7. Let $f: \mathbf{Z}^+ \rightarrow \mathbf{R}$ be defined recursively by $f(1) = 1$ and $f(n + 1) = \sqrt{2 + f(n)}$ for all $n \in \mathbf{Z}^+$. Prove that $f(n) < 2$ for all $n \in \mathbf{Z}^+$.

8. The **Fibonnaci numbers** f_n, $n = 1, 2, 3, \ldots$, are defined recursively by the formulas $f_1 = 1$, $f_2 = 1$, $f_n = f_{n-1} + f_{n-2}$ for $n \geq 3$.
   (a) Write out the first ten Fibonnaci numbers.
   (b) Compute $f_1 + f_2$, $f_1 + f_2 + f_3$, $f_1 + f_2 + f_3 + f_4$.
   (c) Derive a formula for the sum of the first n Fibonnaci numbers and prove it by induction.
   (d) Prove that $f_1^2 + f_2^2 + \ldots + f_n^2 = f_n f_{n+1}$ for all $n \geq 1$.

9. Let H_n be the number of handshakes required if in a group of n people each person shakes with every other person exactly once.
   (a) Compute H_n for $n = 2, \ldots, 6$.
   (b) Find a recursion formula for H_{n+1} in terms of H_n.

(c) Find an explicit formula for H_n.

10. Let $A_1, A_2, \ldots, A_n$ be a collection of finite mutually disjoint sets. Prove that

$$\left| \bigcup_{i=1}^{n} A_i \right| = \sum_{i=1}^{n} |A_i|.$$

(This is Corollary 2.3.5 of Section 2.3. You may assume Theorem 2.3.4 given in that section.)

11. Let $*$ be an associative binary operation on a set A with identity element e. Let B be a subset of A that is closed under $*$. Let $b_1, b_2, \ldots b_n \in B$. Prove that $b_1 * b_2 * \ldots * b_n \in B$.

12. Let $*$ be an associative binary operation on a set A with identity element e. In Exercise 32 of Section 4.1, you were asked to prove that if B and C are subsets of A that are closed under $*$, then $B \cap C$ is closed under $*$. Prove by induction that if $B_1, B_2, \ldots, B_n$ are subsets of A that are all closed under $*$, then $B_1 \cap B_2 \cap \ldots \cap B_n$ is closed under $*$.

13. (a) Let n be an integer. Prove by induction that if n is even, then n^k is even for all $k \in \mathbf{Z}^+$.
    (*Note:* the induction should be done with the variable k, not the variable n.)
    (b) State the converse of part (a). Prove or disprove.

14. Prove by induction that if $n_1, n_2, \ldots, n_t$ are even integers, then $n_1 + n_2 + \ldots + n_t$ is even.

15. Let $A, B_1, B_2, \ldots, B_n$ be sets. Generalize part 2 of Theorem 2.2.3 by proving that

$$A \cup (B_1 \cap B_2 \cap \ldots \cap B_n) = (A \cup B_1) \cap (A \cup B_2) \cap \ldots \cap (A \cup B_n).$$

16. Let A be a nonempty set. Let $f_1, f_2, \ldots, f_n \in \mathbf{F}(A)$. Prove:
    (a) If $f_1, f_2, \ldots, f_n$ are surjective, then the composition $f_1 f_2 \ldots f_n$ is surjective.
    (b) If $f_1, f_2, \ldots, f_n$ are injective, then the composition $f_1 f_2 \ldots f_n$ is injective.
    (c) If $f_1, f_2, \ldots, f_n$ are invertible, then the composition $f_1 f_2 \ldots f_n$ is invertible and $(f_1 f_2 \ldots f_n)^{-1} = f_n^{-1} f_{n-1}^{-1} \ldots f_2^{-1} f_1^{-1}$.

17. Let $P(n)$ be a statement about the positive integer n. Using the negations of conditions 1 and 2 in Theorem 5.2.1, complete the following sentence: $P(n)$ is false for some positive integer n if $\ldots$.

18. Prove Theorem 5.2.2.

19. Prove Theorem 5.2.3.

20. State a modified form of the Second Principle of Induction similar to the modified form of the First Principle of Induction.

21. Prove Theorem 5.2.4.

22. Restate Theorem 5.2.3 in set theory language.

23. Let A be a finite set with n elements. Use Corollary 5.2.8 to prove that $\mathbf{P}(A)$, the power set of A, has 2^n elements.

24. Prove that if n and k are integers such that $1 \le k \le n$, then

$$\binom{n}{k} + \binom{n}{k-1} = \binom{n+1}{k}.$$

25. Prove that if n and k are integers such that $1 \le k \le n$, then $\binom{n}{k}$ is an integer.

26. Prove that $\displaystyle\sum_{k=0}^{n} (-1)^k \binom{n}{k} = 0$.

27. Use the definition of Bernoulli numbers to compute B_5, B_6, B_7, B_8.

28. Verify Bernoulli's Theorem for $k = 1, 2, 3$.

29. Use Bernoulli's Theorem to obtain formulas for $1^k + 2^k + 3^k + \ldots + (n-1)^k$ for $k = 4, 5, 6, 7$.

30. In this problem n and k are fixed positive integers. Let $S_k(n) = 1^k + 2^k + 3^k + \ldots + (n-1)^k$. The purpose of this problem is to give a recursion formula for $S_k(n)$. This will give an alternative to Bernoulli's Theorem for computing $S_k(n)$.
    (a) Prove that if l is a positive integer, then

$$(l+1)^{k+1} - l^{k+1} = 1 + \binom{k+1}{1} l + \binom{k+1}{2} l^2$$

$$+ \binom{k+1}{3} l^3 + \ldots + \binom{k+1}{k} l^k.$$

    (b) Prove that $n^{k+1} = \displaystyle\sum_{l=0}^{n-1} [(l+1)^{k+1} - l^{k+1}]$.

    (c) Use (a) and (b) to prove that

$$n^{k+1} = n + \binom{k+1}{1} S_1(n) + \binom{k+1}{2} S_2(n)$$

$$+ \ldots + \binom{k+1}{k-1} S_{k-1}(n) + \binom{k+1}{k} S_k(n).$$

    (d) Use this recursion formula and *not* Bernoulli's Theorem to compute $S_4(n)$.

31. Use Euler's Theorem to calculate $\varsigma(s)$ for $s = 6$ and $s = 8$.

32. This problem shows the connection between addition and multiplication in $\mathbf{Z}$. Let $t \in \mathbf{Z}$. Use induction to prove that $nt = t + t + t + \ldots + t$ (n times) for all $n \in \mathbf{Z}^+$.

33. Suppose that **P** is a subset of **Z** with the following properties: (1) **P** is closed with respect to addition and multiplication; and (2) for every element x of **Z**, exactly one of the following properties holds: $x \in$ **P**, $x = 0$, or $-x \in$ **P**. Prove that **P** = **Z**⁺.

    Hint: First prove that $1 \in$ P and then using induction, prove that Z⁺ ⊆ P. Finally, prove that P ⊆ Z⁺.

34. Let A be a set with n elements and B a set with m elements. Prove that $A \times B$ has nm elements.

    Hint: With m fixed, prove the result by induction with respect to n.

35. Let A be a set with n elements. Let **S**(A) be the set of all permutations of A; that is, **S**(A) is the set of all bijective functions from A to A. Prove by induction that **S**(A) contains $n!$ elements.

36. Let $f: \mathbf{Z} \to \mathbf{Z}$ be a function with the property that $f(x + y) = f(x) + f(y)$ for all x and $y \in \mathbf{Z}$.
    (a) Prove that $f(0) = 0$.
    (b) Prove that $f(-x) = -f(x)$ for all $x \in \mathbf{Z}$.
    (c) Prove that $f(x - y) = f(x) - f(y)$ for all $x, y \in \mathbf{Z}$.
    (d) Prove that if $x_1, x_2, \ldots, x_n \in \mathbf{Z}$, then $f(x_1 + x_2 + \ldots + x_n) = f(x_1) + f(x_2) + \ldots + f(x_n)$.
    (e) Prove that $\exists\, a \in \mathbf{Z} \ni f(n) = na$ for all $n \in \mathbf{Z}$.

37. A polygon with n sides is called an *n-gon*. An n-gon is said to be *convex* if the straight line connecting any pair of interior points is contained in the interior. Assuming that the sum of the radian measures of the interior angles of a triangle is π, prove by induction that the sum of the radian measures of the interior angles of a convex n-gon is $\pi(n - 2)$.

38. Assuming the Sum Rule,

$$\frac{d}{dx}(f(x) + g(x)) = f'(x) + g'(x),$$

prove by induction that

$$\frac{d}{dx}(f_1(x) + f_2(x) + \ldots + f_n(x)) = f_1'(x) + f_2'(x) + \ldots + f_n'(x)$$

for $n \in \mathbf{Z}^+$.

39. Assuming that the indefinite integral is additive:

$$\int (f(x) + g(x))\, dx = \int f(x)\, dx + \int g(x)\, dx,$$

prove that

$$\int (f_1(x) + f_2(x) + \ldots + f_n(x))\, dx = \int f_1(x)\, dx$$
$$+ \int f_2(x)\, dx + \ldots + \int f_n(x)\, dx$$

for $n \in \mathbf{Z}^+$.

40. Assuming the Product Rule,

$$\frac{d}{dx}(f(x)g(x)) = f(x)g'(x) + f'(x)g(x),$$

prove by induction that

$$\frac{d}{dx}(f(x))^n = n(f(x))^{n-1}f'(x)$$

for $n \in \mathbf{Z}^+$.

41. Prove that for every real number $x \geq 0$ and every positive integer n,

$$e^x \geq 1 + x + \frac{x^2}{2!} + \frac{x^3}{3!} + \ldots + \frac{x^n}{n!}.$$

42. Assuming that $\left(1 + \frac{1}{n}\right)^n < e$, for all $n \in \mathbf{Z}^+$, prove by induction that $n! > \left(\frac{n}{e}\right)^n$.

Hint: This problem can also be done without induction by using the previous problem.

Discussion and Discovery Exercises

D1. Induction proofs are often unsatisfactory in the sense that they do not tell us *why* something is true. Give a proof of the formula

$$1 + 2 + 3 + \ldots + n = \frac{n(n+1)}{2}$$

(see Example 1 of this section) that is not an induction proof. Explain why this may be a better proof than the induction one.

D2. The numbers $1, 1 + 2, 1 + 2 + 3, 1 + 2 + 3 + 4, \ldots$ are called *triangular numbers* because they can be arranged in triangular form:

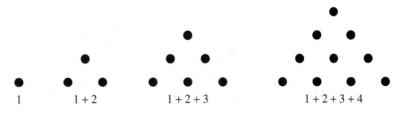

$$\begin{array}{cccc} 1 & 1+2 & 1+2+3 & 1+2+3+4 \end{array}$$

If we let T_n denote the nth triangular number, then $T_1 = 1$, $T_2 = 3$,

$T_3 = 6$, $T_4 = 10$, and so on. In Example 1 of this section, we proved by induction a formula for the nth triangular number; namely,

$$T_n = \frac{n(n+1)}{2}.$$

(a) Prove that the sum of two consecutive triangular numbers is a perfect square.

(b) Let T_n be the nth triangular number. Prove that $8T_n + 1$ is a perfect square.

(c) Prove the following converse of part (b): If t is an integer and $8t + 1$ is a perfect square, then t is a triangular number.

(d) Prove that the difference of the squares of two consecutive triangular numbers is a cube.

(e) Prove that the infinite series $\sum_{n=1}^{\infty} \frac{1}{T_n}$ converges to 2.

(f) Find 2 triangular numbers that are also squares.

(g) Prove that if there exist integers x and y such that $x^2 - 8y^2 = 1$, then y^2 is a number that is both a square and a triangular number.

(h) Let $\alpha = 3 + \sqrt{8}$. It can be proved, although it is not easy to do so, that if one takes powers of α and writes them in the form $x + y\sqrt{8}$, then y^2 is both a square and a triangular number. Use this result to find four numbers that are both square and triangular. Show that the numbers you find are indeed triangular.

D3. Similar to the triangular numbers are the numbers 1, 1 + 3, 1 + 3 + 5, 1 + 3 + 5 + 7, ... and so on. These numbers are called *square* numbers because they are all perfect squares. In fact, if S_n denotes the nth square number, then $S_n = n^2$. (See Exercise 1(a).)

(a) Draw a diagram that justifies the word *square* for this sequence of numbers.

(b) Draw a picture that illustrates the fact that if $n > 1$, then the nth square number is the sum of two consecutive triangular numbers.

D4. Now consider the sequence of numbers 1, 1 + 4, 1 + 4 + 7, 1 + 4 + 7 + 10, ..., and so on. These numbers are called *pentagonal* numbers.

(a) Draw a diagram that justifies the word pentagonal for this sequence of numbers.

(b) If P_n denotes the nth pentagonal number, find a formula for P_n and use induction to prove it.

(c) Define a sequence of numbers that goes the next step beyond pentagonal numbers and that might be called *hexagonal* numbers. Can you draw a picture to justify the name hexagonal? Find and prove a formula for these numbers.

D5. Consider the sequence of numbers: 1, 3 + 5, 7 + 9 + 11, 13 + 15 + 17 + 19, ... Is there a pattern to this sequence that leads to a formula? If so, find the formula and prove it.

5.3 # THE DIVISION ALGORITHM AND GREATEST COMMON DIVISORS

As we all learn in grade school arithmetic, dividing an integer by a positive integer gives an integer plus a remainder that is positive or zero and is less than the original divisor. For example, $\frac{17}{5} = 3 + \frac{2}{5}$, so the remainder is 2. The following theorem, known as the Division Algorithm, is a formal statement of this fact.

Theorem 5.3.1   **Division Algorithm** Let a, b be integers with $b > 0$. Then there exist unique integers q and r such that $a = bq + r$ where $0 \leq r < b$.

PROOF: Let $S = \{n \in \mathbf{Z} \mid n = a - bx, \text{ for some } x \in \mathbf{Z}\}$ and let $S_0 = \{n \in S \mid n \geq 0\}$; that is, S_0 is the set of nonnegative elements of S.

We claim that S_0 is nonempty. Note that $a \in S$ since $a = a - b \cdot 0$. So if $a \geq 0$, $a \in S_0$. If $a < 0$, then $a - ba \in S$. But $a - ba = a(1 - b) \geq 0$ since $a < 0$ and $1 - b \leq 0$. Hence $a - ba \in S_0$. In any case, S_0 is nonempty.

Now let r be the smallest element of S_0. If $0 \in S_0$, then $r = 0$. If $0 \notin S_0$, then r exists by the Well-Ordering Principle.

We have $r = a - bq$ for some $q \in \mathbf{Z}$ and $r \geq 0$. Suppose that $r \geq b$. Then $r - b = a - b(q + 1) \in S_0$. But this contradicts the fact that r is the least element of S_0. Hence $0 \leq r < b$.

To prove uniqueness, suppose there also exist q_1 and r_1 in $\mathbf{Z}$ such that $a = bq_1 + r_1$ where $0 \leq r_1 < b$. Without loss of generality, we can assume that $r \geq r_1$.

Since $a = bq + r$, we have

$$b(q_1 - q) = r - r_1 \geq 0.$$

Therefore, $q_1 - q \geq 0$. If $q_1 - q \neq 0$, then $r - r_1 \geq b$, a contradiction since $0 \leq r < b$ and $0 \leq r_1 < b$. It follows immediately that $q_1 = q$ and $r_1 = r$. ●

Divisors and Greatest Common Divisors

When the remainder in the Division Algorithm is zero, we say that a is divisible by b or b divides a. More formally, we have:

Definition 5.3.2 Let a, b be integers. We say b **divides** a, written $b \mid a$, if there is an integer c such that $bc = a$. We say that b and c are **factors** of a or that a is **divisible** by b and c.

EXAMPLE 1 -4 divides 12, since $(-4)(-3) = 12$.

EXAMPLE 2   3 does not divide 7, since $7 = (3)(2) + 1$.

EXAMPLE 3   We prove by induction that 3 divides $4^n - 1$ for every positive integer n.

Let $P(n)$ be the statement "3 divides $4^n - 1$." Because $4 - 1 = 3$ and $3 \mid 3$, we have that $P(1)$ is true.

Suppose that k is a positive integer for which $P(k)$ is true. Then 3 divides $4^k - 1$. Therefore $4^k - 1 = 3t$ for some $t \in \mathbf{Z}$. Hence $4^{k+1} - 1 = 4 \cdot 4^k - 1 = 4(4^k - 1) + 3 = 4(3t) + 3 = 3(4t + 1)$. This proves that 3 divides $4^{k+1} - 1$ and so $P(k + 1)$ is true.

Therefore $P(n)$ is true for all positive integers n.

Proposition 5.3.3   Let $a, b, c \in \mathbf{Z}$.

1. If $a \mid 1$ then $a = \pm 1$.
2. If $a \mid b$ and $b \mid a$, then $a = \pm b$.
3. If $a \mid b$ and $a \mid c$, then $a \mid bx + cy$ for any $x, y \in \mathbf{Z}$.
4. If $a \mid b$ and $b \mid c$, then $a \mid c$.

PROOF: 1. If $a \mid 1$, then there exists $x \in \mathbf{Z}$ such that $ax = 1$. Hence a has a multiplicative inverse in $\mathbf{Z}$. Thus by Corollary 5.1.8, $a = \pm 1$.

The proofs of 2, 3, and 4 are left as exercises. ●

Given a pair of integers a and b, we can consider all integers that divide both a and b, called common divisors of a and b. For example, if $a = 12$ and $b = 18$, then $\pm 1, \pm 2, \pm 3, \pm 6$ are all the common divisors of a and b. The largest of these is 6 and is divisible by all of the others.

> **Definition 5.3.4** Let a, b be integers, not both zero. A positive integer d is called a **greatest common divisor** of a and b if
>
> 1. d divides a and d divides b; that is, d is a **common divisor** of a and b
> 2. whenever an integer c divides both a and b, then c divides d.

The next theorem shows that every pair of integers a and b (not both zero) has a greatest common divisor and that it is unique. We will then show how the Division Algorithm can be used to compute the greatest common divisor of a pair of integers.

Theorem 5.3.5   Let $a, b \in \mathbf{Z}$, not both zero. Then a greatest common divisor d of a and b exists and is unique. Moreover, there exist integers x and y such that $d = ax + by$.

PROOF: Let $S = \{n \in \mathbf{Z} \mid n = ax + by,$ for some $x, y \in \mathbf{Z}\}$. S is a subset of $\mathbf{Z}$ that contains both a and b since $a = a \cdot 1 + b \cdot 0$ and $b = a \cdot 0 + b \cdot 1$.

Similarly $-a$ and $-b$ are in S, so S must contain positive integers. By the Well-Ordering Principle, S has a smallest positive element. Call it d. We now show that d is a greatest common divisor of a and b.

First, notice that $d \in S$, so there exist integers x and y such that $d = ax + by$. This proves the second statement of the theorem. Next, applying the Division Algorithm to a and d, there exist $q, r \in \mathbf{Z}$ such that $a = dq + r$ where $0 \le r < d$.

But

$$r = a - dq = a - (ax + by)q = a(1 - xq) + b(-yq),$$

which implies that $r \in S$.

Since d is the least positive integer in S, we must have $r = 0$. Therefore $d \mid a$. Similarly, we can show that $d \mid b$.

Finally, suppose that c is a common divisor of a and b. Then there exist integers u and v such that $cu = a$ and $cv = b$. Therefore, $d = ax + by = c(ux + vy)$, and we have that $c \mid d$. This proves that d is a greatest common divisor of a and b.

We leave the proof of uniqueness as an exercise. ●

Since a greatest common divisor (abbreviated g.c.d.) of two integers a and b, not both zero, exists and is unique by the theorem just proved, we will refer to it as *the* greatest common divisor of a and b, and we will denote it by (a, b). In context, there should be no confusion between this notation and the notation for ordered pairs of real numbers.

Euclidean Algorithm

We now give a procedure, using the Division Algorithm, for computing the g.c.d. of two integers. The following lemma will be helpful. We leave the proof as an exercise.

Lemma 5.3.6 Let $a, b \in \mathbf{Z}$, not both zero. Suppose that there exist integers q and r such that $a = bq + r$. Then $(a, b) = (b, r)$.

Let a, b be two positive integers with $a > b$. By repeated application of the Division Algorithm, we get

$$
\begin{aligned}
a &= bq_1 + r_1, & \text{where } q_1, r_1 \in \mathbf{Z}, 0 \le r_1 < b, \\
b &= r_1 q_2 + r_2, & \text{where } q_2, r_2 \in \mathbf{Z}, 0 \le r_2 < r_1, \\
r_1 &= r_2 q_3 + r_3, & \text{where } q_3, r_3 \in \mathbf{Z}, 0 \le r_3 < r_2,
\end{aligned}
$$
$$\vdots$$

Because each remainder is strictly less than the previous one, we eventually get

$$r_{n-2} = r_{n-1}q_n + r_n, \qquad \text{where } q_n, r_n \in \mathbf{Z}, 0 \le r_n < r_{n-1},$$
$$r_{n-1} = r_n q_{n+1} + r_{n+1}, \qquad \text{where } q_{n+1} \in \mathbf{Z} \quad \text{and} \quad r_{n+1} = 0.$$

By Lemma 5.3.6, we have

$$(a, b) = (b, r_1) = (r_1, r_2) = (r_2, r_3) = \ldots = (r_{n-1}, r_n) = (r_n, 0) = r_n.$$

Thus the g.c.d. of a and b is the last nonzero remainder that occurs in this procedure. (If $r_1 = 0$, then $(a, b) = b$). This procedure for finding the g.c.d. of two integers is called the **Euclidean Algorithm**.

EXAMPLE 4   Let $a = 9180$ and $b = 1122$.

$$9180 = (1122)(8) + 204$$
$$1122 = (204)(5) + 102$$
$$204 = (102)(2) + 0.$$

Therefore, $(9180, 1122) = 102$.

EXAMPLE 5   The previous procedure can also be used to find integers x and y such that $(a, b) = ax + by$.

Using Example 4, we get

$$102 = 1122 + 204(-5)$$
$$= 1122 + (9180 + 1122(-8))(-5)$$
$$= 1122 + 9180(-5) + 1122(40)$$
$$= 1122(41) + 9180(-5).$$

Hence $x = -5$ and $y = 41$.

Relatively Prime Integers

Often a pair of integers will have no common divisors except for the obvious ones of 1 and -1.

Definition 5.3.7 Two integers a and b, not both zero, are said to be **relatively prime** if $(a, b) = 1$.

For example, 12 and 35 are relatively prime, as are 14 and 15.

The following result is a useful characterization of relatively prime integers.

Theorem 5.3.8 Let $a, b \in \mathbf{Z}$. Then a and b are relatively prime if and only if there exist integers x and y such that $ax + by = 1$.

PROOF: Suppose a and b are relatively prime. Then by Theorem 5.3.5, $1 = (a, b) = ax + by$ for some $x, y \in \mathbf{Z}$.

Conversely, suppose there exist $x, y \in \mathbf{Z}$ such that $ax + by = 1$. Let $d = (a, b)$. Since d divides both a and b, then by Proposition 5.3.3, part 3, d also divides $ax + by$.

Therefore $d \mid 1$, and by Proposition 5.3.3, part 1, $d = 1$. ●

It is possible for an integer a to divide the product of two integers without a dividing either of the factors. For example, $6 \mid 4 \cdot 3$ but 6 does not divide either 4 or 3. However, if a is relatively prime to one of the factors, then it must divide the other factor. A proof of this fact follows and is our first important application of Theorem 5.3.8.

Theorem 5.3.9 Let $a, b, c \in \mathbf{Z}$. Suppose $a \mid bc$ and $(a, b) = 1$. Then $a \mid c$.

PROOF: Since $a \mid bc$, there is a z in $\mathbf{Z}$ such that $az = bc$. Also, by Theorem 5.3.8, there are x, y in $\mathbf{Z}$ such that $ax + by = 1$. Then $c = c(ax + by) = cax + cby = cax + azy = a(cx + zy)$. Hence $a \mid c$. ●

● HISTORICAL COMMENTS: PYTHAGOREAN TRIPLES

The Pythagorean Theorem gives rise to the equation $x^2 + y^2 = z^2$. Integer solutions of this equation are called *Pythagorean triples*. It is not hard to show that if (x, y, z) is a Pythagorean triple, then any multiple (tx, ty, tz) where t is an integer is also a Pythagorean triple. If (x, y, z) is a Pythagorean triple and the integers $x, y,$ and z have no common factor, we say that (x, y, z) is a *primitive* Pythagorean triple. Some examples are: (3, 4, 5), (5, 12, 13), (8, 15, 17), and (20, 21, 29). The following theorem, due to the Greek mathematician **Diophantus**, characterizes the primitive Pythagorean triples.

Theorem 5.3.10 Let $x, y,$ and z be positive integers. Then (x, y, z) is a primitive Pythagorean triple if and only if there exist relatively prime positive integers a and b, $a > b$, and a and b having opposite parity, (meaning one of them is even and one odd) such that $x = 2ab$, $y = a^2 - b^2$, $z = a^2 + b^2$. ●

By taking $a = 3$ and $b = 2$, for example, we get the triple (12, 5, 13).

Diophantus, who lived in the third century A.D., is famous for having written *Arithmetica*, the earliest known treatise on algebra. It consists of thirteen books, only six of which have survived. Diophantus was interested in methods for finding integer and rational solutions of equations and *Arithmetica* is filled with examples and techniques for solving such equations, which have become known as *Diophantine equations*.

Exercises 5.3

---

1. (a) Prove that every integer n is of the form $3t$, $3t + 1$ or $3t + 2$ for some $t \in \mathbf{Z}$.
   (b) Prove that every integer n is of the form $4t$, $4t + 1$, $4t + 2$ or $4t + 3$ for some $t \in \mathbf{Z}$.

2. Prove that $n^2 - n$ is divisible by 2 for every integer n.

3. Prove that if $n \in \mathbf{Z}$, then $n^2 + 2$ is not divisible by 4.

4. Prove by induction that for every positive integer n:
   (a) 4 divides $5^n - 1$.
   (b) 7 divides $2^{3n} - 1$.

5. Generalize Problem 4 by proving that if $a \in \mathbf{Z}$, then for every positive integer n, $a - 1$ divides $a^n - 1$.

6. (a) Prove part 2 of Proposition 5.3.3.
   (b) Prove part 3 of Proposition 5.3.3.
   (c) Prove part 4 of Proposition 5.3.3.
   (d) State the contrapositive of each part of Proposition 5.3.3.
   (e) State the converse of each part of Proposition 5.3.3. Prove or disprove each converse.

7. Let $a, b \in \mathbf{Z}$, not both zero. Complete the proof of Theorem 5.3.5 by showing that a g.c.d. of a and b is unique.

8. Let $a, b \in \mathbf{Z}$, not both zero. Let $S = \{n \in \mathbf{Z} \mid n = ax + by,$ for some $x, y \in \mathbf{Z}\}$, the set defined in the proof of Theorem 5.3.5. Let $d = (a, b)$. Prove that S is the set of all integer multiples of d.

9. Prove Lemma 5.3.6.

10. Compute the g.c.d.s of the following pairs of integers. Then find integers x and y such that $(a, b) = ax + by$.
    (a) $a = 1840, b = 1518$
    (b) $a = 1001, b = 3465$
    (c) $a = 255, b = -143$

11. Let $a, b \in \mathbf{Z}$, not both zero, and let $d = (a, b)$. Prove that $\left(\dfrac{a}{d}, \dfrac{b}{d}\right) = 1$.

12. Let $a, b, c \in \mathbf{Z}$. Suppose that $(a, c) = (b, c) = 1$. Prove that $(ab, c) = 1$.

13. Let $a, b, c \in \mathbf{Z}$. Suppose that $a \mid c$ and $b \mid c$ and that $(a, b) = 1$. Prove that $ab \mid c$.

14. Let $n, m \in \mathbf{Z}$ such that $(n, m) = 1$. Prove that $n\mathbf{Z} \cap m\mathbf{Z} = nm\mathbf{Z}$.
    Recall that $n\mathbf{Z}$ is the set of all integer multiples of n.

15. Prove that for every integer n, $(n, n + 1) = 1$.

16. (a) Prove that if n is an odd integer, then $(n, n + 2) = 1$.
    (b) If n is an even integer, what is $(n, n + 2)$? Prove your answer.

17. (a) State the contrapositive of Theorem 5.3.9.
    (b) State the converse of Theorem 5.3.9. Prove or disprove.

18. Prove that if (x, y, z) is a Pythagorean triple, then for any integer t, (tx, ty, tz) is a Pythagorean triple.

19. Prove that if (x, y, z) is a Pythagorean triple, then there exists integers d, u, v, w such that $x = du$, $y = dv$, and $z = dw$ where (u, v, w) is a primitive Pythagorean triple.

20. Prove that if (x, y, z) is a primitive Pythagorean triple, then x and y have opposite parity.

21. Prove that if (x, y, z) is a primitive Pythagorean triple, then x or y is divisible by 3.

22. Prove that if (x, y, z) is a primitive Pythagorean triple, then x, y or z is divisible by 5.

23. Let C be the unit circle given by the equation $x^2 + y^2 = 1$. A rational point on C is a point (a, b) where a and b are rational numbers. For example, $(1, 0)$, $(0, 1)$, and $(3/5, 4/5)$ are rational points on C. Prove that there are infinitely many rational points on C.

Discussion and Discovery Exercises

D1. Explain why the restriction that a and b be not both zero in the statement of Theorem 5.3.5 is a necessary condition for that theorem to be true. Also, show where the restriction is actually used in the proof.

D2. Theorem 5.3.5 is an example of what is known as an *existence theorem*. It asserts the existence of the greatest common divisor of two integers.
    (a) What is the key step in the proof of Theorem 5.3.5 that guarantees the existence of the greatest common divisor?
    (b) Have we encountered any other existence proofs in this chapter? If so, what are they?

D3. Let $a, b \in \mathbf{Z}$, not both zero, and suppose there exist integers x and y such that $ax + by = 2$. Does this fact imply that $(a, b) = 2$? If not, what can be said about (a, b)?

D4. In Exercises 15 and 16, we looked at the greatest common divisors $(n, n + 1)$ and $(n, n + 2)$ for all integers n.
    (a) Formulate a similar result that characterizes $(n, n + 3)$ for all integers n.
    (b) Generalize part (a) so that for a fixed positive integer k, you can characterize $(n, n + k)$ for all integers n.

D5. Pythagorean triples give examples of the sum of two squares being itself a square. Can you find positive integers x and y such that $x^2 + y^2$ is a cube? Does it seem likely that there are infinitely many triples (x, y, z) such that $x^2 + y^2 = z^3$?

5.4   PRIMES AND UNIQUE FACTORIZATION

Prime Numbers

Every positive integer greater than 1 has at least two positive divisors: itself and 1. Some positive integers greater than 1 have the property that these are the *only* positive divisors. The integers 2, 3, 5, and 7 have this property but 4 and 6 do not; 6, for example, is divisible by 2 and 3. In order to study the set of integers in any depth, it is necessary to discuss the numbers with this property.

> **Definition 5.4.1** An integer p greater than 1 is called a **prime** number if the only divisors of p are ± 1 and $\pm p$. If an integer greater than 1 is not prime, it is called **composite**.

Prime numbers are the building blocks of the integers since every positive integer can be written as a product of primes in an essentially unique way. This important fact, called the *Unique Factorization Theorem*, will be proven in this section.

The following characterization of composite numbers we leave as an exercise.

Lemma 5.4.2   Let $n \in \mathbf{Z}^+$, $n > 1$. Then n is composite if and only if there exist integers a and b such that $n = ab$ where $1 < a < n$ and $1 < b < n$.

Although not every integer is prime, it is true that every integer is divisible by a prime number. The proof, which is a nice application of the Well-Ordering Principle, is given next.

Proposition 5.4.3   Every integer greater than 1 is divisible by a prime number.

PROOF: Let $T = \{\, n \in \mathbf{Z}^+ \mid n > 1$ and n is not divisible by a prime number$\}$. It suffices to prove that $T = \varnothing$. Suppose, on the contrary, that $T \neq \varnothing$. Then by the Well-Ordering Principle, T has a smallest element. Call it n_0.

Now n_0 itself cannot be prime, otherwise it would be divisible by a prime, namely itself. So by Lemma 5.4.2, there exist integers a and b such that $n_0 = ab$ where $1 < a < n_0$ and $1 < b < n_0$.

Since a is smaller than n_0, a is not in T. Hence a is divisible by some prime p. Since p divides a and clearly a divides n_0, by Proposition 5.3.3, p divides n_0. This is a contradiction to the fact that n_0 is in T. Thus $T = \varnothing$. ●

We saw at the end of the last section that an integer x could divide the product of two numbers a and b without dividing either a or b. This will not happen, however, if x is prime, as we see in the next proposition.

Proposition 5.4.4   If a prime number p divides a product ab, where $a, b \in \mathbf{Z}$, then p divides a or p divides b.

PROOF: Suppose p does not divide a. Then $a \neq 0$ and so $(p, a) = 1$, since the only positive divisors of p are 1 and p. It now follows from Theorem 5.3.9 that p divides b. ●

The following two corollaries should be proved by induction.

Corollary 5.4.5   Let p be a prime and let $a_1, a_2, \ldots, a_m \in \mathbf{Z}$. If p divides the product $a_1 \cdot a_2 \cdot \ldots \cdot a_m$, then p divides a_i for some $i = 1, 2, \ldots, m$.

PROOF: We leave it as an exercise. ●

Corollary 5.4.6   Let p be a prime and let $a \in \mathbf{Z}$. If p divides a^m for some positive integer m, then p divides a.

PROOF: We leave it as an Exercise. ●

EXAMPLE 1   We will use Proposition 5.4.4 to prove that $\sqrt{2}$ is not a rational number. Suppose, on the contrary, that $\sqrt{2}$ is rational. Then we can write $\sqrt{2} = a/b$ where $a, b \in \mathbf{Z}$ and a/b is reduced to lowest terms. (Recall that being reduced to lowest terms is equivalent to saying that a and b have no common factors or in other words that a and b are relatively prime.) Hence $a^2 = 2b^2$, which implies that 2 divides a^2. It follows from Proposition 5.4.4 that 2 divides a. So $a = 2c$, for some c in $\mathbf{Z}$. We get $a^2 = 4c^2 = 2b^2$ or $2c^2 = b^2$, which implies, again by Proposition 5.4.4, that 2 divides b. This is a contradiction to the fact that a and b have no common factors.

Unique Factorization

We are now in a position to prove one of the most important theorems about the integers, the **Unique Factorization Theorem**. The theorem has two parts: an existence part, which says that every integer greater than 1 is a product of primes, and a uniqueness part, which states that this product is unique except for the order in which the primes occur. We will use the Second Principle of Mathematical Induction to prove the existence part and rely heavily on Corollary 5.4.5 to prove the uniqueness part.

Theorem 5.4.7   **Unique Factorization Theorem** Let $n \in \mathbf{Z}, n > 1$. Then n is a prime number or can be written as a product of prime numbers. Moreover, the product is unique, except for the order in which the factors appear.

PROOF: To prove the existence part of the theorem, namely, that every integer greater than 1 is a prime or a product of primes, we use the Second Principle of Mathematical Induction. (See Theorem 5.2.3.)

Let $P(n)$ be the statement: $n = 1$, n is prime or n is a product of primes. Trivially, $P(1)$ is true.

Suppose that $k \in \mathbf{Z}^+$ and that $P(i)$ is true for all i such that $1 \le i \le k$. We will show that $P(k + 1)$ is true.

If $k = 1$, then $P(k + 1) = P(2)$ is true since 2 is prime.

Now suppose that $k > 1$. The induction hypothesis tells us that every integer i such that $1 < i \le k$ is prime or a product of primes. If $k + 1$ is prime, then clearly $P(k + 1)$ is true. If $k + 1$ is not prime, then by Lemma 5.4.2, $k + 1 = ab$, where $1 < a < k + 1$ and $1 < b < k + 1$.

By the induction hypothesis, both a and b are primes or products of primes. Therefore $k + 1 = ab$ is a product of primes, again implying $P(k + 1)$ is true. It now follows by induction that $P(n)$ is true for all integers $n \ge 1$. Therefore, every integer $n > 1$ is a prime or a product of primes.

To prove uniqueness, suppose that $n = p_1 \cdot p_2 \cdot \ldots \cdot p_s$ and that $n = q_1 \cdot q_2 \cdot \ldots \cdot q_t$, where $p_1, p_2, \ldots, p_s, q_1, q_2, \ldots, q_t$ are primes.

We will show that $s = t$ and that after suitable renumbering, $p_i = q_i$ for all $i = 1, 2, \ldots, s$. We can assume, without loss of generality, that $s \le t$.

Since p_1 divides n, p_1 divides $q_1 \cdot q_2 \cdot \ldots \cdot q_t$. By Corollary 5.4.5, p_1 divides q_j for some j. We now renumber the q_is so that $q_j = q_1$.

Because p_1 and q_1 are primes, we must have $p_1 = q_1$. It follows from Exercise 16 of Section 5.1 that $p_2 \cdot p_3 \cdot \ldots \cdot p_s = q_2 \cdot q_3 \cdot \ldots \cdot q_t$.

Now p_2 divides $q_2 \cdot q_3 \cdot \ldots \cdot q_t$, so again by Corollary 5.4.5, p_2 divides q_j for some j. Letting $q_j = q_2$, we have $p_2 = q_2$.

Continuing this process, after s steps, we get $p_i = q_i$ for $i = 1, 2, \ldots, s$. If $s < t$, we would have the equation $1 = q_{s+1} \cdot q_{s+2} \cdot \ldots \cdot q_t$, a contradiction.

Therefore $s = t$ and $p_i = q_i$ for all i. This completes the proof of the theorem. ●

EXAMPLE 2   The integer $n = 22540 = 2 \cdot 2 \cdot 5 \cdot 7 \cdot 7 \cdot 23$ and any factorization of this number must be the same except for a rearrangement of the factors.

Let $n \in \mathbf{Z}^+$, $n > 1$. In writing n as a product of primes, if we group together the primes that appear more than once, then we can write n in the form

$$n = p_1^{m_1} \cdot p_2^{m_2} \cdot \ldots \cdot p_r^{m_r}$$

where $p_1, p_2, \ldots, p_r$ are distinct primes, $p_1 < p_2 < \ldots < p_m$, and the exponents $m_1, m_2, \ldots m_r$ are positive integers. We say that n is written in *standard form*.

EXAMPLE 3   From the previous example, we can write

$$22540 = 2^2 \cdot 5 \cdot 7^2 \cdot 23.$$

● MATHEMATICAL PERSPECTIVE: EUCLID'S THEOREM AND SOME GENERALIZATIONS

There are many interesting and challenging problems involving prime numbers. For example, we show here that there are infinitely many prime numbers. The proof is essentially the same as the one given by Euclid in the third century B.C.

Theorem 5.4.8   **Euclid's Theorem** There are infinitely many prime numbers.

PROOF: Suppose that, on the contrary, there are only finitely many primes $p_1, p_2, \ldots, p_r$. Let $m = p_1 \cdot p_2 \cdot \ldots \cdot p_r + 1$. Let p be a prime that divides m. (Why must such a prime exist?) Because $p = p_i$ for some $i = 1, 2, \ldots, r$, p also divides $p_1 \cdot p_2 \cdot \ldots \cdot p_r$. Therefore, p divides $m - p_1 \cdot p_2 \cdot \ldots \cdot p_r = 1$, a contradiction. Thus there must in fact be infinitely many primes. ●

There are many generalizations of Euclid's Theorem. For example, it is known that there are infinitely many primes in the set $\{1, 5, 9, 13, 17, \ldots\}$. More generally, a famous theorem due to **Peter Gustav Lejeune Dirichlet** (1805–1859) states that there are infinitely many primes in any arithmetic progression of the form $\{an + b \mid n = 1, 2, 3, \ldots\}$ where $a, b \in \mathbf{Z}^+$ and $(a, b) = 1$. The set $\{1, 5, 9, 13, 17, \ldots\}$ is an example with $a = 4$ and $b = 1$. Although the proof of Dirichlet's Theorem is beyond the scope of this book, a special case is proven next.

Theorem 5.4.9   There are infinitely many primes of the form $4n + 3$.

PROOF: Suppose that, on the contrary, there are only finitely many primes $p_1, p_2, \ldots, p_r$ of the form $4n + 3$. Since 3 is such a prime, we will assume that $p_1 = 3$. Let $N = 4p_2p_3 \ldots p_r + 3$. Since N is of the form $4n + 3$, there must be at least one prime divisor of N of the form $4n + 3$. (See Exercise 18 following.) Call this prime p. If $p = 3$, then we would have that $3 \mid 4p_2p_3 \ldots p_r$, implying that $3 \mid p_i$ for some i, a contradiction. On the other hand, if $p \neq 3$, then $p = p_i$ for some i since $p_1, p_2, \ldots, p_r$ are, by assumption, all of the primes of the form $4n + 3$. But since $p \mid N$ and $p \mid 4p_2p_3 \ldots p_r$, we get that $p \mid 3$, again a contradiction. Therefore, there must in fact be infinitely many primes of the form $4n + 3$. ●

Some deceptively simple problems concerning primes remain unsolved, even after centuries of effort by mathematicians. For example, two primes that differ by 2 are called **twin primes**: 3 and 5, 5 and 7, 11 and 13, and 17 and 19 are such pairs. Remarkably, it is unknown whether there are infinitely many pairs of twin primes.

Another unsolved problem involving primes is the assertion, known as the Goldbach Conjecture, that every even integer greater than 4 is the sum of two odd prime numbers. This conjecture has been verified for all even integers less than 10^{10}. So although just about everyone believes that it is true, no general proof seems forthcoming in the immediate future.

Exercises 5.4

1. Prove Lemma 5.4.2.

2. Let $n \in \mathbf{Z}$, $n > 1$. Prove that if n is not divisible by any prime number less than or equal to $\sqrt{n}$, then n is a prime number.

3. Let n be a positive integer greater than 1 with the property that whenever n divides a product ab where a, $b \in \mathbf{Z}$, then n divides a or n divides b. Prove that n is a prime number.

4. Prove Corollary 5.4.5.

5. Prove Corollary 5.4.6.

6. (a) Prove that $\sqrt{3}$ is irrational.
   (b) Prove that $\sqrt[3]{2}$ is irrational.
   (c) Prove that $\sqrt[n]{2}$ is irrational for every $n \in \mathbf{Z}$, $n \geq 2$.
   (d) Prove that if p is a prime number, then $\sqrt[n]{p}$ is irrational for every $n \in \mathbf{Z}$, $n \geq 2$.
   (e) Let n, $m \in \mathbf{Z}$, $n \geq 2$. Prove that if m is not the nth power of an integer, then $\sqrt[n]{m}$ is irrational.

7. (a) Prove that $\log_{10} 3$ is irrational.
   (b) Prove that if r is a rational number such that $r > 1$ and $r \neq 10^n$ for any positive integer n, then $\log_{10} r$ is irrational.

8. Write the following integers in standard form:
   (a) 594   (b) 1,400
   (c) 42,750   (d) 191,737

9. Let $n \in \mathbf{Z}$, $n \geq 1$. Prove that n is a perfect square if and only if, when n is written in standard form, all of the exponents are even.

10. Let a, $b \in \mathbf{Z}$.
    (a) Prove that if $a^2 \mid b^2$, then $a \mid b$.
    (b) Prove that if $a^n \mid b^n$ for some positive integer n, then $a \mid b$.

11. (a) Let a, $b \in \mathbf{Z}$ such that $(a, b) = 1$. Suppose that $ab = x^2$ for some x in $\mathbf{Z}$. Prove that $a = y^2$ and $b = z^2$ for some y and z in $\mathbf{Z}$.
    (b) Show that part (a) is false without the assumption that a and b are relatively prime.
    (c) Let a, $b \in \mathbf{Z}$ such that $(a, b) = 1$. Suppose that $ab = x^n$ for some x in $\mathbf{Z}$ and some positive integer n. Prove that $a = y^n$ and $b = z^n$ for some y and z in $\mathbf{Z}$.
    (d) Let a, b, $c \in \mathbf{Z}$ such that $(a, b) = (a, c) = (b, c) = 1$. Suppose that $abc = x^2$ for some x in $\mathbf{Z}$. Prove that a, b, and c are all squares in $\mathbf{Z}$.

(e) Let $a_1, a_2, \ldots, a_n \in \mathbf{Z}$ such that $(a_i, a_j) = 1$ if $i \neq j$. Suppose that $a_1 a_2 \ldots a_n = x^2$ for some x in $\mathbf{Z}$. Prove that each a_i is a square in $\mathbf{Z}$.

12. Prove that if a positive integer of the form $2^m + 1$ is prime, then m is a power of 2.

13. Prove that 2 is the only prime of the form $n^3 + 1$.

14. Prove that if $2^n - 1$ is prime, then n is prime.

15. Investigate the following statement:

    If n is any positive integer, then $n^2 + n + 41$ is always a prime number.

    If you think it is true, give a proof; if false, give a counterexample.

16. Let $a, b \in \mathbf{Z}^+$, $a > 1$, $b > 1$. Let $a = p_1^{m_1} \cdot p_2^{m_2} \cdot \ldots \cdot p_r^{m_r}$ and $b = p_1^{n_1} \cdot p_2^{n_2} \cdot \ldots \cdot p_r^{n_r}$, where $p_1, p_2, \ldots, p_r$ are primes and m_i and n_i are nonnegative integers, for $i = 1, 2, \ldots, r$. Let $l_i = \min(m_i, n_i)$. Prove that $(a, b) = p_1^{l_1} \cdot p_2^{l_2} \cdot \ldots \cdot p_r^{l_r}$.

17. Use Exercises 8 and 16 to find the greatest common divisor of 1,400 and 42,750.

18. Prove that if a is a positive integer of the form $4n + 3$, then at least one prime divisor of a is of the form $4n + 3$.

19. Prove that if a is a positive integer of the form $3n + 2$, then at least one prime divisor of a is of the form $3n + 2$.

20. Prove that there are infinitely many primes of the form $3n + 2$, $n \in \mathbf{Z}^+$.

21. Prove that there are infinitely many primes of the form $6n + 5$, $n \in \mathbf{Z}^+$.

22. Let n be a positive integer. Prove that the binomial coefficient $\dbinom{2n}{n}$ is divisible by every prime p such that $n < p \leq 2n$ but is not divisible by p^2.

23. Let p be a prime number and t a positive integer. Let $a \in \mathbf{Z}$. Suppose that a divides p^t. Prove that $a = p^k$ for some $k \in \mathbf{Z}$, $1 \leq k \leq t$.

24. Let $n, m \in \mathbf{Z}$, $(n, m) = 1$. Suppose that d is a positive divisor of nm. Prove that there exist positive integers d_1 and d_2 such that $d = d_1 d_2$ where d_1 divides n and d_2 divides m.

25. If n is a positive integer, let $\tau(n)$ denote the number of positive divisors of n. So, for example, $\tau(1) = 1$, $\tau(2) = 2$, $\tau(3) = 2$, $\tau(4) = 3$, $\tau(5) = 2$, $\tau(6) = 4$.
    (a) Prove that if p is a prime number and t is a positive integer, then $\tau(p^t) = t + 1$.
    (b) Let $n, m \in \mathbf{Z}$, $(n, m) = 1$. Prove that $\tau(nm) = \tau(n)\tau(m)$.
    (c) Let $n \in \mathbf{Z}$. Let $n = p_1^{a_1} p_2^{a_2} \ldots p_r^{a_r}$ be the prime factorization of n. Prove that $\tau(n) = (a_1 + 1)(a_2 + 1) \ldots (a_r + 1)$.

(d) Prove that an integer n has an odd number of divisors if and only if n is a perfect square.

Discussion and Discovery Exercises

D1. In Example 1, we proved that $\sqrt{2}$ is irrational. This fact was also proved in Example 5 of Section 1.4. Are the two proofs the same? If not, how do they differ?

D2. Let p_n denote the nth prime number; that is, $p_1 = 2$, $p_2 = 3$, $p_3 = 5$, $p_4 = 7, \ldots$, and so on. Let $q_n = p_1 p_2 p_3 \ldots p_n + 1$. So, for example, $q_1 = 2 + 1 = 3$, $q_2 = 2 \cdot 3 + 1 = 7$, $q_3 = 2 \cdot 3 \cdot 5 + 1 = 31, \ldots$, and so on.
(a) Prove that any prime divisor of q_n must be greater than p_n.
(b) It seems that the q_ns themselves are primes. Prove or disprove.

D3. The modern discipline of cryptography is concerned with the development of secret codes. These codes depend for their effectiveness on the use of large primes. One source of large primes comes from the so-called Mersenne primes. Mersenne primes, named after the 17th century mathematician and monk **Marin Mersenne**, are primes of the form $2^n - 1$. Of course not all numbers of this form are prime, and in fact if $2^n - 1$ is prime, then the exponent n must itself be prime. Some Mersenne primes are: 3, 7, 31, and 127.
(a) Prove that $2^{13} - 1$ is a Mersenne prime.
   Hint: Exercise 2 of this section will help.
(b) Prove that $2^{11} - 1$ is not a Mersenne prime.

D4. Pythagoras and his followers were intrigued by the properties of integers. Certain numbers they designated as *perfect* numbers. These were numbers whose proper divisors added up to the number itself. For example, 6 is a perfect number because the proper divisors of 6 (that is, all divisors except 6), are 1, 2, and 3 and $1 + 2 + 3 = 6$. The next perfect number is 28. Euclid was able to prove that n is an even perfect number if and only if $n = 2^{p-1}(2^p - 1)$ where $2^p - 1$ is a Mersenne prime. No odd perfect numbers have been found and it is conjectured that there are none.
(a) Verify by direct calculation that 28 is a perfect number.
(b) Use Euclid's Theorem to find two more perfect numbers.
(c) Prove that every even perfect number is a triangular number. (See the Discussion and Discovery Exercises in Section 5.2.)

D5. Conjecture a formula for the sum of the first n consecutive odd cubes. Your formula should be of the form $1^3 + 3^3 + 5^3 + \ldots + (2n - 1)^3 = p(n)$ where $p(n)$ is some polynomial expression in the variable n. Then prove your formula.

D6. Using the formula you proved in the previous exercise, prove that every even perfect number is the sum of the first n consecutive odd cubes for some positive integer n.

5.5   CONGRUENCES

Congruences and Their Properties

Consider the following algebra problem: are there any integers x and y such that $x^2 = 4y + 3$? One could plug in integers for x and y to see if a solution with small values exists. But this approach would be futile if in fact, as is the case here, there is no solution. To see that there is no solution, we can argue as follows.

Since $4y + 3$ is odd, x^2 is odd, and by Theorem 1.4.6, x must be odd also. Then we can write $x = 2t + 1$ for some t in $\mathbf{Z}$. So $x^2 = 4t^2 + 4t + 1$. It then follows that $4(t^2 + t - y) = 2$ or that $t^2 + t - y = \frac{1}{2}$, which is impossible, since $t^2 + t - y$ is an integer. Note that here we use the fact that only ± 1 have multiplicative inverses in $\mathbf{Z}$. (See Corollary 5.1.8.)

If we analyze this proof, we see that it can be more succinctly stated as follows: if the square of an integer is divided by 4, the remainder will be 0 or 1 and hence cannot be written in the form $4y + 3$.

In this section we introduce the language of congruences so that proofs like the one above can be written in a more systematic and efficient way. Recall that in Section 4.2, we defined the notion of congruence modulo n, where n is a fixed positive integer; that is, *two integers a and b are* **congruent modulo n** *if the difference $a - b$ is divisible by n*. We write $a \equiv b(\mathrm{mod}\ n)$.

In the exercises of Section 4.2, you were asked to prove that congruence mod n defines an equivalence relation on $\mathbf{Z}$. The following proposition is equivalent to that exercise.

Proposition 5.5.1   Let $a, b, c \in \mathbf{Z}$ and let $n \in \mathbf{Z}^+$.

1. $a \equiv a(\mathrm{mod}\ n)$
2. If $a \equiv b(\mathrm{mod}\ n)$, then $b \equiv a(\mathrm{mod}\ n)$.
3. If $a \equiv b(\mathrm{mod}\ n)$ and $b \equiv c(\mathrm{mod}\ n)$, then $a \equiv c(\mathrm{mod}\ n)$.

Many of the algebraic properties of congruences are similar to corresponding properties of equalities. More specifically, we have:

Proposition 5.5.2   Let $a, b, c, d \in \mathbf{Z}$ and let $n \in \mathbf{Z}^+$.

1. If $a \equiv b(\mathrm{mod}\ n)$ and $c \equiv d(\mathrm{mod}\ n)$, then $a + c \equiv b + d(\mathrm{mod}\ n)$ and $ac \equiv bd(\mathrm{mod}\ n)$.
2. If $ab \equiv ac(\mathrm{mod}\ n)$ and $(a, n) = 1$, then $b \equiv c(\mathrm{mod}\ n)$.

PROOF: 1. Suppose that $a \equiv b$ and $c \equiv d(\mathrm{mod}\ n)$. Then $n \mid a - b$ and $n \mid c - d$, so there exist integers s, t such that $a - b = ns$ and $c - d = nt$.

Hence $(a + c) - (b + d) = (a - b) + (c - d) = n(s + t)$, which implies that $a + c \equiv b + d(\mathrm{mod}\ n)$.

Similarly, $ac - bd = ac - bc + bc - bd = (a - b)c + b(c - d) = n(sc + bt)$, so we have that $ac \equiv bd \pmod n$.

The proof of 2 is left as an exercise. ●

Notice that the cancellation property in 2 of Proposition 5.5.2 does not necessarily hold if a and n are not relatively prime. For example, $(3)(5) \equiv (3)(13) \pmod{24}$ but 5 and 13 are not congruent mod 24.

The following corollary of Proposition 5.5.2 can be proved by induction.

Corollary 5.5.3   If $a \equiv b \pmod n$, then $a^k \equiv b^k \pmod n$ for every positive integer k.

PROOF: We leave it as an exercise. ●

The following proposition is very useful for doing problems involving congruences.

Proposition 5.5.4   Let $n \in \mathbf{Z}, n > 1$. If $a \in \mathbf{Z}$, then a is congruent mod n to exactly one of the integers $0, 1, 2, \ldots, n - 1$.

PROOF: Let $a \in \mathbf{Z}$. By the Division Algorithm (Theorem 5.3.1), there exist q, r in $\mathbf{Z}$ such that $a = nq + r$, where $0 \leq r < n$.

Thus $a - r$ is divisible by n, so $a \equiv r \pmod n$. Hence a is congruent to one of the integers $0, 1, 2, \ldots, n - 1$.

If a is also congruent to some s where $0 \leq s < n$, then $r \equiv s \pmod n$. But since r and s are both less than n, their difference cannot be divisible by n unless $r = s$. Therefore a is congruent mod n to exactly one of the integers $0, 1, 2, \ldots, n - 1$. ●

EXAMPLE I   We will show that the square of any integer is congruent to 0 or 1 mod 4. Let $x \in \mathbf{Z}$. By Proposition 5.5.4, $x \equiv 0, 1, 2$, or $3 \pmod 4$. By Corollary 5.5.3, $x^2 \equiv 0^2, 1^2, 2^2$, or $3^2 \pmod 4$. But $0^2 \equiv 0, 1^2 \equiv 1, 2^2 \equiv 0, 3^2 \equiv 1 \pmod 4$. Therefore, by the transitive property of congruences, $x^2 \equiv 0$ or $1 \pmod 4$. One way to interpret this result is that the square of any integer leaves a remainder of 0 or 1 when divided by 4.

The importance of the congruence approach is that the proof reduces to considering just four cases. In general, working with a congruence mod n will involve working with only the n integers $0, 1, 2, \ldots, n - 1$.

EXAMPLE 2   We can now use congruences to solve the problem stated at the beginning of this section; namely, to show that there are no integers x and y such that $x^2 = 4y + 3$.

Suppose such integers x and y exist. Then $x^2 - 3$ is divisible by 4 or, in other words, $x^2 \equiv 3 \pmod 4$. But the previous example shows that $x^2 \equiv 0$ or $1 \pmod 4$. Hence we have a contradiction. So we can conclude that no such integers exist.

For each $a \in \mathbf{Z}$, we let $[a]$ be the equivalence class of a with respect to the equivalence relation of congruence mod n: $[a] = \{x \in \mathbf{Z} \mid x \equiv a(\bmod\ n)\}$.

We will call $[a]$ the **congruence class** of a mod n.

By Proposition 5.5.4, if $a \in \mathbf{Z}$, then $a \equiv r$ for exactly one integer r such that $0 \leq r \leq n - 1$. Therefore $a \in [r]$. Thus there are exactly n distinct congruence classes; namely, $[0], [1], [2], \ldots, [n - 1]$. To see which equivalence class a given integer a is in, we just compute the remainder when a is divided by n.

EXAMPLE 3 When $n = 2$, there are two congruence classes: $[0] = \{x \in \mathbf{Z} \mid x \equiv 0(\bmod\ 2)\}$ is the set of even integers and $[1] = \{x \in Z \mid x \equiv 1(\bmod\ 2)\}$ is the set of odd integers.

EXAMPLE 4 When $n = 3$, there are three congruence classes:

$[0] = \{x \in \mathbf{Z} \mid x \equiv 0(\bmod\ 3)\}$ is the set of multiples of 3;
$[1] = \{x \in \mathbf{Z} \mid x \equiv 1(\bmod\ 3)\}$ is the set of integers that leave a positive remainder of 1 when divided by 3; and
$[2] = \{x \in \mathbf{Z} \mid x \equiv 2(\bmod\ 3)\}$ is the set of integers that leave a positive remainder of 2 when divided by 3.

Notice that these three classes are distinct and that every integer is in exactly one of them.

EXAMPLE 5 Let $n = 6$. Then there are six congruence classes: $[0], [1], [2], [3], [4]$, and $[5]$. Let $a = 74$. Since the remainder when 74 is divided by 6 is 2, it follows that $74 \in [2]$. By the transitive property of congruences, any integer congruent to 74 mod 6 will be congruent to 2 mod 6. Similarly, any integer congruent to 2 mod 6 will be congruent to 74 mod 6. Hence $[74] = [2]$.

The previous example shows that there is more than one way to represent a congruence class. For $n = 6$, $[2] = [74]$. Similarly, $[2] = [8] = [14] = [20]$ and $[3] = [9] = [15] = [21]$. For $n = 13$, $[4] = [17] = [30] = [43]$.

In general, for congruences mod n, $[a] = [b]$ if and only if $a \equiv b(\bmod\ n)$. We leave the proof of this fact as an exercise.

Reminders:

1. For any integer a, $[a]$ is a set, not an integer.
2. If $0 \leq a < n$, then $[a]$ can be described as the set of integers that give a remainder of a when divided by n.
3. If $[a] = [b]$, it does not mean that $a = b$, only that $a \equiv b(\bmod\ n)$ or that a and b give the same remainder when divided by n.

The Set of Congruence Classes

For $n \in \mathbf{Z}$, $n > 1$, we will let $\mathbf{Z}_n$ denote the set of congruence classes mod n. Thus $\mathbf{Z}_n = \{[0], [1], [2], \ldots, [n - 1]\}$.

In Chapter 4 we discussed the notion of a binary operation on a set; that is, a rule for combining two elements of a set to give a third element of the set. There are two important binary operations on the set $\mathbf{Z}_n$. We define them next. The first one will be called addition because it depends on ordinary addition of integers and the second will be called multiplication because it depends on ordinary multiplication of integers.

First we define addition on $\mathbf{Z}_n$ using the symbol $+$ as follows: if $[a]$, $[b] \in \mathbf{Z}_n$, then $[a] + [b] = [a + b]$, for any $a, b \in \mathbf{Z}$.

So, for example, if $n = 7$, $[4] + [2] = [6]$ and $[5] + [4] = [9] = [2]$.

An apparent problem with this definition is that it seems to depend on the choice of the integers a and b to represent the equivalence classes $[a]$ and $[b]$. If this definition gives a real binary operation, then the sum of the classes $[a]$ and $[b]$ should be the same if a and b are replaced by other integers representing the same classes. In that case we can say that addition is *well defined*.

For example, suppose $n = 6$. Then $[3] + [5] = [8] = [2]$. But we can also write $[3] = [15]$ and $[5] = [47]$ so adding $[15]$ and $[47]$ should give the same answer as adding $[3]$ and $[5]$. We check: $[15] + [47] = [62] = [2]$.

More generally, we must show that if $[a] = [a']$ and $[b] = [b']$, then $[a + b] = [a' + b']$. But if $[a] = [a']$ and $[b] = [b']$, then $a \equiv a'$ and $b \equiv b' (\mathrm{mod}\ n)$. Now applying Proposition 5.5.2, we get that $a + b \equiv a' + b' (\mathrm{mod}\ n)$, which implies that $[a + b] = [a' + b']$. We have now proved that addition on $\mathbf{Z}_n$ is well defined.

The next theorem gives some important algebraic properties of addition in $\mathbf{Z}_n$.

Theorem 5.5.5   Let $n \in \mathbf{Z}$, $n > 1$.

1. Addition in $\mathbf{Z}_n$ is commutative: $[a] + [b] = [b] + [a]$ for all $a, b \in \mathbf{Z}$.
2. Addition in $\mathbf{Z}_n$ is associative: $([a] + [b]) + [c] = [a] + ([b] + [c])$ for all $a, b, c \in \mathbf{Z}$.
3. $[0]$ is the identity element of $\mathbf{Z}_n$ with respect to addition: $[a] + [0] = [a] = [0] + [a]$ for all $a \in \mathbf{Z}$.
4. Every element of $\mathbf{Z}_n$ has an inverse with respect to addition. For any a in $\mathbf{Z}$, the additive inverse of $[a]$ is $[-a]$.

PROOF: 1. Let $a, b \in \mathbf{Z}$. Then $[a] + [b] = [a + b] = [b + a] = [b] + [a]$. Note that in the step $[a + b] = [b + a]$, we use the fact that addition in $\mathbf{Z}$ is commutative.

The proofs of 2, 3, and 4 are left as exercises. ●

Next we define multiplication on $\mathbf{Z}_n$ by letting $[a][b] = [ab]$, for all $a, b \in \mathbf{Z}$.

As with addition, it is necessary to show that this binary operation is well defined. We leave the proof as an exercise.

Theorem 5.5.6 Let $n \in \mathbf{Z}$, $n > 1$.

1. Multiplication in $\mathbf{Z}_n$ is commutative: $[a][b] = [b][a]$ for all $a, b \in \mathbf{Z}$.
2. Multiplication in $\mathbf{Z}_n$ is associative: $([a][b])[c] = [a]([b][c])$ for all $a, b, c \in \mathbf{Z}$.
3. $[1]$ is the multiplicative identity of $\mathbf{Z}_n$: $[a][1] = [a] = [1][a]$ for all $a \in \mathbf{Z}$.
4. The following distributive laws hold: $[a]([b] + [c]) = [a][b] + [a][c]$ and $([a] + [b])[c] = [a][c] + [b][c]$ for all $a, b, c \in \mathbf{Z}$.

PROOF: We prove 2 and leave the rest as exercises. If $a, b, c \in \mathbf{Z}$, $([a][b])[c] = [ab][c] = [(ab)c] = [a(bc)] = [a][bc] = [a]([b][c])$.

Because $\mathbf{Z}_n$ is a finite set, it is possible to give addition and multiplication tables for $\mathbf{Z}_n$ when n is small. Following are the tables for $\mathbf{Z}_5$ and $\mathbf{Z}_6$:

$$\mathbf{Z}_5$$

+	[0]	[1]	[2]	[3]	[4]
[0]	[0]	[1]	[2]	[3]	[4]
[1]	[1]	[2]	[3]	[4]	[0]
[2]	[2]	[3]	[4]	[0]	[1]
[3]	[3]	[4]	[0]	[1]	[2]
[4]	[4]	[0]	[1]	[2]	[3]

·	[0]	[1]	[2]	[3]	[4]
[0]	[0]	[0]	[0]	[0]	[0]
[1]	[0]	[1]	[2]	[3]	[4]
[2]	[0]	[2]	[4]	[1]	[3]
[3]	[0]	[3]	[1]	[4]	[2]
[4]	[0]	[4]	[3]	[2]	[1]

$$\mathbf{Z}_6$$

+	[0]	[1]	[2]	[3]	[4]	[5]
[0]	[0]	[1]	[2]	[3]	[4]	[5]
[1]	[1]	[2]	[3]	[4]	[5]	[0]
[2]	[2]	[3]	[4]	[5]	[0]	[1]
[3]	[3]	[4]	[5]	[0]	[1]	[2]
[4]	[4]	[5]	[0]	[1]	[2]	[3]
[5]	[5]	[0]	[1]	[2]	[3]	[4]

·	[0]	[1]	[2]	[3]	[4]	[5]
[0]	[0]	[0]	[0]	[0]	[0]	[0]
[1]	[0]	[1]	[2]	[3]	[4]	[5]
[2]	[0]	[2]	[4]	[0]	[2]	[4]
[3]	[0]	[3]	[0]	[3]	[0]	[3]
[4]	[0]	[4]	[2]	[0]	[4]	[2]
[5]	[0]	[5]	[4]	[3]	[2]	[1]

Unlike addition, not every element of $\mathbf{Z}_n$ has an inverse with respect to multiplication. An element $[a]$ has an inverse with respect to multiplication if there exists $[x]$ in $\mathbf{Z}_n$ such that $[a][x] = [1]$.

From the multiplication table for $\mathbf{Z}_5$, we see that each element except $[0]$ has a multiplicative inverse. For example, $[2]^{-1} = [3]$ and $[4]^{-1} = [4]$.

In $\mathbf{Z}_6$, however, only $[1]$ and $[5]$ have multiplicative inverses and $[5]^{-1} = [5]$.

The following theorem gives an explicit criterion for an element of $\mathbf{Z}_n$ to have a multiplicative inverse.

Theorem 5.5.7   Let $[a] \in \mathbf{Z}_n$. Then $[a]$ has a multiplicative inverse if and only if a and n are relatively prime.

PROOF: If $[a]$ has a multiplicative inverse, then there exists an element $[x] \in \mathbf{Z}_n$ such that $[a][x] = [1]$. It follows that $[ax] = [1]$ and hence $ax \equiv 1(\bmod\ n)$. (Why?)

This implies that n divides $ax - 1$, which means that $ax - 1 = nt$ for some $t \in \mathbf{Z}$. Hence $ax + n(-t) = 1$. By Theorem 5.3.8, a and n are relatively prime.

Conversely, if a and n are relatively prime, again by Theorem 5.3.8, there exist integers x and y such that $ax + ny = 1$. Hence $[1] = [ax + ny] = [ax] + [ny] = [a][x] + [n][y]$.

But $[n] = [0]$, so we have $[a][x] = [1]$, and therefore $[x]$ is the multiplicative inverse of $[a]$. ●

The proof of Theorem 5.5.7 tells us how to find the multiplicative inverse of an element of $\mathbf{Z}_n$; namely, $[x]$ is the inverse of $[a]$ if and only if x is a solution of the congruence $ax \equiv 1(\bmod\ n)$. The next theorem gives us a procedure for solving such congruences.

Theorem 5.5.8   Let $a, b, n \in \mathbf{Z}$, $n > 1$, and suppose that $(a, n) = 1$. Then the congruence $ax \equiv b(\bmod\ n)$ has a unique solution mod n; that is, there is an integer x such that $ax \equiv b(\bmod\ n)$, and if y is another integer such that $ay \equiv b(\bmod\ n)$, then $y \equiv x(\bmod\ n)$.

PROOF: Since $(a, n) = 1$, there are integers t and s such that $as + nt = 1$. Multiplying by b, we get $asb + ntb = b$. Letting $x = sb$, we have $ax \equiv b(\bmod$

n). If $ay \equiv b(\bmod\, n)$ for some y in $\mathbf{Z}$, then $ax \equiv ay(\bmod\, n)$, and by Proposition 5.5.2, $x \equiv y(\bmod\, n)$. ⬤

Notice that if x is a solution to the congruence $ax \equiv b(\bmod\, n)$, then the set of all solutions to the congruence is the congruence class $[x]$.

EXAMPLE 6 We solve the congruence $11x \equiv 9(\bmod\, 29)$. The proof of Theorem 5.5.8 tells us how to proceed. We first must find integers s and t such that $as + nt = 1$. The procedure for finding such s and t was described in Example 5 of Section 5.3. In this case, we get $s = 8$ and $t = -3$. Therefore, $x = sb = (8)(9) = 72$, and so any integer congruent to $72(\bmod\, 29)$ is a solution of the congruence.

To find the least positive solution, we merely compute the remainder when 72 is divided by 29. Thus the least positive solution to the congruence $11x \equiv 9(\bmod\, 29)$ is 14.

The set of all solutions to the congruence is the congruence class $[14]$ in $\mathbf{Z}_{29}$.

EXAMPLE 7 In $\mathbf{Z}_{12}$, the elements with multiplicative inverses are $[1]$, $[5]$, $[7]$, and $[11]$. To find the inverse of $[7]$, for example, we must solve the congruence $7x \equiv 1(\bmod\, 12)$. Since $1 = 12(3) + 7(-5)$, $x = -5 \equiv 7(\bmod\, 12)$ is a solution. Hence $[7]^{-1} = [7]$.

⬤ MATHEMATICAL PERSPECTIVE: FERMAT'S LITTLE THEOREM

One of the most famous theorems on congruences was proven by **Pierre de Fermat** (1601–1665). Fermat was a French judge who worked as a Councilor in the parliament at Toulouse. But his great love was mathematics, especially number theory. Although Fermat was only an "amateur" mathematician, he formulated some very sophisticated theorems. With one exception, he did not give proofs of his theorems but he undoubtedly knew how to prove many of them. However, some of his results were so difficult that they were only proved years later by mathematicians such as Euler. He also made some claims that he was not able to prove or that were false. For example, Fermat claimed that integers of the form $2^{2^n} + 1$ were always prime. Numbers of this form have come to be known as *Fermat numbers.* It is easy to check that $2^{2^n} + 1$ is prime for $n = 0, 1, 2, 3, 4$. But $2^{2^5} + 1$ is *not* prime since it is divisible by 641. In fact it is known that $2^{2^n} + 1$ is not prime for $5 \le n \le 22$. Moreover, there are no known values of $n > 4$ for which $2^{2^n} + 1$ is prime.

Fermat's theorem on congruences, sometimes referred to as Fermat's Little Theorem, says that if n is any integer and p is a prime number, then $n^p - n$ is divisible by p. In congruence notation, it says that $n^p \equiv n(\bmod\, p)$.

For example, $5^7 - 5$ must be divisible by 7. Indeed, $5^7 - 5 = 78120 = (7)(11160)$.

A corollary of Fermat's Little Theorem is that if n is relatively prime to p, then $n^{p-1} \equiv 1 \pmod{p}$. This corollary can be used to find the remainder when certain large powers are divided by a prime.

EXAMPLE 8   Find the remainder if 5^{213} is divided by 11. By the corollary to Fermat's Little Theorem, $5^{10} \equiv 1 \pmod{11}$. Since $213 = (10)(21) + 3$, we get $5^{213} = (5^{10})^{21}5^3 \equiv 5^3 = 125 \equiv 4 \pmod{11}$. Therefore, the remainder is 4.

The proof of Fermat's Little Theorem can be done by induction. We leave the proof and the proof of the corollary as exercises.

An interesting problem is to consider the converse of Fermat's Little Theorem. Its statement would be: if k is a positive integer and if $n^k \equiv n \pmod{k}$ for all integers n, then k is a prime. Equivalently, this could be stated in its contrapositive form: if k is not a prime, then there exists an integer n such that n^k is not congruent to n mod k.

How does one go about proving or disproving this converse? If no proof presents itself immediately, one might try different cases. This would mean that for a given composite number k, we would need to find an integer n such that n^k is not congruent to n mod k. Let's try some.

$k = 4$: 2^4 is not congruent to 2 mod 4.
$k = 6$: 2^6 is not congruent to 2 mod 6.
$k = 8$: 2^8 is not congruent to 2 mod 8.
$k = 9$: 2^9 is not congruent to 2 mod 9.
$k = 10$: 2^{10} is not congruent to 2 mod 10.

We see from these examples that not only does the converse of Fermat's Little Theorem hold in these cases but we don't have to look far to find the integer n. In all of these cases, 2^k is not congruent to 2 mod k. Might it be the case that for *every* composite number k, 2^k is not congruent to 2 mod k? Unfortunately, this is not the case since $2^{341} \equiv 2 \pmod{341}$ and $341 = (11)(31)$ is not prime. From Exercise 29 of this section, to see that $2^{341} \equiv 2 \pmod{341}$, it suffices to show that $2^{341} \equiv 2 \pmod{11}$ and $2^{341} \equiv 2 \pmod{31}$. To prove each of these congruences Fermat's Little Theorem and its corollary can be used. We leave the details as an exercise (Exercise 30(a)).

But 341 does not give a counterexample to the converse of Fermat's Little Theorem since 3^{341} is not congruent to 3 mod 341. It is not until we consider $561 = (3)(11)(17)$ that we arrive at the correct counterexample. It can be shown that for every integer n, $n^{561} \equiv n \pmod{561}$. So indeed the converse of Fermat's Little Theorem is false.

Numbers k with the property that $n^k \equiv n \pmod{k}$ for all integers n are called **Carmichael numbers**. Other Carmichael numbers are 1105, 1729, and 2821. Carmichael numbers occur infrequently. There are only 43 of them less than one million. However, it was proved in 1992 that there are infinitely many Carmichael numbers. A description of this last result can be found in *What's Happening in the Mathematical Sciences 1993*. [2]

Exercises 5.5

1. Let $x \in \mathbf{Z}$.
   (a) Prove that if x is even, then $x^2 \equiv 0 (\mathrm{mod}\ 4)$.
   (b) Prove that if x is odd, then $x^2 \equiv 1 (\mathrm{mod}\ 4)$.
   (c) State the converses of parts (a) and (b). Prove or disprove each.

2. Let $x \in \mathbf{Z}$.
   (a) Prove that if x is even, then $x^2 \equiv 0$ or $4 (\mathrm{mod}\ 8)$.
   (b) Prove that if x is odd, then $x^2 \equiv 1 (\mathrm{mod}\ 8)$.

3. (a) Prove that there are no integers x and y such that $x^2 = 8y + 3$.
   (b) Prove that there are no integers x and y such that $x^2 = 5y + 2$.

4. Prove that there are no integers x, y, z such that $x^2 + y^2 = 8z + 3$.

5. Prove that there are no integers x and y such that $x^2 + 2y^2 = 805$.

6. Prove that if $n \in \mathbf{Z}$ and $n \equiv 2 (\mathrm{mod}\ 3)$, then there are no integers x and y such that $n = x^2 + 3y^2$.

7. (a) Let x be an odd positive integer. Prove that if $x \equiv 7 (\mathrm{mod}\ 8)$, then x cannot be written as the sum of three integer squares.

   (b) State the converse of the implication in part (a). This converse is true but its proof is difficult.

8. Find the congruence class in $\mathbf{Z}_{17}$ of each of the following integers. Write your answer in the form $[x]$ where $0 \leq x \leq 16$.
   (a) 29   (b) 166   (c) 10,227

9. Prove part 2 of Proposition 5.5.2.

10. Prove Corollary 5.5.3.

11. Prove that in $\mathbf{Z}_n$, $[a] = [b]$ if and only if $a \equiv b (\mathrm{mod}\ n)$.

12. (a) Prove that addition in $\mathbf{Z}_n$ is associative.
    (b) Prove that $[0]$ is the additive identity of $\mathbf{Z}_n$.
    (c) Prove that every element of $\mathbf{Z}_n$ has an additive inverse.

13. (a) Prove that multiplication in $\mathbf{Z}_n$ is well defined.
    (b) Prove that multiplication in $\mathbf{Z}_n$ is commutative.
    (c) Prove that $[1]$ is the multiplicative identity of $\mathbf{Z}_n$.

14. Prove the distributive laws in $\mathbf{Z}_n$, as stated in part 4 of Theorem 5.5.6.

15. Compute the following for the given $\mathbf{Z}_n$. Write your answer in the form $[a]$ where $0 \leq a \leq n - 1$.
    (a) $[21] + [19]$ in $\mathbf{Z}_{12}$
    (b) $[34][27]$ in $\mathbf{Z}_{45}$
    (c) $[33]([83] - [67])$ in $\mathbf{Z}_{100}$

    (d) The additive inverse of [18] in $\mathbf{Z}_{26}$

16. Find the least positive integer x that satisfies the following congruences:
    (a) $6x \equiv 57(\text{mod } 23)$
    (b) $90x \equiv 41(\text{mod } 73)$
    (c) $11x \equiv 211(\text{mod } 571)$

17. Find the multiplicative inverses of the following elements:
    (a) [8] in $\mathbf{Z}_{17}$ (b) [11] in $\mathbf{Z}_{54}$
    (c) [9] in $\mathbf{Z}_{16}$ (d) [201] in $\mathbf{Z}_{674}$

18. Write out the addition and multiplication tables for $\mathbf{Z}_7, \mathbf{Z}_8$, and $\mathbf{Z}_9$. Then list the elements in each of these sets that have multiplicative inverses.

19. Let $n \in \mathbf{Z}, n > 1$. Prove that every nonzero element of $\mathbf{Z}_n$ has a multiplicative inverse if and only if n is prime.

20. Let $n \in \mathbf{Z}, n > 1$. Let $a \in \mathbf{Z}$ such that $a^2 \equiv 1(\text{mod } n)$. Prove that $[a]^{-1} = [a]$.

21. Let $a, b, n \in \mathbf{Z}, n > 1$. Prove that if $a \equiv b(\text{mod } n)$, then $a \equiv b(\text{mod } m)$ for every m in $\mathbf{Z}, m > 1$, such that m divides n.

22. Let $a, x, y, n \in \mathbf{Z}, n > 1$. Prove that if $ax \equiv ay(\text{mod } n)$, then $x \equiv y\left(\text{mod }\dfrac{n}{d}\right)$, where $d = (a, n)$.

    Hint: See Exercise 11 of Section 5.3.

23. Let p be a prime number and let $a, b \in \mathbf{Z}$.
    (a) Suppose that $a^2 \equiv b^2(\text{mod } p)$. Prove that $a \equiv \pm b(\text{mod } p)$.
    (b) Does the result of part (a) hold if p is not prime? Prove your answer.

24. Let $n \in \mathbf{Z}, n > 1$. Without using any results on equivalence classes from Chapter 4, prove that
    (a) $\mathbf{Z}$ is the union of the distinct congruence classes mod n.
    (b) Any two distinct congruence classes mod n are disjoint.

25. Let p be a prime. Prove that the binomial coefficient $\dbinom{p}{k}$ is divisible by p for every integer k such that $1 \leq k \leq p - 1$.

26. Prove Fermat's Little Theorem; that is, if p is a prime, then for every integer n, $n^p - n$ is divisible by p.
    Hint: First use induction to prove it for all positive integers n. The Binomial Theorem and the previous exercise will be helpful.

27. Prove the corollary to Fermat's Little Theorem: if p is a prime and n is an integer relatively prime to p, then $n^{p-1} \equiv 1(\text{mod } p)$.

28. (a) Find the remainder if 19^{55} is divided by 13.
    (b) Find the remainder if 11^{21} is divided by 35.

29. Let $n, m \in \mathbf{Z}^+$ and $a, b \in \mathbf{Z}$. Suppose that $a \equiv b(\text{mod } n)$ and $a \equiv b(\text{mod } m)$ and that $(n, m) = 1$. Prove that $a \equiv b(\text{mod } nm)$.
    Hint: See Exercise 13 of Section 5.3.

30. (a) Prove that $2^{341} \equiv 2 \pmod{341}$.
    (b) Prove that 3^{341} is not congruent to 3 mod 341.

31. Suppose that $k = p_1 p_2 \dots p_r$ where the p_is are distinct primes and $r > 1$. Suppose also that for each $i = 1, 2, \dots, r$, $p_i - 1$ divides $k - 1$. Prove that k is a Carmichael number.

32. Verify that 1105 and 1729 are Carmichael numbers.

33. Let $f_n = 2^{2^n} + 1$ be the nth **Fermat number**.
    (a) Prove by induction that for all $n \geq 1$, $f_0 f_1 f_2 \dots f_{n-1} = f_n - 2$.
    (b) Prove that if n and m are distinct positive integers, then f_n and f_m are relatively prime.
    (c) Use part (b) to give another proof that there are infinitely many primes.

Discussion and Discovery Exercises

D1. Back in the 17th century, Fermat considered the problem of determining which positive integers can be written as the sum of two squares. For example, $2 = 1^2 + 1^2$, $5 = 2^2 + 1^2$, $8 = 2^2 + 2^2$, $10 = 3^2 + 1^2$, and so forth. But not all integers can be written as the sum of two squares. It is easy to see, for example, that 3 and 7 cannot. In a letter to Mersenne, Fermat claimed that he had a proof that any prime congruent to 1 mod 4 is a sum of two squares. The first published proof of Fermat's Theorem was given by Euler about one hundred years later.
    (a) Verify Fermat's Theorem for primes up to 37.
    (b) Prove the converse of Fermat's Theorem; that is, if p is an odd prime and p is a sum of two squares, then $p \equiv 1 \pmod{4}$.
    (c) List the first 20 integers, not necessarily primes, that are the sum of two squares.
    (d) Based on your answer to part (c), state a theorem that characterizes all integers that are the sum of two squares.
    (e) Find a positive integer that can be written as the sum of two squares in two different ways. (Note: they must truly be different ways. For example, if $n = x^2 + y^2$, then don't consider the equations $n = y^2 + x^2$ or $n = (\pm x)^2 + (\pm y)^2$ as new representations.)

D2. Explain in your own words the meaning of the set $\mathbf{Z}_n$. Make your explanation clear enough to be understood by a group of freshman math majors who know what the set of integers is but not much else about it. You should be somewhat informal in your writing, but above all be clear and write only complete sentences. You may use $\mathbf{Z}_4$ as an example (but only as an example) if you feel it will help to make your explanation clearer.

D3. Following the general instructions of Exercise D2, explain what it means for addition in $\mathbf{Z}_n$ to be well defined.

GENERALIZING A THEOREM

One way that mathematicians have been able to discover new facts is by extending or generalizing the results of a known theorem by conjecturing and then proving a result that verifies more cases of the known theorem or that proves a theorem of which the known theorem is a special case. For example, Euclid's Theorem that there exist infinitely prime numbers was generalized by Dirichlet when he proved that there are infinitely many primes in any arithmetic progression. (See the comments at the end of Section 5.4.)

In this section, we illustrate this approach to discovering new facts by looking at Theorem 1.4.4, which says that for every integer n, n is even if and only if n^2 is even.

One way to extend this result would be to consider other exponents. For example, is it true that for every integer n, n is even if and only if n^3 is even? Let's try a proof. It would be helpful to reread the proof of Theorem 1.4.4, since it seems reasonable to try a similar proof to that one.

We will first prove that if n is even, then n^3 is even. So let n be an even integer. Then $n = 2t$ for some integer t. Thus

$$n^3 = (2t)^3 = 8t^3 = 2(4t^3),$$

which shows that n^3 is even.

We now wish to prove (or disprove) the converse; namely, if n^3 is even, then n is even. Recall that in Section 1.4, when we proved that if n^2 is even then n is even, we actually proved the contrapositive of the converse; namely, if n is odd, then n^2 is odd. It makes sense to try that approach again. We will prove that if n is odd, then n^3 is odd.

Suppose that n is an odd integer. Then $n = 2t + 1$ for some integer t. We get

$$n^3 = (2t + 1)^3 = 2(4t^3 + 6t^2 + 3t) + 1.$$

Hence n^3 is odd. We have now established that if n^3 is even, then n is even. We may summarize our results in a proposition.

Proposition 5.6.1   For every integer n, n is even if and only if n^3 is even.

It is possible to prove that n^3 even implies n even directly without proving the contrapositive. However, such a proof does require using some results from Section 5.4. We leave the solution as an exercise.

The next logical extension of Theorem 1.4.4 is to exponent 4 and indeed it is true that for every integer n, n is even if and only if n^4 is even. The proof is similar to the proof of Proposition 5.6.1 and is left as an exercise.

Now instead of trying more exponents, suppose we try to state and prove a general theorem that includes the ones we have already done. The obvious generalization would be the following: if k is a positive integer, then for every integer n, n is even if and only if n^k is even.

A natural way to start a proof is to try to mimic the proof of Theorem 1.4.4 or Proposition 5.6.1. We first prove that if n is an even integer, then n^k is even. If $n = 2t$ for some integer t, then $n^k = (2t)^k = 2 \cdot 2^{k-1} t^k$, an even number.

To prove the converse, we will again mimic the proof of Theorem 1.4.4 and prove its contrapositive; namely, if n is odd, then n^k is odd. Let n be an odd integer. Then we can write $n = 2t + 1$, where $t \in \mathbf{Z}$. So $n^k = (2t + 1)^k$.

Here we pause because our exponent k is unspecified. If we are to follow the proof of Theorem 1.4.4, we need to expand $(2t + 1)^k$. Fortunately, the Binomial Theorem (Theorem 5.2.7) will allow us to do that. We get

$$(2t + 1)^k = \sum_{i=0}^{k} \binom{k}{i} (2t)^{k-i}.$$

Each term of this sum except the last one has a factor of 2 in it and the last term is 1. Thus $(2t + 1)^k$ can be expressed as $2m + 1$ for some integer m, proving that n^k is odd.

Our proof is now complete and we have the following result that generalizes Theorem 1.4.4 and Proposition 5.6.1.

Theorem 5.6.2 Let k be a positive integer. Then for every integer n, n is even if and only if n^k is even.

In proving Theorem 1.4.4 and Proposition 5.6.1, we needed to use only simple algebraic properties of the integers. But in proving the more general result in Theorem 5.6.2, we needed to use the Binomial Theorem. It is usually the case in mathematics that extending a theorem often requires using more sophisticated methods or additional theorems and lemmas.

Theorem 5.6.2 can also be proved by induction. The details are left as an exercise.

There is another way to extend the result of Theorem 1.4.4. To say that an integer n is even is equivalent to saying that it is divisible by 2. So it is natural to ask whether Theorem 1.4.4 can be extended to divisibility by 3; that is, *is it true that for every integer n, n is divisible by 3 if and only if n^2 is divisible by 3*? (We will stick with n^2 for now and consider higher powers of n later.)

To start out at least, let's try following the proof of Theorem 1.4.4 since that approach has proved successful so far. The proof that if n is divisible by 3, then n^2 is divisible by 3 is straightforward and is left as an exercise.

Next we want to show that if n^2 is divisible by 3, then n is divisible by 3. As in the proof of Theorem 1.4.4, we will prove the contrapositive; namely, if n is not divisible by 3, then n^2 is not divisible by 3.

In the case of divisibility by 2, to say n is not divisible by 2 means that n is odd or, equivalently, $n = 2t + 1$ for some integer t. What is a comparable statement for n to be *not divisible* by 3? Here the Division Algorithm helps. One consequence of it is that for every integer n, there exist integers t and r such that $n = 3t + r$, where $0 \le r < 3$. In other words, we can write $n = 3t$, $n = 3t + 1$, or $n = 3t + 2$, for some integer t. So if n is not divisible by 3, then $n = 3t + 1$ or $n = 3t + 2$ for some integer t.

We have two cases to consider then. If $n = 3t + 1$, then

$$n^2 = (3t + 1)^2 = 9t^2 + 6t + 1 = 3(3t^2 + 2t) + 1$$

and if $n = 3t + 2$, then

$$n^2 = (3t + 2)^2 = 9t^2 + 12t + 4 = 3(3t^2 + 4t + 1) + 1.$$

In both cases, we see that n^2 is not divisible by 3. Thus we have proved the following:

Proposition 5.6.3   For every integer n, n is divisible by 3 if and only if n^2 is divisible by 3.

Note: It is possible to prove directly, without using the contrapositive, that if n^2 is divisible by 3, then n is divisible by 3. (See Exercise 5.)

The next case to consider is divisibility by 4. A quick look at examples and we see that the statement "For every integer n, n is divisible by 4 if and only if n^2 is divisible by 4" is false. Let $n = 2$. Then $n^2 = 4$ is divisible by 4 but n is not.

So we are led to the following question: For what positive integers k is it true that for every integer n, n is divisible by k if and only if n^2 is divisible by k?

It is easy to see that one of the implications is true for every positive integer k; namely, for every integer n, if n is divisible by k, then n^2 is divisible by k. We leave the proof as an exercise.

It is the converse that is interesting. Let $S(k)$ be the statement "For every integer n, if n^2 is divisible by k, then n is divisible by k." We would like to know for which positive integers k the statement $S(k)$ is true.

We know that $S(k)$ is true for $k = 1$ (trivially), for $k = 2$ (Theorem 1.4.4), and for $k = 3$ (Proposition 5.6.3). It is false for $k = 4$.

Since 2 and 3 are primes, we might conjecture that $S(k)$ is true if k is prime and false if k is not prime. The following proposition states that $S(k)$ is true if k is a prime. The proof is left as an exercise. A word of caution first. Unlike the proofs of Theorem 1.4.4 and Proposition 5.6.3, it is best to prove Proposition 5.6.4 directly and not prove its contrapositive. One or more results on primes from Section 5.4 will be helpful here.

Proposition 5.6.4   If k is a prime number, then for every integer n, if n^2 is divisible by k, then n is divisible by k.

Now what about composite values of k? We have only shown that $S(k)$ is false for one value of k, $k = 4$. It would help to reinforce our conjecture by looking at one or two more examples. Let's try $k = 6$. If we compute the squares of all positive integers up to 50, we see that the only ones divisible by 6 are 36, 144, 324, 576, 900, 1296, and 1764. Each one of these squares is the square of a number divisible by 6. It would seem then that $S(6)$ may very well be true. Let's see if we can prove it.

One approach is to prove the contrapositive of $S(6)$ as was done in the proofs of Theorem 1.4.4 and Proposition 5.6.3. We leave this method of proof as an exercise and instead prove $S(6)$ directly. Suppose that n is an integer such that n^2 is divisible by 6. It follows that n^2 is divisible by 2 and by 3. Now since $S(2)$ and $S(3)$ are true, it follows that n is divisible by 2 and by 3. Because 2 and 3 are relatively prime, we can use Exercise 13 of Section 5.3 to conclude that n is divisible by 6. So now we have the following:

Proposition 5.6.5 For every integer n, if n^2 is divisible by 6, then n is divisible by 6.

So now we have to revise our conjecture. We know that $S(k)$ is true if k is prime but not always false if k is not prime. The next step might be to generalize Proposition 5.6.5. If we study the proof of Proposition 5.6.5, we see that the roles played by 2 and 3 could be replaced by any pair of distinct primes. Hence, we have the following result and again we leave the proof for the exercises.

Proposition 5.6.6 Let p and q be distinct primes and let $k = pq$. For every integer n, if n^2 is divisible by k, then n is divisible by k.

Next, let's consider how we may generalize Proposition 5.6.6 even more. A reasonable generalization would be to look at the product of more than two distinct primes. In fact, $S(k)$ is true whenever k is a product of distinct primes. A rigorous proof will require a suitable generalization of Exercise 13 of Section 5.3. In fact, a generalization of Exercise 12 of Section 5.3 is required first. We state these generalizations as lemmas and of course leave the details to you. Both proofs should be done by induction.

Lemma 5.6.7 Let $a_1, a_2, \ldots, a_m, b \in \mathbf{Z}$ and suppose that $(a_1, b) = (a_2, b) = \ldots = (a_m, b) = 1$. Then $(a_1 a_2 \ldots a_m, b) = 1$.

Lemma 5.6.8 Let $a_1, a_2 \ldots, a_m, b \in \mathbf{Z}$ and suppose that $a_i \mid b$ for all $i = 1, 2, \ldots, m$. Suppose also that $(a_i, a_j) = 1$ if $i \neq j$. Then $a_1 a_2 \ldots a_m \mid b$.

Now you should be able to prove the following. (See Exercise 12.)

Proposition 5.6.9 Let $p_1, p_2, \ldots, p_m$ be distinct prime numbers. Let $k = p_1 p_2 \ldots p_m$. For every integer n, if n^2 is divisible by k, then n is divisible by k.

Is it possible that $S(k)$ is true for other k besides those k that are the product of distinct primes? We'll try some examples.

$S(8)$ is false since 4^2 is divisible by 8 but 4 is not. Similarly, $S(9)$ is false since 3^2 is divisible by 9 but 3 is not. In fact, $S(k)$ is false for $k = 12, 16,$ and 18 as well. (See Exercise 13.)

Perhaps then Proposition 5.6.9 gives all values of $k \geq 2$ for which $S(k)$ is true. This is indeed the case.

Proposition 5.6.10 $S(k)$ is false if k is not the product of distinct primes.

> **PROOF:** We leave this as an exercise. (You may find it helpful to prove special cases of this result first. For example, first prove that $S(k)$ is false if $k = p^2$, where p is a prime, and then if $k = p^m$, $m \geq 2$.) ●

We summarize these results in the following theorem.

Theorem 5.6.11   If $k \geq 2$, then $S(k)$ is true if and only if k is a product of distinct primes.

We conclude with one more generalization. If k and t are positive integers, let $S(k, t)$ be the statement "for every integer n, n is divisible by k if and only if n^t is divisible by k." In Theorem 5.6.3, we proved that $S(2, t)$ is true for all positive integers t. In Theorem 5.6.11, we proved (or rather *you* will prove) that $S(k, 2)$ is true if and only if k is a product of distinct primes.

As a final exercise, formulate and prove the appropriate theorem for $S(k, t)$ in general (Exercise 15).

● HISTORICAL COMMENTS: FERMAT'S LAST THEOREM

One of the most famous generalizations in all of mathematical history concerns Pythagorean triples. As we saw in Section 5.3, the equation $x^2 + y^2 = z^2$ has infinitely many solutions in positive integers. One generalization of this result would be to look at the equation $x^3 + y^3 = z^3$ and ask whether there are any solutions to this equation in positive integers. That equation and the more general one $x^n + y^n = z^n$ have a fascinating history.

A Latin translation of Diophantus' *Arithmetica* was published by **Claude Gaspard de Bachet** (1581–1638) in 1621. A copy of this translation became available to Fermat. In a page margin of his copy of Bachet's translation, Fermat claimed that the equation $x^n + y^n = z^n$ has *no* solution in positive integers. He wrote that he had a wonderful proof of this fact but that the margin was too small to contain it. No record of Fermat's "proof" has ever been found. Fermat's claim became known as Fermat's Last Theorem and its history comprises one of the most interesting stories in all of mathematics.

With one exception, Fermat never published any proofs of his work. That exception was a proof that the equation $x^4 - y^4 = z^2$ has no solutions in positive integers. An easy corollary of this result is the $n = 4$ case of Fermat's Last Theorem. Fermat's proof is elementary enough to be included in most introductory books on number theory. The case $n = 3$ of Fermat's Last

Theorem was essentially proved by Euler. Euler's proof is not elementary in the sense that it requires more sophisticated techniques than the $n = 4$ case requires.

One can show that if Fermat's Last Theorem is true for odd prime exponents p and also for $n = 4$, then it is true for every exponent n. (See Exercise 17 following.) So the challenge was to prove that if p is an odd prime then the equation $x^p + y^p = z^p$ has no positive integer solutions.

In the 1820s, independent proofs of Fermat's Last Theorem for $p = 5$ were given by both Dirichlet and **Adrien-Marie Legendre** (1752–1833) and in 1839, **Gabriel Lame** (1795–1870) gave a proof for $p = 7$.

The most significant work on Fermat's Last Theorem in the 19th century was done by **Ernst Kummer** (1810–1893) and the Bernoulli numbers that we discussed in Section 5.2 played an important role. In 1847, Kummer proved that if p is an odd prime and p does not divide the numerator of any of the Bernoulli numbers $B_2, B_4, \ldots, B_{p-3}$, then Fermat's Last Theorem is true for the prime p.

Primes that satisfy the conditions of Kummer's Theorem are called *regular* primes. A prime that is not regular is called *irregular*. It is easy to check that 3, 5, and 7 are regular primes. In fact, the first irregular prime is 37 since the numerator of B_{32} is 7709321041217, which is divisible by 37. In a random list of primes, regular primes seem to occur roughly three-fifths of the time. But, oddly enough, although it is known that there are infinitely many irregular primes, it is not known if there are infinitely many regular primes.

Over the next 100 years or so various methods were developed to prove special cases of Fermat's Last Theorem, so that by 1975 it was established to be true for all prime exponents less than 125,000. These methods centered on analyzing what are called *cyclotomic integers,* which are complex numbers involving roots of unity. (Complex numbers and roots of unity are discussed in Section 7.3.)

Also over the years, many mathematicians, famous and amateur alike, thought that they had proven Fermat's Last Theorem. But until 1994, every one of these proofs had a flaw in it.

Then in 1986 the history of Fermat's Last Theorem took a surprising turn. A German mathematician named **Gerhard Frey** published a paper in which he claimed that Fermat's Last Theorem would follow from a conjecture about *elliptic curves*. Elliptic curves are given by equations of the form $y^2 = ax^3 + bx^2 + cx + d$ where the cubic polynomial on the right-hand side has no repeated roots. The conjecture that Frey was referring to was the Shimura–Taniyama–Weil conjecture, a statement about elliptic curves with rational coefficients that is very difficult to state in simple terms so we will make no attempt here. What was stunning to the mathematical community was that Frey's approach was new and totally unexpected. However, there were few details in Frey's paper, so work had to be done to give a rigorous proof of his claim. This was done by **Ken Ribet**, who proved in 1990 that if the Shimura–Taniyama–Weil conjecture was true for so-called semistable elliptic curves, then Fermat's Last Theorem was also true. Then in the summer of 1993, **Andrew Wiles** announced that he had a proof of the Shimura–Taniyama–Weil conjecture for semistable elliptic curves and hence a proof of Fermat's Last Theorem. The proof was long and difficult, so it was necessary for mathematicians to check Wiles' work. Alas a mistake, a serious one, was found in subsequent months. Wiles worked for a year on the gap in his proof and with the help of **Richard Taylor** was able to clear up all difficulties. Finally, after more than 350 years, the first correct proof of Fermat's Last Theorem was given!

Exercises 5.6

1. Prove the statement "For every integer n, if n^3 is even, then n is even" without proving its contrapositive.

2. Prove that for every integer n, n is even if and only if n^4 is even.

3. Use induction to give another proof of Theorem 5.6.2.

4. Prove that for every integer n, if n is divisible by 3, then n^2 is divisible by 3.

5. Prove the statement "For every integer n, if n^2 is divisible by 3, then n is divisible by 3" without proving its contrapositive.

6. Let k be a positive integer. Prove that for every integer n, if n is divisible by k, then n^2 is divisible by k.

7. Prove Proposition 5.6.4.

8. Prove $S(6)$ by proving its contrapositive.

9. Prove Proposition 5.6.6.

10. Prove Lemma 5.6.7.

11. Prove Lemma 5.6.8.

12. Prove Proposition 5.6.9.

13. Prove that $S(k)$ is false for $k = 12$, 16, and 18.

14. (a) Prove that $S(k)$ is false if $k = p^2$ where p is a prime.
    (b) Prove that $S(k)$ is false if $k = p^m$ where p is a prime and $m \geq 2$.
    (c) Prove Proposition 5.6.10.

15. State and prove a theorem that classifies all k and t for which $S(k, t)$ is true.

16. Prove that if the equation $x^4 - y^4 = z^2$ has no solutions in positive integers, then Fermat's Last Theorem is true for $n = 4$.

17. Prove that if Fermat's Last Theorem is true for all odd prime exponents p and also for $n = 4$, then it is true for every exponent n.

18. Prove that if the equation $x^n + y^n = z^n$ has no solutions in positive integers, then it has no solutions in positive rational numbers.

19. Verify that 3, 5, 7, and 11 are regular primes.

20. Prove that there is no right triangle with integer sides whose area is a perfect square.

    Hint: First prove that if such a triangle exists, then there must be such a triangle whose sides form a primitive Pythagorean triple. Then use Theorem 5.3.10 and Exercise 11(d) of Section 5.4 together with Fermat's result that the equation $x^4 - y^4 = z^2$ has no solutions in positive integers to derive a contradiction.

INFINITE SETS

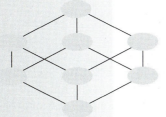

6

Introduction

We have studied many different sets in this text. Some of them have a finite number of elements; the set $\{1, 2, \ldots , 10\}$ for example. But most of the interesting sets we have looked at have an infinite number of elements, the sets $\mathbf{Z}$, $\mathbf{Q}$, and $\mathbf{R}$ being the most important examples. In the 19th century, mathematicians attempted to introduce rigor into calculus. In doing so, they undertook a comprehensive study of the properties of the real numbers. As the theory advanced, it soon became clear that infinite sets in general had to be studied. But the very idea of a set with an infinite number of elements was resisted for a long time, mainly because it seemed to lead to contradictions.

If A and B are finite sets, it is usually easy to determine which of the sets has the greater number of elements; we just count the number of elements in each set. Now suppose that A and B are infinite sets. Is there a sense in which one of these sets is "bigger" than the other? Clearly, counting elements is not a viable method in this case. However, there is another way to show that two finite sets A and B have the same number of elements: find a bijective function with domain A and codomain B. It follows from Exercise 27(a) of Section 3.2 that if such a function exists, then $|A| = |B|$. By analogy, we could say that two infinite sets have the "same size" if there is a bijection between them.

Here are two examples that illustrate the fact that the notion of two infinite sets having the same size can run counter to our intuition.

Let C_1 and C_2 be two concentric circles. We can establish a bijection between C_1 and C_2 by associating a point p_1 on C_1 with a point p_2 on C_2 if p_1 and p_2 lie on the same radial line from the common center of the circles.

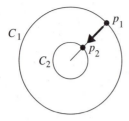

By our definition, then, we would conclude that C_1 and C_2 have the "same size." However, one of the concentric circles will clearly have a smaller circumference than the other.

As another example, consider the set $\mathbf{Z}$ of integers and the set $\mathbf{E}$ of even integers. The function $f: \mathbf{Z} \to \mathbf{E}$ defined by $f(n) = 2n$ is a bijection between $\mathbf{Z}$ and $\mathbf{E}$.

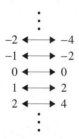

Thus we have the seeming contradiction of a set having the same size as one of its proper subsets.

Gradually, as the 19th century came to a close, these examples came to be seen as **paradoxes** rather than actual mathematical contradictions. Paradoxes, in this context, are seemingly contradictory statements that are actually true. In the last example, if we accept the fact that two sets having the same size *means* that there is a bijection between them, then it is a true statement to say that $\mathbf{Z}$ and $\mathbf{E}$ have the same size.

It was the mathematician Georg Cantor (1845–1918) who in a series of papers put the theory of infinite sets on a firm logical foundation. We will discuss many of Cantor's ideas in this chapter.

6.1   COUNTABLE SETS

Numerically Equivalent Sets

Cantor distinguished infinite sets by their "size" in the same way we discussed, by labeling sets as having the "same size" if there is a bijection between them. We will call such sets *numerically equivalent*. The formal definition is given next. As a result of this definition we will be able to give a formal definition of what it means for a set to be finite.

> **Definition 6.1.1** Let A and B be sets contained in some universal set U. We say that A and B are **numerically equivalent** if there exists a bijection $f: A \to B$. We say that A and B have the same **cardinality**.

If A and B are numerically equivalent, we write $A \approx B$.

The definition of numerical equivalence actually defines an equivalence relation on the power set of U. The proof of this fact is left as an exercise. Specifically, we have the following:

1. For every set A, $A \approx A$.
2. For all sets A and B, if $A \approx B$, then $B \approx A$.
3. For all sets, A, B, and C, if $A \approx B$ and $B \approx C$, then $A \approx C$.

> **Definition 6.1.2** Let A be a set contained in a universal set U. We say that A is a **finite** set if $A = \varnothing$ or if $A \approx \{1, 2, \ldots, n\}$ for some positive integer n. The integer n is the number of elements of A and is denoted by $|A|$.
>
> A set that is not finite is called an **infinite** set.

Note that if A and B are finite sets, then $A \approx B$ if and only if $|A| = |B|$. We leave the formal proof as an exercise.

The cardinality of a set is a measure of the size of a set. So, for example, two finite sets have the same cardinality if and only if they have the same number of elements. The cardinality or **cardinal number** of a finite set is the number of elements in the set. Every nonnegative integer then is a cardinal number. (0 is the cardinality of $\varnothing$.)

Countable Sets

The most natural example of an infinite set, the one we first encounter in arithmetic, is the set $\mathbf{Z}^+$ of positive integers. So we will first look at sets that are numerically equivalent to $\mathbf{Z}^+$. Note that since $\mathbf{Z}^+$ has no largest element, it must be an infinite set. (See Exercise 18 of Section 5.1.)

> **Definition 6.1.3** Any set A that is numerically equivalent to $\mathbf{Z}^+$ is called a **countably infinite** set. A set that is either finite or countably infinite is called **countable**. A set that is not countable is called **uncountable**.

Any two countably infinite sets then have the same cardinality. We denote this cardinality by $\aleph_0$ (read "aleph zero" or "aleph naught"). $\aleph_0$ is our first example of an infinite cardinal number.

EXAMPLE 1   The set $\mathbf{Z}$ of all integers is countably infinite. To prove this we must find a bijective function $f: \mathbf{Z} \to \mathbf{Z}^+$. One possible way to do this is to map the positive integers to the even positive integers and map the nonpositive integers to the odd positive integers. The function defined by

$$f(n) = \begin{cases} 2n, & \text{if } n > 0 \\ -2n + 1, & \text{if } n \leq 0 \end{cases}$$

is such a function. We leave as an exercise the fact that f is bijective.

EXAMPLE 2   In light of the previous example and the discussion at the beginning of this chapter, it should seem quite plausible that the set $\mathbf{E}$ of even integers is countably infinite. To prove this fact, it would seem necessary to find a bijective function $f: \mathbf{E} \to \mathbf{Z}^+$. But since $\mathbf{Z}^+ \approx \mathbf{Z}$ by Example 1, and since numerical equivalence is an equivalence relation, it suffices to show that there is a bijection function $f: \mathbf{Z} \to \mathbf{E}$. The function $f(n) = 2n$ is such a function. Therefore $\mathbf{E}$ is countably infinite.

The previous examples suggest that any infinite set of integers is countably infinite. In fact, *any* infinite subset of a countably infinite set is countably infinite.

Theorem 6.1.4   Let A be a countably infinite set and let B be an infinite subset of A. Then B is countably infinite.

PROOF: Since A is countably infinite, there exists a bijection $f: \mathbf{Z}^+ \to A$. To prove that B is countably infinite, we will define a bijective function $g: \mathbf{Z}^+ \to B$. To do this, we first inductively define a collection of nonempty subsets of $\mathbf{Z}^+$.

Let $S_1 = \{i \in \mathbf{Z}^+ \mid f(i) \in B\}$. S_1 is the set of positive integers that get mapped to B by the function f. Since f is surjective, S_1 is an infinite subset of $\mathbf{Z}^+$. Next we define S_2 as follows: by the Well-Ordering Principle, S_1 has a smallest element that we will call k_1. So we let $S_2 = S_1 - \{k_1\}$. Now suppose that we have defined $n - 1$ nonempty subsets of $\mathbf{Z}^+$, $S_1, S_2, \ldots, S_{n-1}$. By the Well-Ordering Principle, each S_i has a smallest element. Call it k_i. We define $S_n = S_1 - \{k_1, k_2, \ldots, k_{n-1}\}$. Then S_n is nonempty since B is an infinite set. Let k_n be the smallest element of S_n. Now we define $g: \mathbf{Z}^+ \to B$ by $g(n) = f(k_n)$.

Note that the sequence of integers $k_1, k_2, \ldots$ has the property that if $n < m$, then $k_n < k_m$. We now show that g is a bijective function. To see that g is injective, let $n, m \in \mathbf{Z}^+$ and suppose that $g(n) = g(m)$. Then

$f(k_n) = f(k_m)$. Since f is injective, we have $k_n = k_m$. It now follows from the above remark that $n = m$. Therefore, g is injective.

To prove that g is surjective, let $b \in B$. We must find $n \in \mathbf{Z}^+$ such that $g(n) = b$. Since f is surjective, there exists $t \in \mathbf{Z}^+$ such that $f(t) = b$. Note that $t \in S_1$ since $f(t) \in B$. Let m be the number of integers in S_1 that are less than t. If there are no such integers, that is, if $m = 0$, then $t = k_1$. Otherwise the integers that are less than t are $k_1, k_2, \ldots, k_m$. Since $S_{m+1} = S_1 - \{k_1, k_2, \ldots, k_m\}$, t must be the smallest element of S_{m+1} and hence $t = k_{m+1}$. Thus if $m = 0$ or not, $g(m + 1) = f(t) = b$. This proves that g is surjective.

Since g is a bijection, it follows that B is countably infinite. ⬤

Corollary 6.1.5 Every subset of $\mathbf{Z}$ is countable.

We have seen that a set A is countably infinite if there is a bijection from $\mathbf{Z}^+$ to A. Sometimes it may be difficult or cumbersome to define such a function. In our next result, we prove that we only have to find a surjective function from $\mathbf{Z}^+$ to A.

Theorem 6.1.6 Let A be an infinite set. Suppose there exists a surjection $f: \mathbf{Z}^+ \to A$. Then A is countably infinite.

PROOF: We will inductively define a bijective function $g: \mathbf{Z}^+ \to A$.

Let $g(1) = f(1)$. Let n_2 be the smallest positive integer such that $f(n_2) \neq g(1)$. Let $g(2) = f(n_2)$. Now suppose that $k \in \mathbf{Z}^+$ and $g(1), g(2), \ldots, g(k - 1)$ have been defined. Let n_k be the smallest positive integer such that $f(n_k) \in A - \{g(1), g(2), \ldots, g(k - 1)\}$. We define $g(k) = f(n_k)$. This definition gives a function from $\mathbf{Z}^+$ to A. From the definition of g, $g(k) \neq g(i)$ for any $i = 1, 2, \ldots, k - 1$. Thus g is injective.

We now prove that g is surjective. Let $a \in A$. Since f is surjective, a is in the image of f. Let t be the smallest positive integer such that $f(t) = a$. Then $f(t) \neq f(i)$ for any $i = 1, 2, \ldots, t - 1$. Let r be the number of distinct elements from among $f(1), f(2), \ldots, f(t - 1)$. Then $g(1), g(2), \ldots, g(r)$ are equal to $f(1), f(2), \ldots, f(t - 1)$ in some order.

Hence t is the smallest positive integer such that $f(t) \in A - \{g(1), g(2), \ldots, g(r)\}$. It follows from the definition of g that $g(r + 1) = f(t) = a$, proving that g is surjective.

We have thus constructed a bijection from $\mathbf{Z}^+$ to A and so A is countably infinite. ⬤

We next relate the idea of a countable set with the notion of a sequence.

Definition 6.1.7 Let U be some universal set. A **sequence** of elements of U is a function $f: \mathbf{Z}^+ \to U$.

If $f: \mathbf{Z}^+ \to U$ is a sequence and we write $a_n = f(n)$ for each $n \in \mathbf{Z}^+$, then the sequence can be expressed in the form $\{a_1, a_2, a_3, \ldots\}$ or more simply $\{a_n\}_{n=1}^{\infty}$. Note that the elements of the sequence are not necessarily distinct. In fact, they would all assume the same value if f is a constant function.

Note however that the image of f is a countable set. That fact follows from Theorem 6.1.6.

Now let A be a countably infinite set and let $f: \mathbf{Z}^+ \to A$ be a bijection. Let $a_n = f(n)$ for each $n \in \mathbf{Z}^+$. Then the elements of A can be identified with the sequence $\{a_1, a_2, a_3, \ldots\}$ with no elements repeated; that is, $a_i \neq a_j$ if $i \neq j$.

Conversely, if the elements of a set can be written as a sequence, then that set is finite or countably infinite.

EXAMPLE 3 Let $A = \left\{1, \frac{1}{2}, \frac{1}{3}, \ldots\right\} = \left\{\frac{1}{n} \mid n \in \mathbf{Z}^+\right\}$. The elements of A form a sequence and therefore A is countably infinite. Notice that this gives us an example of a countably infinite set that is also a *bounded* set: every element of A is in the closed interval $[0, 1]$.

Unions of Countable Sets

We now consider the question of whether there are "bigger" sets than $\mathbf{Z}$ that are countable. More specifically, are there countable sets of real numbers that contain the set $\mathbf{Z}$ as a proper subset? For example, our intuition would tell us that the set $A = \mathbf{Z} \cup \{\sqrt{2}\}$ is a countable set. In fact, if S is any finite set of real numbers, then the set $\mathbf{Z} \cup S$ is countable. We leave the proof as an exercise. (See Exercise 7.)

Now what about the union of $\mathbf{Z}$ with another countably infinite set? Such a union will be countably infinite. More generally, we have the following result:

Theorem 6.1.8 Let $A_1, A_2, \ldots, A_k$ be a finite collection of countably infinite sets, all contained in some universal set U. Then $A_1 \cup A_2 \cup \ldots \cup A_k$ is a countably infinite set.

PROOF: Since each A_i is countably infinite, there is a bijection $f_i: \mathbf{Z}^+ \to A_i$ for each $i = 1, 2, \ldots, k$. Let $A = A_1 \cup A_2 \cup \ldots \cup A_k$. We will define a surjective function $f: \mathbf{Z}^+ \to A$. It will then follow from Theorem 6.1.6 that A is countably infinite.

We let

$$f(1) = f_1(1)$$
$$f(2) = f_2(1)$$
$$\vdots$$
$$f(k) = f_k(1)$$
$$f(k+1) = f_1(2)$$
$$\vdots$$
$$f(2k) = f_k(2)$$
$$\vdots$$

An explicit formula for f can be given as follows. Let $n \in \mathbf{Z}^+$. Using a modification of the Division Algorithm, we can write $n = mk + r$ where m is a nonnegative integer and r is an integer such that $1 \leq r \leq k$. Moreover, r and m are unique. Then $f(n) = f(mk + r) = f_r(m + 1)$.

Now f is not necessarily injective but it is surjective. To see this, let $a \in A$. Then $a \in A_i$ for some i. Since f_i is surjective, there exists $n \in \mathbf{Z}^+$ such that $f_i(n) = a$. Then $f((n - 1)k + i) = f_i(n) = a$. Therefore f is surjective and our proof is complete. ●

The Rationals Are Countable

We now consider the set $\mathbf{Q}$ of rational numbers. At first glance it seems to be a "much bigger" set than $\mathbf{Z}$. As we will see in Chapter 7, between any two real numbers there are infinitely many rational numbers. This means that any interval on the real line, no matter how small, will contain infinitely many rational numbers. Despite this fact, the rational numbers *do* form a countable set.

Theorem 6.1.9   The set $\mathbf{Q}$ of rational numbers is countably infinite.

PROOF: We will show that the set $\mathbf{Q}^+$ of positive rationals is countably infinite. Since it is easy to define a bijection from $\mathbf{Q}^+$ to $\mathbf{Q}^-$, it will follow that the set $\mathbf{Q}^-$ of negative rationals is also countably infinite. Since $\mathbf{Q} = \mathbf{Q}^+ \cup \mathbf{Q}^- \cup \{0\}$, Theorem 6.1.8 implies that $\mathbf{Q}$ is countably infinite.

Any element of $\mathbf{Q}^+$ can be written as a/b where $a, b \in \mathbf{Z}^+$. So we can write the elements of $\mathbf{Q}^+$ in a rectangular array where the elements in a given row all have the same denominator; that is, the numbers in row 1 have denominator 1, the numbers in row 2 have denominator 2, the numbers in row 3 have denominator 3, ... , and so on. It will look like this:

$$
\begin{array}{cccc}
\dfrac{1}{1} & \dfrac{2}{1} & \dfrac{3}{1} & \dfrac{4}{1}\cdots \\[2ex]
\dfrac{1}{2} & \dfrac{2}{2} & \dfrac{3}{2} & \dfrac{4}{2}\cdots \\[2ex]
\dfrac{1}{3} & \dfrac{2}{3} & \dfrac{3}{3} & \dfrac{4}{3}\cdots \\[2ex]
\dfrac{1}{4} & \dfrac{2}{4} & \dfrac{3}{4} & \dfrac{4}{4}\cdots \\[2ex]
\vdots & \vdots & \vdots & \vdots
\end{array}
$$

Note that elements of $\mathbf{Q}^+$ appear more than once in this listing. In fact each element appears an infinite number of times. We define $f: \mathbf{Z}^+ \to \mathbf{Q}^+$ by going through this array diagonally:

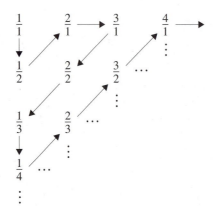

That is, we can define:

$$f(1) = \frac{1}{1}, \qquad f(2) = \frac{1}{2}, \qquad f(3) = \frac{2}{1}, \qquad f(4) = \frac{3}{1}, \qquad f(5) = \frac{2}{2},$$

$$f(6) = \frac{1}{3}, \qquad f(7) = \frac{1}{4}, \qquad f(8) = \frac{2}{3}. \ldots$$

We will forego giving an explicit formula for f since it is complicated. But it is clear from the construction of f that every positive rational is in the image of f. It is now a consequence of Theorem 6.1.6 that $\mathbf{Q}^+$ is countably infinite. We can now conclude that $\mathbf{Q}$ is a countably infinite set. ⬤

Corollary 6.1.10   Any subset of $\mathbf{Q}$ is countable.

Cartesian Products of Countable Sets

Another set that is "bigger" than $\mathbf{Z}$ is the Cartesian product $\mathbf{Z} \times \mathbf{Z}$, the set of all ordered pairs of integers. In view of the examples we have done so far, it should not be surprising that $\mathbf{Z} \times \mathbf{Z}$ is a countable set. In fact if A and B are any countably infinite sets, then the Cartesian product $A \times B$ is countably infinite. One proof of that fact uses an argument similar to the diagonal argument of Theorem 6.1.9. We leave the details of this proof as an exercise. (See Exercise 12.) We give another proof next that uses the Unique Factorization Theorem for Integers.

Theorem 6.1.11   Let A and B be countably infinite sets contained in some universal set U. Then $A \times B$ is countably infinite.

PROOF: Since A and B are countably infinite, there exist bijections $g: \mathbf{Z}^+ \rightarrow A$ and $h: \mathbf{Z}^+ \rightarrow B$. For each $i \in \mathbf{Z}^+$, let $a_i = g(i)$ and $b_i = h(i)$. Since g is injective and surjective, every element of A can be written uniquely in the form a_i for some $i \in \mathbf{Z}^+$. Similarly, every element of B can be written uniquely in the form b_j for some $j \in \mathbf{Z}^+$.

Define $f: A \times B \to \mathbf{Z}^+$ by $f(a_i, b_j) = 2^i 3^j$. We first show that f is injective. Suppose that (a_i, b_j) and (a_k, b_l) are in $A \times B$ and that $f(a_i, b_j) = f(a_k, b_l)$. Then $2^i 3^j = 2^k 3^l$. It follows from the Unique Factorization Theorem for Integers that $i = k$ and $j = l$. Hence $(a_i, b_j) = (a_k, b_l)$ and we can conclude that f is injective.

Clearly f is *not* surjective because the image of f consists of only those positive integers that can be written as a power of 2 times a power of 3. But the image of f is an infinite subset of $\mathbf{Z}$, so by Corollary 6.1.5, Im f is countably infinite. Therefore, $A \times B$ is numerically equivalent to a countably infinite set, namely Im f, and is thus itself countably infinite. ◐

Note that using the primes 2 and 3 in this proof was not essential. Any pair of primes would suffice. Similarly, by using three primes, one can show that the Cartesian product of three countably infinite sets is countably infinite. In general, we have the following result, whose proof is left as an exercise.

Theorem 6.1.12   Let $A_1, A_2, \ldots, A_k$ be a finite collection of countably infinite sets, all contained in some universal set U. Then the Cartesian product $A_1 \times A_2 \times \ldots \times A_k$ is a countably infinite set.

So far we have only looked at infinite sets that are countable. There are indeed infinite sets that are not countable: the set $\mathbf{R}$ of real numbers is the outstanding example. In the next section, we will prove that $\mathbf{R}$ is not countable and see other examples of uncountable sets.

◐   HISTORICAL COMMENTS: CANTOR AND THE INFINITE

It might be said that Georg Cantor introduced the infinite to mathematics. Certainly, mathematicians had considered infinity previously in many different ways; for example, the ancient Greeks had computed areas by exhausting a region with infinitely many simple geometric figures, and by Cantor's time, the notion of limit was being made rigorous. But Cantor treated infinite objects as "completed" (the terminology is due to Gauss), so that their sizes could be measured and compared. In fact, as we will see in the next section, he concluded that different sizes of infinity existed.

This point of view was very controversial in the mathematical community. Among Cantor's many critics were **Henri Poincaré** (1854–1912), who remarked that one should "never introduce objects that one cannot define completely in a finite number of words," and the similarly acid **Leopold Kronecker** (1823–1891). Cantor went so far as to blame Kronecker, a professor at the University of Berlin, for his own failure to rise above his position at the less prestigious University of Halle.

Sadly, Cantor may have let such feelings of persecution dominate his later life. He suffered several breakdowns and died in a hospital in 1918.

His mathematics, however, has stood the test of time, and his place in history was elegantly foretold by **David Hilbert** (1862–1943), who called Cantor's work "the most astonishing product of mathematical thought" and predicted that "No one shall expel us from the paradise which Cantor has created for us."

Exercises 6.1

1. Let U be some universal set. Prove that numerical equivalence, as defined in Definition 6.1.1, defines an equivalence relation on U.

2. Prove that if A and B are finite sets, then $A \approx B$ if and only if $|A| = |B|$.

3. Prove that the function defined in Example 1 is bijective.

4. Prove that $\mathbf{Z} \approx \mathbf{Z}^+$ by finding a bijective function $g\colon \mathbf{Z}^+ \to \mathbf{Z}$.

5. Let $\mathbf{Z}^-$ be the set of negative integers. Prove that $\mathbf{Z}^- \approx \mathbf{Z}^+$ by finding a bijective function $f\colon \mathbf{Z}^+ \to \mathbf{Z}^-$. Prove that your function is bijective.

6. Prove that the following sets S are numerically equivalent to $\mathbf{Z}$ by explicitly defining a bijective function $f\colon \mathbf{Z} \to S$. Then prove that your function is indeed bijective.
   (a) $S = \mathbf{O}$, the set of odd integers
   (b) $S = n\mathbf{Z}$, the set of multiples of a fixed positive integer n
   (c) $S = \{n \in \mathbf{Z} \mid n = 3t + 1 \text{ for some } t \in \mathbf{Z}\}$
   (d) $S = \{x \in \mathbf{Z} \mid x > 100\}$
   (e) $S = \mathbf{Z} \cup \{\sqrt{2}\}$
   (f) $S = \{x \in \mathbf{R} \mid \sin x = 0\}$
   (g) $S = \mathbf{Z} \times \{0\}$
   (h) $S = \mathbf{Z}^+ \times \{-1, 1\}$

7. Prove that if S is any finite set of real numbers, then the set $\mathbf{Z} \cup S$ is countably infinite.

8. (a) Prove that if A is a finite set and $S \subseteq A$, then S is a finite set.
   (b) Prove that if A and B are finite sets, so are $A \cup B$ and $A \cap B$.
   (c) Prove that if A is an infinite set and B is a finite set, then $A - B$ is an infinite set.

9. Let A, B, C, D be subsets of a universal set U. Suppose that $A \approx B$ and $C \approx D$. Prove that $A \times C \approx B \times D$.

10. Compute the values $f(10)$, $f(15)$, and $f(20)$ of the function defined in the proof of Theorem 6.1.9.

11. Prove that each of the following sets is countably infinite. Use the results of this section rather than constructing an explicit bijection to $\mathbf{Z}^+$, and remember to show that the set is not finite.

(a) $\mathbf{Q} \cap [0, 1]$

(b) $\mathbf{Q} - \mathbf{Z}$

(c) $\mathbf{Q}^+ \cup \{e^x \mid x \in \mathbf{Z}\}$

(d) The set of ordered pairs $\left(n, \dfrac{k}{n}\right)$ where $n \in \mathbf{Z}^+$ and $k = 0$ or 1

(e) The set of prime numbers

(f) The rational points on the unit circle:

$$\{(x, y) \mid x^2 + y^2 = 1, x \in \mathbf{Q}, y \in \mathbf{Q}\}$$
Hint: Pythagorean Triples!

12. (a) Give another proof of Theorem 6.1.11 using an argument similar to the one given in the proof of Theorem 6.1.9.

    (b) Give a proof of Theorem 6.1.12 along the lines of the proof of Theorem 6.1.11 given in the text.

    (c) Use induction and Theorem 6.1.11 to give another proof of Theorem 6.1.12.

13. Still another proof of Theorem 6.1.11 goes as follows. Using the notation of the proof of Theorem 6.1.11, define $f_1: A \times B \to \mathbf{Z}^+$ by $f_1(a_i, b_j) = 2^{i-1}(2j - 1)$. Verify that f_1 is a bijection.

14. Let A and B be sets contained in a universal set U. Suppose there exists an injection $f: A \to B$. Prove that if B is countable, then A is countable.

15. Let A and B be sets contained in a universal set U. Suppose there exists a surjection $f: A \to B$. Prove that if A is countably infinite, then B is countable.

16. Prove that $\mathbf{Q}$ is countably infinite by finding a surjection $f: \mathbf{Z} \times \mathbf{Z} \to \mathbf{Q}$ and applying Exercise 15.

17. Prove that the set of all closed intervals $[a, b]$ where $a, b \in \mathbf{Q}$ is a countably infinite set.

Discussion and Discovery Exercises

D1. (a) Imagine that you owned a hotel with a countably infinite number of rooms. Suppose that on a given day, every room was taken. A new guest arrives. Explain how you could provide the new guest with a room of her own while still providing the guests already there with rooms of their own.

    (b) Then suppose that a countably infinite number of new guests arrive. How could the owner find a room for each new guest while again making sure that all of the current guests have their own rooms?

6.2   UNCOUNTABLE SETS, CANTOR'S THEOREM, AND THE SCHROEDER–BERNSTEIN THEOREM

In this section we will study infinite sets, such as the real numbers, that are uncountable. Before beginning, though, we must recall some simple facts about the decimal expansion of real numbers.

We will assume that any nonnegative real number x can be written in decimal form

$$x = a_0.a_1a_2a_3\ldots = a_0 + \frac{a_1}{10} + \frac{a_2}{10^2} + \frac{a_3}{10^3} + \ldots$$

where $a_0, a_1, a_2, a_3, \ldots$ are nonnegative integers and $a_1, a_2, a_3, \ldots$ are between 0 and 9, and that every such expansion in fact represents a real number. Such an expansion is unique except for expansions like

$$\frac{1}{2} = .5000\ldots = .4999\ldots.$$

That is, the decimal representation is unique unless the expansion ends in an infinite string of 9s. (To see this, notice that if these expansions represent different real numbers, there would be a real number between them. What could its expansion be?)

To ensure uniqueness, then, we will adopt the following convention: *we will not use decimal expansions ending in an infinite string of 9s.*

One more thing: there is nothing special about the decimal expansion. We could expand using any positive integer greater than 1, not just using 10. The most important case is when the base is 2. This is called the **binary** expansion, and it uses only the digits 0 and 1. Of course, the same problem arises: the binary representation of a real number is unique only if we agree not to use expansions ending in an infinite string of 1s.

Uncountable Sets

With all this in hand, we are now able to prove that some very familiar sets are uncountable.

Theorem 6.2.1   The closed interval $[0, 1)$ is uncountable.

PROOF: The proof is by contradiction. We assume that $[0, 1)$ is countable and derive a contradiction. If $[0, 1)$ is countable, then there is a bijection f: $\mathbf{Z}^+ \to [0, 1)$. For each $i \in \mathbf{Z}^+$, let $x_i = f(i)$. Now by the previous remarks, each x_i has a decimal expansion of the form

$$x_i = 0.a_1^i a_2^i \ldots$$

where $a_1^i, a_2^i \ldots$ are integers between 0 and 9.

We will derive our contradiction by constructing a real number x in $[0, 1)$ that is not equal to any x_i.

Let $x = 0.b_1 b_2 b_3 \ldots$ where $b_i = 0$ if $a_i^i = 1$ and $b_i = 1$ if $a_i^i \neq 1$. This means that x differs from x_1 in the first decimal place, differs from x_2 in the second decimal place, differs from x_3 in the third decimal place, and so on. Hence $x \neq x_i$ for all i. In other words, $x \neq f(i)$ for any $i \in \mathbf{Z}^+$. But this contradicts the fact that f is surjective. Thus $[0, 1)$ cannot be countable and therefore must be uncountable. ●

Corollary 6.2.2 The closed interval $[0, 1]$ and the set $\mathbf{R}$ of real numbers are uncountable.

PROOF: If $[0, 1]$ were countable, then by Theorem 6.1.4, any subset of $[0, 1]$ would be countable. In particular, this would make $[0, 1)$ countable, contradicting Theorem 6.2.1. Therefore, $[0, 1]$ is uncountable. Similarly, the set $\mathbf{R}$ is uncountable. ●

Since we now know that there are subsets of $\mathbf{R}$ that are uncountable, we will consider the question of the numerical equivalence of such sets. For example, it is not hard to prove that every closed interval $[a, b]$ is numerically equivalent to $[0, 1]$. It follows that any two closed intervals $[a, b]$ and $[c, d]$ are numerically equivalent. (See Exercise 1.) Similarly, any two bounded open intervals are numerically equivalent. (See Exercise 2.)

EXAMPLE 1 The function $f(x) = \tan x$ is a bijection from the open interval $\left(-\dfrac{\pi}{2}, \dfrac{\pi}{2} \right)$ to the real numbers $\mathbf{R}$. Hence any bounded open interval is numerically equivalent to $\mathbf{R}$.

EXAMPLE 2 Every closed interval $[a, b]$ is numerically equivalent to any open interval (c, d). This fact can be proven fairly easily using the Schroeder–Bernstein Theorem, which will be proved later in this section.

Any two sets that are numerically equivalent to $\mathbf{R}$ have the same cardinality. We denote this cardinality by $\mathbf{c}$, called the **power of the continuum**. $\mathbf{c}$ is our second example of an infinite cardinal number. It follows from the previous remarks that every open or closed interval of real numbers has cardinality $\mathbf{c}$.

Cantor's Theorem

Since there is a natural ordering of the positive integers, it is possible to determine the larger of two finite cardinal numbers. What about the infinite cardinals? Since $\aleph_0$ is the cardinality of $\mathbf{Z}$ and $\mathbf{c}$ is the cardinality of $\mathbf{R}$, it makes sense that $\aleph_0$ be a "smaller" cardinal number than $\mathbf{c}$ because $\mathbf{Z}$ is a "smaller" set than $\mathbf{R}$: $\mathbf{Z}$ is a subset of $\mathbf{R}$ and $\mathbf{Z}$ is not numerically equivalent to **R**. We are led to the following definition of what it means for one set to be "smaller" than another. We will use the familiar "less than" symbol $<$.

> **Definition 6.2.3** Let A and B be sets contained in some universal set U. We say that $A < B$ if there is an injection $f: A \rightarrow B$ but no bijection from A to B.
> We say that $A \preceq B$ if $A < B$ or if A is numerically equivalent to B.

Now if we have two cardinal numbers, say **x** and **y**, we will say that **x** $<$ **y** if **x** is the cardinality of set X, **y** is the cardinality of a set Y, and $X < Y$.

Similarly, we will say that **x** $\preceq$ **y** if **x** is the cardinality of set X, **y** is the cardinality of a set Y, and $X \preceq Y$.

For these definitions to make sense, however, they should not depend on the choice of the sets X and Y. In other words, we need the following lemma.

Lemma 6.2.4   Let A, B, C, D be sets contained in a universal set U. Suppose that $A \approx B$, $C \approx D$. If $A < C$, then $B < D$. If $A \preceq C$, then $B \preceq D$.

PROOF: See Exercise 6 for the proof. ●

So we can now say that $\aleph_0 < \mathbf{c}$.

We might ask if there are any infinite cardinal numbers **x** such that **x** $<$ $\aleph_0$. The answer is no since it can be shown, although it is difficult, that every infinite set contains a countably infinite subset.

Are there other infinite cardinal numbers besides $\aleph_0$ and **c**? The answer is yes, and in fact there are an infinite number of infinite cardinals. The following result, known as Cantor's Theorem, explains why.

Theorem 6.2.5   **Cantor's Theorem** Let A be a set contained in a universal set U. Let $\mathbf{P}(A)$ denote the power set of A. Then $A < \mathbf{P}(A)$.

PROOF: We first need to show that there is an injection $f: A \rightarrow \mathbf{P}(A)$. Such an injection must map elements of A to subsets of A. There is a natural way to do this. Let $a \in A$ and map a to the subset $\{a\}$; that is, $f(a) = \{a\}$. It is easy to see that f is injective: for if $f(a) = f(b)$ for some $a, b \in A$, then $\{a\} = \{b\}$ and hence $a = b$.

Now we will show that there is *no* surjective function from A to $\mathbf{P}(A)$. Suppose, on the contrary, that there *is* a surjective function $g: A \rightarrow \mathbf{P}(A)$. We will derive a contradiction.

If $a \in A$, then $g(a)$ is a subset of A. Now the element a may or may not be an element of this subset; that is, for some a, we may have $a \in g(a)$ and for some a, we may have $a \notin g(a)$.

Let $B = \{a \in A \mid a \notin g(a)\}$. B is the set of elements a of A that get mapped to a subset of A that does not contain a.

Since g is surjective and B is a subset of A, B must be in the image of g. This will be true even if B is the empty set. So there exists an element a_0 in A such that $g(a_0) = B$. Now we consider the question: is $a_0 \in B$ or not? Here is where we will derive a contradiction by showing that either possibility leads to an absurdity.

Suppose $a_0 \in B$. Then, by definition of B, $a_0 \notin g(a_0)$. But $g(a_0) = B$, so we are led to the contradictory statement that if $a_0 \in B$, then $a_0 \notin B$.

Now suppose that $a_0 \notin B$. Then it follows that $a_0 \in g(a_0) = B$. So if $a_0 \notin B$, then $a_0 \in B$! There is no way out of this contradiction except to conclude that B is not in the image of g, contradicting the fact that g is surjective. Therefore, no such surjective function can exist.

We have thus proved that $A < \mathbf{P}(A)$. ●

The Continuum Hypothesis

Question: Is there an infinite cardinal $\mathbf{x}$ such that $\aleph_0 < \mathbf{x} < \mathbf{c}$? This problem has a long and interesting history. Cantor conjectured that no such infinite cardinal $\mathbf{x}$ exists. His conjecture has come to be known as the **Continuum Hypothesis**. One consequence of the Continuum Hypothesis is that any subset of the real numbers is either countable or is numerically equivalent to **R**. At the International Congress of Mathematicians in Paris in 1900, the eminent mathematician **David Hilbert** (1862–1943) gave an address in which he outlined the outstanding mathematical problems that he wished to see solved in the 20th century. The first problem on his list was the Continuum Hypothesis.

The first real progress on the Continuum Hypothesis was made by **Kurt Gödel** (1906–1978) in the 1930s when he proved that the Continuum Hypothesis was consistent with the axioms of set theory. (For a brief discussion of these axioms, see Section 6.3.) In other words, using the commonly accepted axioms on which the theory of sets is based, it is impossible to *disprove* the Continuum Hypothesis. However, in 1963, **Paul Cohen** (1934–   ) proved that the *negation* of the Continuum Hypothesis is also consistent with the axioms of set theory. Thus the Continuum Hypothesis is independent of the axioms of set theory and can be neither proved nor disproved from those axioms!

The Schroeder–Bernstein Theorem

Note that $\preceq$ is a relation on $\mathbf{P}(U)$, where U is a universal set. But what kind of a relation is it?

According to Exercise 8(b), like the corresponding relation on **Z**, $\preceq$ is reflexive and transitive but not symmetric. It is too much to hope that $\preceq$ be antisymmetric: if $A \preceq B$ and $B \preceq A$, there is no reason that A and B must actually be equal. We can prove, though, that A and B must be numerically equivalent.

However, a little thought will convince you that this last statement is not simple to prove. Indeed, if $A \preceq B$, then we have an injection $A \to B$, whereas $B \preceq A$ gives an injection $B \to A$. But these injections are not necessarily inverses of each other, so we cannot immediately conclude that either is a bijection (that is, a surjection as well).

This important result is known as the Schroeder–Bernstein Theorem, and has surprising and interesting consequences.

Theorem 6.2.6   **Schroeder–Bernstein Theorem** Let A and B be sets, and suppose that $A \preceq B$ and $B \preceq A$. Then $A \approx B$.

PROOF: Let $f: A \to B$ and $g: B \to A$ be injections. We need to construct a bijection $h: A \to B$, so let $a \in A$. The key is to trace a back as far as possible through f and g. That is, if $a = a_1 \in g(B)$, let b_1 be the unique element of B such that $a_1 = g(b_1)$; if $b_1 \in f(A)$, let a_2 be the unique element of A such that $b_1 = f(a_2)$; if $a_2 \in g(B)$, let b_2 be the unique element of B such that $a_2 = g(b_2)$; and so on. We might call this tracing the *ancestry* of a, and we can visualize it as follows:

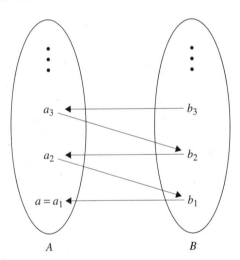

We thus construct a set $\{a = a_1, b_1, a_2, b_2, a_3, b_3, \ldots\}$ of ancestors of a. Now this may be a finite set: we may reach an oldest ancestor $a_i \notin g(B)$ or $b_i \notin f(A)$. Or the process may not stop, yielding an infinite set of ancestors. So let us define the following subsets of A:

$A_A = \{a \in A \mid a \text{ has an oldest ancestor } a_i \in A\}$,
$A_B = \{a \in A \mid a \text{ has an oldest ancestor } b_i \in B\}$,
$A_\infty = \{a \in A \mid a \text{ has no oldest ancestor}\}$.

These sets are clearly pairwise disjoint and $A_A \cup A_B \cup A_\infty = A$; that is, every element of A must satisfy exactly one of the three conditions.

Finally, notice that we could perform the same analysis on B, and arrive at analogous subsets B_B, B_A, and B_∞ of B. In fact, f maps A_A bijectively to B_A and A_∞ to B_∞; similarly, g maps B_B bijectively to A_B. (See the exercises.) Thus we may construct the bijection h as follows:

$$h(a) = \begin{cases} f(a), & \text{if } a \in A_A \cup A_\infty \\ b, & \text{where } a = g(b), \text{if } a \in A_B. \end{cases}$$

One consequence of the Schroeder–Bernstein Theorem and Cantor's Theorem is that there must be an infinite number of infinite cardinals. For if $A \prec \mathbf{P}(A)$ for any set A, then $\mathbf{P}(A) \prec \mathbf{P}(\mathbf{P}(A)) \prec \mathbf{P}(\mathbf{P}(\mathbf{P}(A))) \prec \ldots$. Each one of these sets has a different cardinality, giving us an infinite collection of cardinal numbers. In Exercise D2, you are asked to justify this statement.

The Schroeder–Bernstein Theorem can often be used to give fairly easy proofs of the numerical equivalence of two sets when explicitly constructing a bijection is difficult.

Corollary 6.2.7 $(0, 1) \approx [0, 1]$.

PROOF: By Exercise 9, since $(0, 1)$ is a subset of $[0, 1]$, we know that $(0, 1) \preceq [0, 1]$. On the other hand, by Exercise 1(b), $[0, 1]$ is numerically equivalent to $[1/4, 3/4]$, which is a subset of $(0, 1)$. Therefore, $[0, 1] \preceq (0, 1)$, and the result follows immediately from the Schroeder–Bernstein Theorem.

A similar argument can be used to prove that any two intervals of real numbers are numerically equivalent. See Exercise 14.

An interesting consequence of Theorem 6.2.5 is the fact that since $\mathbf{Z} \prec \mathbf{P}(\mathbf{Z})$, $\mathbf{P}(\mathbf{Z})$ must be an uncountable set. With the Schroeder–Bernstein Theorem, we can now prove that the cardinality of the power set of $\mathbf{Z}$ is in fact $\mathbf{c}$. For convenience, we will actually look at the power set of $\mathbf{Z}^+$; in the exercises, you will be asked to show that the power set of any countable set is numerically equivalent to $\mathbf{P}(\mathbf{Z})$.

Corollary 6.2.8 $\mathbf{P}(\mathbf{Z}^+) \approx \mathbf{R}$.

PROOF: We will show that $\mathbf{P}(\mathbf{Z}^+) \approx [0, 1)$; this is sufficient since we know that $\mathbf{R}$ and $[0, 1)$ are numerically equivalent.

First, let us define an injection $f \colon \mathbf{P}(\mathbf{Z}^+) \to [0, 1)$. For a subset A of $\mathbf{Z}^+$, let $f(A) = 0.a_1a_2a_3 \ldots$ (a *decimal* expansion) where

$$a_i = \begin{cases} 0, & \text{if } i \in A \\ 1, & \text{if } i \notin A. \end{cases}$$

Since no $f(A)$ ends with an infinite string of 9's, $f(A)$ is always an element of $[0, 1)$. Moreover, if $f(A) = 0.a_1a_2a_3 \ldots = 0.b_1b_2b_3 \ldots = f(B)$ for subsets A and B of $\mathbf{Z}^+$, these decimal expansions are unique, and so $a_i = b_i$, for all

$i \in \mathbf{Z}^+$. That is, $i \in A$ if and only if $i \in B$ or $A = B$. We conclude that f is an injection, and so $\mathbf{P}(\mathbf{Z}^+) \preceq [0, 1)$.

To define an injection $g: [0, 1) \to \mathbf{P}(\mathbf{Z}^+)$, we will use the *binary* expansion of elements of $[0, 1)$: let

$$g(0.a_1a_2a_3 \ldots) = \{i \in \mathbf{Z}^+ \mid a_i = 0\}.$$

Recall that the binary expansion is unique because we do not allow an infinite string of 1s. Therefore, there is no ambiguity in the definition of this function.

If $A = g(0.a_1a_2a_3 \ldots) = g(0.b_1b_2b_3 \ldots)$, then $a_i = b_i = 0$ if $i \in A$, and otherwise $a_i = b_i = 1$. Thus, $0.a_1a_2a_3 \ldots = 0.b_1b_2b_3 \ldots$ proving that g is an injection. Therefore, $[0, 1) \preceq \mathbf{P}(\mathbf{Z}^+)$. ●

A closing comment may help you to understand this last proof. In constructing f, we did not really need to use the decimal expansion. The ternary, or base 3, expansion would have sufficed, *but the binary would not*! However, the binary expansion is necessary to define g, whereas *the decimal would not work*. Can you see why these statements are true? You are asked to carefully explain them in the exercises.

● MATHEMATICAL PERSPECTIVE: THE PERILS OF INTUITION

Cantor's Theorem and its consequences are certainly surprising and to some extent counterintuitive. This is undoubtedly why the mathematical community found them so difficult to accept. However, there are many examples of startling results that defy our intuition.

For example, mathematicians in the 19th century attempted to give a rigorous definition of the "obvious" notion of a continuous curve, such as the following:

Camille Jordan (1838–1922) gave a perfectly reasonable definition: a curve is a set of points $(f(t), g(t))$, where f and g are continuous functions of $t \in [0, 1]$. Amazingly, **Giuseppe Peano** (1858–1932) found a function that meets this definition yet whose image is a square, the region $[0, 1] \times [0, 1]$:

This "space-filling" curve is not at all what Jordan had in mind. Peano's function is not a bijection; **Eugen Netto** (1846–1919) showed that a continuous bijection from an interval to a square is impossible. Of course, the fact that a bijection exists is a consequence of the fact that the interval and the square have cardinality **c**.

But perhaps the most unbelievable result came in 1924. The Banach–Tarski Theorem states that a ball can be broken up into finitely many pieces, and then reassembled to form a ball twice as big as the original! This remarkable discovery was made in the field of *measure theory*, the attempt to generalize the notion of the length of an interval (or the volume of a ball) to arbitrary sets, and clearly revealed the need to understand the foundations of mathematical logic as well as the axioms of set theory.

Exercises 6.2

1. (a) Prove that every closed interval $[a, b]$ is numerically equivalent to $[0, 1]$.
   (b) Prove that any two closed intervals $[a, b]$ and $[c, d]$ are numerically equivalent.

2. Prove that any two open intervals (a, b) and (c, d) are numerically equivalent.

3. Let $f: [0, 1] \rightarrow (0, 1)$ be the function defined by $f(0) = \frac{1}{2}, f\left(\frac{1}{n}\right) = \frac{1}{n + 2}$ for any $n \in \mathbf{Z}^+$ and $f(x) = x$ otherwise. Prove that f is a bijection.

4. Let $[a, b]$ be a closed interval and (c, d) an open interval. Prove that $[a, b]$ and (c, d) are numerically equivalent.

5. (a) Prove that the function $f(x) = \frac{1}{1 + x}$ defines a bijection from $\mathbf{R}^+$, the set of positive real numbers, to the open interval $(0, 1)$.
   (b) Prove that $\mathbf{R}^+$ is numerically equivalent to any bounded open or closed interval.
   (c) Prove that any unbounded interval of the form (a, ∞) is numerically equivalent to $\mathbf{R}$.
   (d) Prove that any unbounded interval of the form $(-\infty, a)$ is numerically equivalent to $\mathbf{R}$.

6. Prove Lemma 6.2.4.

7. Determine the cardinality of each of the following sets. Prove your answer.
   (a) $[0, 1] \cup [2, 3]$
   (b) $[0, 2] \cap (1, 3)$

(c) The set of irrational numbers

(d) $\mathbf{R} \times \mathbf{R}$

(e) $\{(x, p) \in \mathbf{R} \times \mathbf{R} \mid x \in \mathbf{R}, p \text{ is a prime}\}$

(f) $\left\{\dfrac{x}{p} \in \mathbf{R} \,\middle|\, x \in \mathbf{R}, p \text{ is a prime}\right\}$

(g) $\mathbf{F}(\mathbf{Z}^+)$, the set of all functions $f: \mathbf{Z}^+ \to \mathbf{Z}^+$

(h) the set of real numbers whose decimal expansions do not contain the digit 5

8. Let A, B, C be sets contained in a universal nonempty set U. The symbols $<$ and $\preceq$ define relations on $\mathbf{P}(U)$.
   (a) Prove that $<$ is transitive but not reflexive or symmetric.
   (b) Prove that $\preceq$ is reflexive and transitive but not symmetric.

9. Let A and B be sets such that $A \subseteq B$. Prove that $A \preceq B$.

10. Let $A, B, C,$ and D be sets contained in a universal set U such that $A \preceq B$ and $C \preceq D$. Prove that $A \times C \preceq B \times D$.

11. Let A, B be sets contained in a universal set U. Suppose that $A \approx B$. Prove that $\mathbf{P}(A) \approx \mathbf{P}(B)$.
    Hint: Exercises 21 and 22 of Section 3.2 will be helpful for this problem.

12. Let A, B be sets contained in a universal set U such that $A \subseteq B$. Suppose that A is countable and B is uncountable. Prove that $B - A$ is uncountable.

13. Prove or disprove: the relation $<$ is antisymmetric on $\mathbf{P}(U)$, where U is a universal set.

14. Using the Schroeder–Bernstein Theorem, prove that any two intervals of real numbers are numerically equivalent.
    Hint: The possibilities are: $(a, b), [a, b], (a, b], [a, b), (a, \infty), [a, \infty), (-\infty, b), (-\infty, b], and (-\infty, \infty).$

15. Fill in the details of the proof of the Schroeder–Bernstein Theorem by showing:
    (a) f maps A_A bijectively to B_A.
    (b) f maps A_∞ bijectively to B_∞.
    (c) g maps B_B bijectively to A_B.
    (d) h is a bijection.

16. Consider the proof of Corollary 6.2.8.
    (a) Explain why the proof that f is an injection will fail if the binary expansion is used, and why it would be valid for any $n > 2$ if the base n expansion is used.
    (b) Explain why the proof that g is an injection will fail if the base n expansion $(n > 2)$ is used.

17. Prove that there is no universal set of all sets.
    Hint: Let U be the set of all sets, and use Cantor's Theorem to derive a contradiction.

Discussion and Discovery Exercises

D1. Do you think that the set of all permutations of $\mathbf{Z}^+$ is countable? Since this is a subset of the set of all functions from $\mathbf{Z}^+$ to $\mathbf{Z}^+$, its cardinality can be no greater than your answer to Exercise 7(g). Also, it's obviously infinite. (Proof?) For guidance, you might want to consider the finite case: the set of permutations of $\{1, 2, \ldots, n\}$ has cardinality $n!$, and $n < n!$ as long as n is at least 3.

D2. Prove the statement given after the proof of the Schroeder–Bernstein Theorem: that there are an infinite number of infinite cardinals, by showing that, for an infinite set A, the sets $\mathbf{P}(A)$, $\mathbf{P}(\mathbf{P}(A))$, $\mathbf{P}(\mathbf{P}(\mathbf{P}(A)))$, $\ldots$ all have different cardinality. Indicate where you actually use the Schroeder–Bernstein Theorem.

6.3 COLLECTIONS OF SETS

Russell's Paradox

In Section 2.3 we discussed arbitrary collections of sets and considered their union and intersection. Such a collection of sets or *set of sets* is usually described by an indexing set and this indexing set can be infinite. For example, for each positive integer i, let $A_i = (-i, i) = \{x \in \mathbf{R} \mid -i < x < i\}$ as in Example 3 of Section 2.3. Then the set $S = \{A_1, A_2, A_3, \ldots\}$ is a collection of open intervals whose indexing set is $\mathbf{Z}^+$. Using the terminology of this chapter, we would say that S is a *countably infinite collection* of open intervals since the indexing set is countably infinite. Similarly, if r is a real number, let $A_r = (r, \infty) = \{x \in \mathbf{R} \mid x > r\}$ as in Example 5 of Section 2.3. Here the indexing set is uncountable, so we say that the collection $\{A_r \mid r \in \mathbf{R}\}$ is an *uncountable collection* of sets. So there are many ways of constructing sets by taking sets of sets, sets of sets of sets, and so on.

However, there are many pitfalls and surprises inherent in any casual approach to building sets. For example, we know (Exercise 17 of the previous section) that there is no object as enormous as "the set of *all* sets."

The mathematician and philosopher **Bertrand Russell** (1872–1970) discovered that there are severe logical problems with just assuming that any combination of sets results in another set, and his discovery shocked the mathematical world. He asked the following question: can a set be an element of itself? Certainly, most sets are not: $\{1\} \notin \{1\}$, since that set contains only one element, namely 1. On the other hand, if there were such a thing as the set of all sets, then as a set, it would have to be an element of itself.

So Russell constructed $\mathscr{A} = \{A \mid A \notin A\}$, the set of all sets that are not elements of themselves. Then $\{1\} \in \mathscr{A}$, but the set of all sets is not an element of $\mathscr{A}$. So far, so good. But then Russell asked: what about $\mathscr{A}$ itself? If $\mathscr{A} \in$

$\mathscr{A}$, then by its very definition, $\mathscr{A} \notin \mathscr{A}$! And if $\mathscr{A} \notin \mathscr{A}$, then this must mean that $\mathscr{A} \in \mathscr{A}$.

This argument has come to be known as **Russell's Paradox**. It is not a semantic trick; it means that not every collection can, logically, be considered a set. In other words, the study of sets should properly begin (just like our study of the integers in Chapter 5 did) with axioms that specify how sets can be constructed. Many different lists of such axioms have been studied, and although we will not examine them in detail here (you may some day in a course in mathematical logic), we will make an assumption that is common to most: if all the sets we consider are subsets of one larger set—a universal set—then the operations of union, intersection, and Cartesian product generate sets.

With this in mind, we can now begin to extend some of the earlier results of this chapter to arbitrary collections of sets. As before, in many examples, a universal set such as **C**, **R**, **Q**, or **Z** is clear, but its existence is always assumed, even if not explicitly stated.

Countable Unions of Countable Sets

We have seen that the finite union of countably infinite sets is countable. What about infinite unions?

Theorem 6.3.1   For every $i \in \mathbf{Z}^+$, let A_i be a countably infinite set. Then $\bigcup_{i \in \mathbf{Z}^+} A_i$ is also countably infinite.

PROOF: For every $i \in \mathbf{Z}^+$, we have a bijection $f_i \colon Z^+ \to A_i$. Write $f_i(j) = a^i_j$, for every $j \in \mathbf{Z}^+$, so that $\bigcup_{i \in \mathbf{Z}^+} A_i = \{a^i_j \mid i, j \in \mathbf{Z}^+\}$. We may arrange the elements of the union in a table and then enumerate them by tracing the diagonals, much as we did in proving that **Q** is countably infinite.

Explicitly, let us define the *diagonal height* of a^i_j to be the integer $i + j$. Then there are $m - 1$ elements in the union that have diagonal height m.

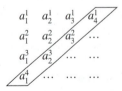

Diagonal height $m = 5$

Thus, if we enumerate the elements of the union by their diagonal height (eliminating repeated elements, of course) we will have an enumeration of the entire union:

$$\{a^1_1, a^2_1, a^1_2, a^3_1, a^2_2, a^1_3, a^4_1, a^3_2, a^2_3, a^1_4, \ldots\}.$$

Because $\bigcup_{i \in \mathbf{Z}^+} A_i$ can be written as a sequence, it follows that $\bigcup_{i \in \mathbf{Z}^+} A_i$ is countably infinite. ●

Notice that the *uncountable* union of countably infinite sets may well be uncountable. In fact, this may be true of the uncountable union of finite sets: $\bigcup_{r \in \mathbf{R}} \{r\} = \mathbf{R}$.

Another way to combine a finite number of sets, say n, is to form the Cartesian product, the set of all n-tuples with coordinates chosen from the respective sets. To do this for infinitely many sets, we must extend the idea of an n-tuple. Let's begin with the easiest case, where we have countably many sets.

> **Definition 6.3.2** Let $\{A_i \mid i \in \mathbf{Z}^+\}$ be a countably infinite collection of sets. The **Cartesian product** of the collection, denoted $\prod_{i \in \mathbf{Z}^+} A_i$, is the set of all sequences of the form $(a_1, a_2, a_3, \dots, a_i, \dots)$ such that $a_i \in A_i$ for all $i \in \mathbf{Z}^+$.

EXAMPLE 1 Let $A = \{0, 1, 2, 3, 4, 5, 6, 7, 8, 9\}$, the set of base 10 digits. Then $\prod_{i \in \mathbf{Z}^+} A$ (here we've chosen all of the sets to be the same) is the set of all sequences $(a_1, a_2, a_3, \dots)$ of the ten digits.

This last example is reminiscent of the proof that $[0, 1)$ is an uncountable set. In fact, we can mimic that proof to determine the size of many Cartesian products.

Theorem 6.3.3 Let $\{A_i \mid i \in \mathbf{Z}^+\}$ be a countably infinite collection of sets, and suppose that each A_i contains at least two elements. Then $\prod_{i \in \mathbf{Z}^+} A_i$ is uncountable.

PROOF: Suppose, to the contrary, that $\prod_{i \in \mathbf{Z}^+} A_i$ is countable. Then there is a bijection $f \colon \mathbf{Z}^+ \to \prod_{i \in \mathbf{Z}^+} A_i$, and we can write $f(i) = \mathscr{A}_i$, for $i \in \mathbf{Z}^+$. Now the elements of the Cartesian product are sequences, so we can write $\mathscr{A}_i = (a_1^i, a_2^i, a_3^i, \dots)$, where $a_j^i \in A_j$. That is,

$$\mathscr{A}_1 = (a_1^1, a_2^1, a_3^1, \dots)$$
$$\mathscr{A}_2 = (a_1^2, a_2^2, a_3^2, \dots)$$
$$\mathscr{A}_3 = (a_1^3, a_2^3, a_3^3, \dots)$$
$$\vdots$$

Now since each A_i contains at least 2 elements, we can choose elements $b_i \in A_i$ such that $b_i \neq a_i^i$, for every $i \in \mathbf{Z}^+$. But then $\mathscr{B} = (b_1, b_2, b_3, \ldots) \in \prod_{i \in \mathbf{Z}^+} A_i$, yet $\mathscr{B} \neq \mathscr{A}_i$, for any $i \in \mathbf{Z}^+$, a contradiction. ●

We have until now required our index set in forming a Cartesian product to be $\mathbf{Z}^+$. The reason is not hard to see. We were trying to extend n-tuples $(a_1, a_2, a_3, \ldots, a_n)$ to infinite lists; naturally, we looked at sequences $(a_1, a_2, a_3, \ldots)$ indexed by all positive integers.

But what if the index set to our collection is uncountable? Can we still form a sensible product? The key lies in remembering that although our notation for a sequence, $(a_1, a_2, a_3, \ldots)$, resembles that for an n-tuple, the actual definition of a sequence is that it is a function with domain the index set $\mathbf{Z}^+$. This definition can easily be extended to any index set—finite, countably infinite, or uncountable.

> **Definition 6.3.4** Let $\{A_i \mid i \in I\}$ be a collection of sets. The **Cartesian product** of the collection, denoted $\prod_{i \in I} A_i$, is the set of all functions $f: I \to \bigcup_{i \in I} A_i$ such that $f(i) \in A_i$ for all $i \in I$.

Before concluding, we must say a word about the significance of these last two definitions. It may seem intuitively clear that if $A_i \neq \varnothing$, for all $i \in I$, then certainly $\prod_{i \in I} A_i$ must also be nonempty. However, this innocuous statement turns out to be very deep.

To be specific, one of the most common foundations for set theory is a system known as the *Zermelo–Frankel Axioms*. These axioms assure, among other things, the existence (as sets!) of the empty set, the power set of a set, the union of a set of sets, and the like. The consequences of these axioms are collectively referred to as Zermelo–Frankel Set Theory, or simply ZF. Much of mathematics, and almost all of what we have done, can be carried out in ZF.

However, the above statement cannot; that is, the statement

$$\text{If } I \neq \varnothing \text{ and } A_i \neq \varnothing, \text{ for all } i \in I, \text{ then } \prod_{i \in I} A_i \neq \varnothing$$

known as the Axiom of Choice, cannot be proven or disproven from the Zermelo–Frankel Axioms. (Gödel proved that the Axiom of Choice is consistent with the Zermelo–Frankel Axioms and Cohen proved that the negation of the Axiom of Choice is also consistent with the Zermelo–Frankel Axioms.) If the Axiom of Choice is added to the list and assumed in addition to the other axioms, the resulting system is referred to as ZFC.

Many interesting and useful results can be proven in ZFC using the Axiom of Choice, and it is sometimes surprising when it is required. For example, it is needed to prove that the cardinal numbers are linearly ordered: if A and B are sets, then either $A \preceq B$ or $B \preceq A$. We used it, in fact, in our proof of Theorem 6.3.1. (See the exercises.)

It is safe to say that the vast majority of mathematicians today work in ZFC, and it has been our context from the beginning.

● HISTORICAL COMMENTS: MATHEMATICS IS LOGIC

Bertrand Russell's discovery of the paradox mentioned at the beginning of this section sprung from his ambitious goal to describe a complete logical foundation for mathematics. This enormous effort bore fruit: *Principia Mathematica,* three volumes written with **Alfred North Whitehead** (1861–1947). The *Principia* develops logic in a formal way (using notation similar to ours from Chapter 1 but much more strictly) and set theory, but with a specific caveat: a collection cannot be a member of itself. Thus, Russell and Whitehead explicitly avoid the paradox.

Interestingly, another mathematician was not so clearsighted. **Gottlob Frege** (1848–1925) spent much of his career on a project similar to Russell's: his two-volume *Fundamental Laws of Mathematics.* Shortly before the second volume was to go to press, he received a letter from Russell containing an exposition of the paradox. Frege's dismay is best described by his own words: "A scientist can hardly meet with anything more undesirable than to have the foundation give way just as the work is finished."

Exercises 6.3

1. Let $\{A_i \mid i \in I\}$ be a collection of subsets of a set A and $\{B_j \mid j \in J\}$ be a collection of subsets of a set B. Let $f: A \to B$.

   (a) Prove that $f\left(\bigcup_{i \in I} A_i\right) = \bigcup_{i \in I} f(A_i)$.

   (b) Prove that $f^{-1}\left(\bigcup_{j \in J} B_j\right) = \bigcup_{j \in J} f^{-1}(B_j)$.

   (c) Prove that $f\left(\bigcap_{i \in I} A_i\right) \subseteq \bigcap_{i \in I} f(A_i)$.

   (d) Prove that $f^{-1}\left(\bigcap_{j \in J} B_j\right) = \bigcap_{j \in J} f^{-1}(B_j)$.

2. Suppose we define an order on $\mathbf{R} \times \mathbf{R}$ as follows:

$$(a, b) < (c, d) \Leftrightarrow a < c \quad \text{or} \quad (a = c \text{ and } b < d).$$

This is called the *lexicographic* order. We define $(a, b) \leq (c, d)$ in the obvious way.

(a) Let (a, b), $(c, d) \in \mathbf{R} \times \mathbf{R}$. Prove that if $(a, b) \leq (c, d)$ and $(c, d) \leq (a, b)$, then $(a, b) = (c, d)$.

(b) Let (a, b), (c, d), $(e, f) \in \mathbf{R} \times \mathbf{R}$. Prove that if $(a, b) \leq (c, d)$ and $(c, d) \leq (e, f)$, then $(a, b) \leq (e, f)$.

(c) Let $(a, b) \in \mathbf{R} \times \mathbf{R}$ and let $A_{(a,b)} = \{(x, y) \in \mathbf{R} \times \mathbf{R} \mid (x, y) > (a, b)\}$. Prove that the collection $\{A_{(a,b)} \mid (a, b) \in \mathbf{R} \times \mathbf{R}\}$ is a decreasing chain. (See the definition of decreasing chain in Section 2.3.)

3. Let $\{A_i \mid i \in I\}$ be a collection of sets and suppose that $\bigcup_{i \in I} A_i$ is countably infinite. Must at least one of the A_is be countably infinite? Prove or disprove.

4. Let $A = \{0, 1\}$. Prove that the set $\prod_{i \in \mathbf{Z}^+} A$ is numerically equivalent to $\mathbf{R}$.

5. Let $A = \{0, 1\}$. Prove that the set $\prod_{r \in \mathbf{R}} A$ is numerically equivalent to $\mathbf{P}(\mathbf{R})$, the power set of $\mathbf{R}$.

6. Let I be a set. Then $\prod_{i \in I} \mathbf{Z}$ is the set of all functions $f: I \to \mathbf{Z}$. We define the **direct sum** over I of $\mathbf{Z}$, denoted $\bigoplus_{i \in I} \mathbf{Z}$, to be the set of all functions $f: I \to \mathbf{Z}$ such that $f(i) = 0$, for all but a finite number of elements $i \in I$.

(a) Prove that $\prod_{i \in I} \mathbf{Z} = \bigoplus_{i \in I} \mathbf{Z} \Leftrightarrow I$ is a finite set.

(b) Prove that if I is countable, so is $\bigoplus_{i \in I} \mathbf{Z}$.

7. Give a different proof of Theorem 6.3.1 by constructing a bijection $f: \mathbf{Z}^+ \times \mathbf{Z}^+ \to \bigcup_{i \in \mathbf{Z}^+} A_i$.

Hint: Think of $\bigcup_{i \in \mathbf{Z}^+} A_i$ *as a matrix with infinitely many rows and columns.*

Discussion and Discovery Exercises

D1. Precisely explain how the Axiom of Choice is used in the proof of Theorem 6.3.1.

D2. Let $\{A_i \mid i \in \mathbf{Z}^+\}$ be a countably infinite collection of sets. What conditions on the sets A_i will guarantee that $\prod_{i \in \mathbf{Z}^+} A_i$ is countable? Try to be as general as possible.

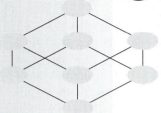

THE REAL AND COMPLEX NUMBERS

7

7.1 FIELDS

In Chapter 5 we studied the set of integers from the point of view of their axioms and the results that can be derived from these axioms. In this chapter we consider two other important sets: the set **R** of *real numbers,* which should be very familiar to you, and the set **C** of *complex numbers,* which you may find less familiar.

Both the real and complex numbers satisfy similar axioms, many of them the same as those satisfied by the integers, but with some important differences. Any set that satisfies these axioms is called a *field.* In this section, we discuss fields in general as well as the notion of an *ordered field.*

Definition 7.1.1 A **field** F is a nonempty set with two binary operations. The first is called **addition** and is denoted by $+$; the second is called **multiplication** and is denoted by $\cdot$. These binary operations satisfy the following axioms:

1. Addition is associative: $(a + b) + c = a + (b + c)$ for all a, b, $c \in F$.
2. Addition is commutative: $a + b = b + a$ for all a, $b \in F$.
3. F has an identity with respect to addition: there exists an element 0 in F such that $a + 0 = 0 + a = a$ for all $a \in F$. 0 is called the **zero element** (or just simply zero) or **additive identity** of F.

4. Every element of F has an inverse with respect to addition: for each a in F, there is an element $-a \in F$ such that $a + (-a) = -a + a = 0$. $-a$ is called the **additive inverse** of a or **negative** a.
5. Multiplication is associative: $(a \cdot b) \cdot c = a \cdot (b \cdot c)$ for all a, b, $c \in F$.
6. Multiplication is commutative: $a \cdot b = b \cdot a$ for all a, $b \in F$.
7. F has an identity with respect to multiplication that is distinct from the additive identity: there is an element 1 in F such that $1 \neq 0$ and $a \cdot 1 = 1 \cdot a = a$ for all $a \in F$. 1 is called the **multiplicative identity** (or just the *identity*) of F and is often just called **one**.
8. Every nonzero element of F has an inverse with respect to multiplication: if $a \in F$ and $a \neq 0$, then there exists an element $a^{-1} \in F$ such that $a \cdot a^{-1} = a^{-1} \cdot a = 1$. a^{-1} is called the **multiplicative inverse** or simply the **inverse** of a.
9. The following distributive law holds: $a \cdot (b + c) = a \cdot b + a \cdot c$ for all a, b, $c \in F$.

Note that all of the axioms except 8 are also satisfied by the set $\mathbf{Z}$ of integers. But since only 1 and -1 have inverses in $\mathbf{Z}$ with respect to multiplication, $\mathbf{Z}$ is not a field.

EXAMPLE 1   The set $\mathbf{R}$ of real numbers is a field with respect to the usual operations of addition and multiplication in $\mathbf{R}$. The additive identity is the real number zero, which we write as 0, and the multiplicative identity is the real number 1.

If $a \in \mathbf{R}$, $a \neq 0$, then we will often write the multiplicative inverse of a as $1/a$, in keeping with standard notation. Similarly, if a, $b \in \mathbf{R}$, $b \neq 0$, then we write ab^{-1} as a/b.

EXAMPLE 2   The set $\mathbf{Q} = \left\{ \dfrac{a}{b} \,\middle|\, a, b \in \mathbf{Z}, b \neq 0 \right\}$ of rational numbers is a field with respect to the usual way of adding and multiplying fractions. Note that if $a/b \in \mathbf{Q}$, $a \neq 0$, then the multiplicative inverse of a/b is b/a.

Let F be a field. Note that we use the same notation for the additive identity of F, namely 0, that we use for the real number zero. The reason is simply because the real number 0 is the identity for addition in $\mathbf{R}$ and it is convenient to keep the same symbol. It should be clear from the context whether we are referring to the real number 0 or to the additive identity of F.

Similarly, we use the symbol 1 to mean the real number one as well as the multiplicative identity of F.

Elementary Properties of Fields

There are many more examples of fields. Since our focus in this chapter is on the real and complex numbers, we will restrict our attention to properties of fields that are most relevant to these examples. Many of these are quite familiar to you as properties of the real numbers, but they can be proved for arbitrary fields using only the axioms and results derived from those axioms.

The first result collects statements identical to those in Propositions 5.1.1 and 5.1.2. We state them without proof, because they are proved in exactly the same way as in Chapter 5, using only the axioms shared by $\mathbf{Z}$ and any field; namely, Axioms 1–7 and 9.

First, some comments on notation: for simplicity and convenience, we will write ab instead of $a \cdot b$ for multiplication in a field. Also, we will use the familiar exponent notation: if $a \in F$ and n is a positive integer, then a^n will mean the product of a with itself n times and a^{-n} will mean the product of a^{-1} with itself n times.

Proposition 7.1.2   Let F be a field and let $a, b, c \in F$.

1. $a0 = 0a = 0$.
2. $(-a)b = a(-b) = -(ab)$.
3. $-(-a) = a$.
4. $(-a)(-b) = ab$.
5. $a(b - c) = ab - ac$.
6. $(-1)a = -a$.
7. $(-1)(-1) = 1$.

The following proposition gives some properties of addition and multiplication in a field which are quite familiar as properties of the real numbers.

Proposition 7.1.3   Let F be a field and let $a, b, c \in F$.

1. If $a + b = a + c$, then $b = c$.
2. If $ab = ac$ and $a \neq 0$, then $b = c$.
3. If $a \neq 0$, then $(a^{-1})^{-1} = a$.

PROOF: The proof is left as an exercise.

Ordered Fields

Recall that in Chapter 5, Axioms 9 and 10 for the integers gave two properties of the set $\mathbf{Z}^+$ of positive integers. These properties are also satisfied by the set $\mathbf{R}^+$ of positive real numbers. Guided by this example, we single out fields that have similar special subsets.

> **Definition 7.1.4** Let F be a field. We say that F is an **ordered** field if there exists a subset P of F satisfying the following two properties:
>
> 1. P is closed with respect to addition and multiplication; that is, if $x, y \in P$, then $x + y \in P$ and $xy \in P$.
> 2. For every element x in F, exactly one of the following statements is true: $x \in P$, $-x \in P$, or $x = 0$.
>
> If F is an ordered field with respect to the subset P, then P is called the set of **positive** elements of F.

EXAMPLE 3   The set $\mathbf{R}$ of real numbers is an ordered field with $P = \mathbf{R}^+$.

EXAMPLE 4 $\mathbf{Q}$ is an ordered field with $P = \mathbf{Q}^+$, the set of positive rationals.

Recall that in Chapter 5, we used Axioms 9 and 10 for $\mathbf{Z}$ to introduce the notion of inequality in $\mathbf{Z}$. We can do the same thing for ordered fields.

> **Definition 7.1.5** Let F be an ordered field with respect to the subset P. If $x, y \in F$, we say $x < y$ or $y > x$ (x is **less than** y or y is **greater than** x) if $y - x \in P$.

As was the case with integers, inequalities in an ordered field satisfy some basic properties. The following proposition is analogous to Propositions 5.1.3 and 5.1.5 and the proofs are the same.

Proposition 7.1.6   Let F be an ordered field with respect to the subset P and let $a, b, c \in F$.

1. Exactly one of the following holds: $a < b$, $b < a$, or $a = b$.
2. If $a > 0$, then $-a < 0$ and if $a < 0$, then $-a > 0$.
3. If $a > 0$ and $b > 0$, then $a + b > 0$ and $ab > 0$.
4. If $a > 0$ and $b < 0$, then $ab < 0$.
5. If $a < 0$ and $b < 0$, then $ab > 0$.
6. If $a < b$ and $b < c$, then $a < c$.
7. If $a < b$, then $a + c < b + c$.
8. If $a < b$ and $c > 0$ then $ac < bc$.
9. If $a < b$ and $c < 0$, then $ac > bc$.
10. If $a \in F$, $a \neq 0$, then $a^2 \in P$.

In $\mathbf{Z}$, the set $\mathbf{Z}^+$ satisfies the Well-Ordering Principle; every nonempty subset of $\mathbf{Z}^+$ has a smallest element. A similar result is *not* true for $\mathbf{R}^+$ or $\mathbf{Q}^+$. In fact there is no smallest element of $\mathbf{R}^+$, for if $x \in \mathbf{R}^+$, then $\frac{1}{2}x < x$, proving that no element of $\mathbf{R}^+$ can be the smallest.

This last statement is true in general: in any ordered field F the subset P has no smallest element. (See the exercises.)

The next result gives some additional properties of the subset P of an ordered field F.

Proposition 7.1.7 Let F be an ordered field with respect to the subset P. Then:

1. $1 \in P$.
2. If $x \in P$, then $x^{-1} \in P$.
3. P is an infinite set.

PROOF: 1. Suppose that $1 \notin P$. Then $-1 \in P$ and since P is closed under multiplication, $(-1)(-1) \in P$. But $(-1)(-1) = 1$ by part 7 of Proposition 7.1.3. So we are led to the contradiction that if $1 \notin P$, then $1 \in P$. Therefore, $1 \in P$.

2. The proof is left as an exercise.

3. Since $1 \in P$ and P is closed under addition, the elements $1 + 1, 1 + 1 + 1, \ldots$ are all in P. Now suppose that two sums of this form are equal, say $1 + 1 + \ldots + 1$ (n times) $= 1 + 1 + \ldots + 1$ (m times) with $n > m$. Then, by part 1 of Proposition 7.1.4, we would have $1 + 1 + \ldots + 1$ ($n - m$ times) $= 0$. But this implies that $0 \in P$, a contradiction. Therefore the sums $1 + 1 + \ldots + 1$ (n times), $n \in \mathbf{Z}^+$, are all distinct, giving us infinitely many elements of P.

Corollary 7.1.8 Every ordered field is infinite.

Finite Fields

In Section 5.5 we discussed the set $\mathbf{Z}_n$ of congruence classes mod n where n is a fixed positive integer greater than 1. Theorems 5.5.5 and 5.5.6 show that $\mathbf{Z}_n$ satisfies Axioms 1–7 and 9 of a field. The additive identity is the congruence class $[0]$ and the multiplicative identity is the class $[1]$.

However, for some n, Axiom 8 will not be satisfied. For example, in $\mathbf{Z}_4$ there is a nonzero element that does not have a multiplicative inverse. (Which one?) So $\mathbf{Z}_4$ is not a field. However, it is easy to see that $\mathbf{Z}_2$ and $\mathbf{Z}_3$ are fields. So exactly which $\mathbf{Z}_n$ are fields and which are not?

Theorem 7.1.9 $\mathbf{Z}_n$ is a field if and only if n is a prime number.

PROOF: First suppose that n is a prime number. As just pointed out, any $\mathbf{Z}_n$ satisfies all of the axioms of a field except possibly Axiom 8. So it suffices to show that any nonzero element of $\mathbf{Z}_n$ has a multiplicative inverse when n is prime.

Let $[a] \in \mathbf{Z}_n$, $[a] \neq [0]$. Since a and 0 are not in the same congruence class, it follows that a is not congruent to $0 \mod n$. Hence n does not divide a and since n is prime, a and n must be relatively prime. Then by Theorem

5.3.8, there exist integers x and y such that $ax + ny = 1$. Hence $[a][x] + [n][y] = [ax] + [ny] = [ax + ny] = [1]$. But $[n] = [0]$ since $n \equiv 0 \pmod{n}$. Thus $[a][x] = [1]$, so $[x]$ is the multiplicative inverse of $[a]$. Therefore, if n is a prime number, $\mathbf{Z}_n$ is a field.

Now we want to prove the converse: if $\mathbf{Z}_n$ is a field, then n is a prime number. We will prove the contrapositive of the converse: if n is not a prime number, then $\mathbf{Z}_n$ is not a field.

Let n be a composite number. Then there exist integers a and b such that $n = ab$ where $1 < a < n$ and $1 < b < n$. We get

$$[a][b] = [ab] = [n] = [0].$$

Now note that $[a] \neq [0]$ since $1 < a < n$ and $[b] \neq [0]$ since $1 < b < n$. We claim that $[a]$ does not have a multiplicative inverse. Suppose, on the contrary, that $[a]$ does have a multiplicative inverse. Then there exists an integer x such that $[x][a] = [1]$. Multiplying both sides of this equation by $[b]$, we get

$$([x][a])[b] = [1][b] = [b].$$

Since multiplication in $\mathbf{Z}_n$ is associative, we have

$$[x]([a][b]) = [b].$$

But $[a][b] = [0]$ from above. Hence $[b] = [x][0] = [x0] = [0]$. This is a contradiction, since $[b] \neq [0]$. Therefore, $[a]$ does not have a multiplicative inverse, proving that $\mathbf{Z}_n$ is not a field. ●

Notice that by Corollary 7.1.8, if p is a prime, then the field $\mathbf{Z}_p$ is not an ordered field.

● MATHEMATICAL PERSPECTIVE: OTHER FIELDS

We have seen relatively few examples of fields in this section: $\mathbf{Q}$, $\mathbf{R}$, $\mathbf{C}$, and the finite fields $\mathbf{Z}_p$. Are there others? In trying to answer a question like this, mathematicians often take a very commonsense approach: use the examples we know to construct new and different ones.

Suppose we start with $\mathbf{Q}$ and a real number not in $\mathbf{Q}$, say $\sqrt{2}$. The field $\mathbf{R}$ contains both $\mathbf{Q}$ and $\sqrt{2}$, of course, but is there a smallest field that does as well? Essentially, we can just build that field F. If $\mathbf{Q} \subseteq F \subseteq \mathbf{R}$, $\sqrt{2} \in F$, and F is a field, then since F is closed, F must also contain, for example, $2\sqrt{2}, 1 - \sqrt{2}, \frac{1}{\sqrt{2}}, (\sqrt{2})^3$, and all elements of this type. However, since $\frac{1}{\sqrt{2}} = \frac{1}{2}\sqrt{2}$ and $(\sqrt{2})^2 = 2$, it is not hard to see that $F = \{r + s\sqrt{2} \mid r, s \in \mathbf{Q}\}$. We denote this field $\mathbf{Q}(\sqrt{2})$. Now obviously $\mathbf{Q}(\sqrt{2}) \neq \mathbf{Q}$, but it is not so clear that $\mathbf{Q}(\sqrt{2}) \neq \mathbf{R}$.

To prove this last statement, we can adapt our proof that $\sqrt{2}$ is irrational and show that $\sqrt{3} \notin \mathbf{Q}(\sqrt{2})$. Suppose to the contrary that $\sqrt{3} = r +$

$s\sqrt{2}$, where $r, s \in \mathbf{Q}$. Then $3 = r^2 + 2s^2 + 2rs\sqrt{2}$, and this would mean that $\sqrt{2}$ is rational. (You should check the special cases $r = 0$ and $s = 0$.)

$\mathbf{Q}(\sqrt{2})$ is called an *algebraic number field,* and is clearly one of a plethora of other fields: we could just as easily construct $\mathbf{Q}(\sqrt{3})$, $\mathbf{Q}(\sqrt[3]{4})$, and infinitely many others. The simple representation we found, however, is due to the fact that $\sqrt{2}$ is a solution of the equation $x^2 - 2 = 0$, in which the coefficients are all rational numbers. No such polynomial equation can be found for π, and so $\mathbf{Q}(\pi)$, the smallest field of real numbers containing $\mathbf{Q}$ and π, has no analogous representation. This very deep fact—that π is a *transcendental number*—was first proved by **C. L. F. Lindemann** (1852–1939). It is an indication of the rich variety of infinite fields.

The idea of adding solutions of equations to a field led **Evariste Galois** (1811–1832) to construct new finite fields from $\mathbf{Z}_p$. The difference was that Galois added solutions α of equations $f(x) = [0]$, where $f(x)$ is a polynomial with coefficients in $\mathbf{Z}_p$. If $f(x)$ is irreducible (that is, does not factor), he showed that $\mathbf{Z}_p(\alpha)$ is a finite field, and that $|\mathbf{Z}_p(\alpha)|$ is a power of p. Later, **E. H. Moore** (1862–1932) showed that any finite field can be constructed in this way.

In short, mathematicians today have a very clear picture of the structure of finite fields, whereas the nature of infinite fields, in particular algebraic number fields, remains an active area of research.

Exercises 7.1

1. Prove Proposition 7.1.3.

2. Let F be a field and let $a, b \in F$.
   (a) Suppose that $ab = 0$. Prove that $a = 0$ or $b = 0$.
   (b) Suppose that $a^2 = b^2$. Prove that $a = \pm b$.

3. Let F be a field. Let $a \in F$, $a \neq 0$, and $n \in \mathbf{Z}^+$.
   (a) Prove that $a^n \neq 0$.
   (b) Prove that $a^{-n} = (a^n)^{-1}$.

4. Let F be a field and let $a, b \in F$, with $a \neq 0$. Prove that the equation $ax = b$ has a unique solution x in F.

5. Let F be a field and let $a, b \in F$.
   (a) Prove that $(ab)^2 = a^2b^2$.
   (b) Prove that $(ab)^n = a^nb^n$ for every positive integer n.
   (c) Prove that if $a, b \neq 0$, then $(ab)^{-n} = a^{-n}b^{-n}$ for every positive integer n.

6. Let F be a field and let $a \in F$, $a \neq 0$. Let n and m be positive integers.
   (a) Prove that $a^na^m = a^{n+m}$.
   (b) Prove that $a^na^{-m} = a^{n-m}$.
   (c) Prove that $a^{-n}a^{-m} = a^{-n-m}$.

7. (a) Prove that between any two rational numbers there is another rational number; that is, if $a, b \in \mathbf{Q}$ and $a < b$, then there exists $z \in \mathbf{Q}$ such that $a < z < b$.

   (b) Prove that between any two rational numbers there are infinitely many rational numbers.

8. Does the set of irrational numbers form a field?

9. Prove that if x is an irrational number, then x^{-1} is irrational.

10. Let F be an ordered field with respect to the subset P.

   (a) Prove part 2 of Proposition 7.1.7; that is, if $x \in P$, then $x^{-1} \in P$.

   (b) Prove that if $x, y \in P$ and $x < y$, then $x^{-1} > y^{-1}$.

   (c) Prove that if $x, y \in F$, then $x^2 + y^2 \geq xy$ and $x^2 + y^2 \geq -xy$.

11. Let F be an ordered field. Let $a, b, c, d \in F$. Prove or disprove: if $a < b$ and $c < d$, then $ac < bd$.

12. Let F be an ordered field with respect to the subset P. Let $x, y \in P$.

   (a) Prove that if $x < y$, then $x^2 < y^2$.

   (b) Prove that if $x^2 < y^2$, then $x < y$.

   (c) Prove that if $x < y$, then for any positive integer n, $x^n < y^n$.

13. Let F be an ordered field and let $a \in F$. Prove that there exists $b \in F$ such that $b > a$.

14. Prove that in any ordered field F, the subset P has no smallest element.

15. Prove that $\mathbf{Q}^+$ is the only subset of $\mathbf{Q}$ that makes $\mathbf{Q}$ an ordered field; that is, if $\mathbf{Q}$ is an ordered field with respect to a subset P, then $P = \mathbf{Q}^+$.

16. Let F be an ordered field with respect to the subset P. Prove that -1 is not a square in F: there does not exist an element x in F such that $x^2 = -1$.

17. Let F be a field with the following two properties:

   (1) -1 is not a square in F.

   (2) the sum of any two nonsquares of F is a nonsquare of F.

   Let P be the set of nonzero squares of F.

   $$P = \{x \in F \mid x = y^2 \text{ for some } y \in F, y \neq 0\}.$$

   Let N be the set of nonsquares of F.

   (a) Prove that if $x \in P$, then $-x \in N$.

   (b) Prove that if $x \in N$, then $-x \in P$.

   (c) Prove that F is an ordered field with respect to the subset P.

18. Do not use Theorem 7.1.9 to do this problem.

   (a) Prove that $\mathbf{Z}_3$ is a field.

   (b) Prove that $\mathbf{Z}_4$ is not a field.

   (c) Prove that $\mathbf{Z}_6$ is not a field.

   (d) Let $F = \{[0], [2], [4]\} \subseteq \mathbf{Z}_6$. Prove that F is a field.

19. (a) Prove that the equation $[2]x = [3]$ has no solutions in $\mathbf{Z}_4$.

   (b) Prove that the equation $[3]x = [3]$ has three solutions in $\mathbf{Z}_6$.

20. Prove that $\mathbf{Q}(\sqrt{2}) = \{r + s\sqrt{2} \mid r, s \in \mathbf{Q}\}$ is a field. (You may assume that addition and multiplication in $\mathbf{Q}(\sqrt{2})$ are associative and commutative since $\mathbf{Q}(\sqrt{2}) \subseteq \mathbf{R}$.)

Discussion and Discovery Exercises

D1. Suppose that F is an ordered field with respect to the subset P. By Exercise 14, we know that P has no smallest element. Use this fact to give a different proof of Corollary 7.1.8, that F is infinite.

D2. $\mathbf{Q}$ is a subset of $\mathbf{R}$ and is also a field. We would say that $\mathbf{Q}$ is a *subfield* of $\mathbf{R}$, since we are using the same addition and multiplication. Are there any subfields of $\mathbf{Q}$? Are there any subfields of $\mathbf{Z}_p$?

7.2   THE REAL NUMBERS

Bounded Sets

We saw in Section 7.1 that both $\mathbf{R}$ and $\mathbf{Q}$ are ordered fields. It is possible to construct the real numbers from the rational numbers, but the construction is long and fairly complicated so we will not attempt it here. However, one property that distinguishes the real numbers from the rationals is an important one that we will take as an axiom, the Least Upper Bound Axiom.

We first need some definitions.

> **Definition 7.2.1** Let S be a nonempty set of real numbers.
>
> 1. We say that S is **bounded above** if there exists a real number x such that $a \leq x$ for all $a \in S$. The number x is called an **upper bound** for the set S.
> 2. We say that S is **bounded below** if there exists a real number y such that $a \geq y$ for all $a \in S$. The number y is called a **lower bound** for the set S.
> 3. If S is bounded above and bounded below, we say that S is a **bounded** set.

EXAMPLE 1   The closed interval $[0, 1]$ is a bounded set. Any number greater than or equal to 1 is an upper bound for S and any number less than or equal to 0 is a lower bound.

EXAMPLE 2   The open interval $(0, 1)$ is a bounded set. As with Example 1, any number greater than or equal to 1 is an upper bound for S and any number less than or equal to 0 is a lower bound.

EXAMPLE 3   The set $\mathbf{Z}^+$ of positive integers is *not* bounded above. It is bounded below and any number less than or equal to 1 is a lower bound.

EXAMPLE 4   The interval $(-\infty, 0]$ is bounded above but not bounded below.

Least Upper and Greatest Lower Bounds

Notice that an upper or lower bound for a set S may or may not be in S. But in both Examples 1 and 2, the number 1 is clearly an important upper bound: it is the *smallest* of all. (We will prove this shortly.) Similarly, 0 is the *largest* lower bound in both examples. The existence of these special bounds in every case is the essence of the axiom we mentioned previously.

> **Definition 7.2.2**
>
> 1. Let S be a set of real numbers that is bounded above. We say that x is the **least upper bound** for S if x is an upper bound for S and if $x \leq z$ whenever z is an upper bound for S.
> 2. Let S be a set of real numbers that is bounded below. We say that y is the **greatest lower bound** for S if y is a lower bound for S and if $y \geq z$ whenever z is a lower bound for S.

EXAMPLE 5   The least upper bound of $[0, 1]$ is 1 and the greatest lower bound is 0.

EXAMPLE 6   We claim that the least upper bound of $(0, 1)$ is 1. Clearly 1 is an upper bound for $(0, 1)$. Now suppose $0 < z < 1$. There exists a real number y between z and 1; namely $y = (z + 1)/2$. Since $y \in (0, 1)$, it follows that z is not an upper bound for $(0, 1)$. This shows that no real number less than 1 is an upper bound for $(0, 1)$. Therefore 1 is the least upper bound of $(0, 1)$.
  Similarly, 0 is the greatest lower bound of $(0, 1)$.

EXAMPLE 7   The set $\mathbf{Z}^+$ does not have a least upper bound since it is not bounded above, and 1 is the greatest lower bound.

EXAMPLE 8   The interval $(-\infty, 0]$ has 0 as its least upper bound and does not have a greatest lower bound.
  We can now state our final axiom for the set of real numbers.

> *Least Upper Bound Axiom:*   Any nonempty subset of $\mathbf{R}$ that is bounded above has a least upper bound.

It seems reasonable to expect that any nonempty subset of **R** that is bounded below has a greatest lower bound. This is true but we need not take it as an axiom, for it can be proved from the Least Upper Bound Axiom. (See Exercise 2.)

The least upper bound and greatest lower bound of a set, if they exist, are unique. (See Exercise 3.) We will denote the least upper bound of a set S by *lub S* and the greatest lower bound by *glb S*.

The Archimedean Principle

An important consequence of the Least Upper Bound Axiom is the following theorem, known as the **Archimedean Principle**.

Theorem 7.2.3   Let a and b be positive real numbers. Then there exists a positive integer n such that $na > b$.

PROOF: The proof is by contradiction. Suppose that $na \leq b$ for all positive integers n. Let $S = \{na \mid n \in \mathbf{Z}^+\}$. Clearly S is a nonempty set, and by our assumption b is an upper bound for S. By the Least Upper Bound Axiom, S has a least upper bound. Call it z.

Since $a > 0$, $z - a < z$. Thus $z - a$ is not an upper bound for S. Consequently, some element of S must be greater than $z - a$; that is, there exists $x \in S$ such that $x > z - a$. Since $x \in S$, $x = ma$ for some positive integer m. Then $ma > z - a$ and it follows that $z < ma + a = (m + 1)a$. But $(m + 1)a$ is in S, giving us a contradiction since z is an upper bound for S. The theorem now follows.   ●

Corollary 7.2.4   If b is any positive real number, then there exists a positive integer n such that $1/n < b$.

PROOF: Applying the Archimedean Principle with $a = 1$ and b replaced by $1/b$, there exists a positive integer n such that $n > 1/b$. It now follows from Exercise 10(b) of Section 7.1 that $1/n < b$.   ●

Another important consequence of the Archimedean Principle is the fact that between any two real numbers there is a rational number. To prove this, we need to prove a lemma first.

Lemma 7.2.5:   Let $x \in \mathbf{R}$. Then there exists an integer m such that $m - 1 \leq x < m$.

PROOF: Suppose first that $x > 0$. By the Archimedean Principle, there exists a positive integer n such that $n > x$. By the Well-Ordering Principle, there must be a smallest such integer; call it m. Then $m > x$, and since $m - 1 < m$, $m - 1 \leq x$.

If $x = 0$, the result is obvious by letting $m = 1$.

Suppose finally that $x < 0$. Then $-x > 0$, so by the Archimedean Principle again, there exists a positive integer n_0 such that $n_0 > -x$. Then $x + n_0 > 0$, so by our first case, there exists an integer m_0 such that $m_0 - 1 \leq x + n_0 < m_0$. Let $m = m_0 - n_0$. Then $m - 1 \leq x < m$. ●

Theorem 7.2.6   Between any two real numbers there is a rational number.

PROOF: Let $x, y \in \mathbf{R}$ such that $x < y$. Then $y - x > 0$, so by the Archimedean Principle, there exists a positive integer n such that $n > 1/(y - x)$. It follows that $ny - nx > 1$ or $nx + 1 < ny$. By the previous lemma, there exists $m \in \mathbf{Z}$ such that $m - 1 \leq nx < m$. We get $m \leq nx + 1 < ny$ and hence $nx < m < ny$. It now follows that the rational number m/n is between x and y. ●

Incompleteness of Q

We mentioned earlier in this section that the Least Upper Bound Axiom distinguishes the real numbers from the rational numbers. To justify that statement, we need to demonstrate that $\mathbf{Q}$ does not satisfy the Least Upper Bound Axiom—that there is a nonempty subset of $\mathbf{Q}$ that is bounded above but does not have a least upper bound in $\mathbf{Q}$.

Let $S = \{x \in \mathbf{Q} \mid x > 0 \text{ and } x^2 < 2\}$. Clearly, $S \neq \varnothing$ and S is bounded above since 2 is an upper bound for S. (Why?) Suppose that S has a least upper bound in $\mathbf{Q}$. Call it z. We will show that $z^2 = 2$. This will be a contradiction since it was proved in Section 1.2 that $\sqrt{2}$ is not a rational number.

Assume that $z^2 < 2$. Then $z < \sqrt{2}$ by Exercise 12(b) of Section 7.1. By Theorem 7.2.6, there is a rational number r such that $z < r < \sqrt{2}$. Hence $z^2 < r^2 < 2$, implying that $r \in S$. This is a contradiction since z is an upper bound for S.

Suppose, on the other hand, that $z^2 > 2$. Then $z > \sqrt{2}$, so there exists a rational number r such that $z > r > \sqrt{2}$. We get $z^2 > r^2 > 2$. Thus if $w \in S$, $w^2 < 2 < r^2$, and therefore $w < r$. Thus r is an upper bound for S, a contradiction since $r < z$ and z is the *least* upper bound for S.

Therefore, we must have $z^2 = 2$, and, as was pointed out previously, this is a contradiction since $\sqrt{2}$ is irrational. We conclude that S does not have a least upper bound in $\mathbf{Q}$.

In this argument, we assumed that the integer 2 has a square root in $\mathbf{R}$. In fact all positive real numbers have a square root in $\mathbf{R}$, and we conclude this section with a proof.

Theorem 7.2.7   If $x \in \mathbf{R}$, $x > 0$, then there exists $y \in \mathbf{R}$, $y > 0$, such that $y^2 = x$.

PROOF: Let $S = \{t \in \mathbf{R} \mid t > 0 \text{ and } t^2 < x\}$. We first show that S is a nonempty set. Let $t = x/(x + 1)$. Then $0 < t < 1$ and $t < x$, so $0 < t^2 < t < x$. Thus $t \in S$, so $S \neq \varnothing$.

We now show that S is bounded above. Specifically, we will show that $x + 1$ is an upper bound for S. Suppose there exists $t \in S$ such that $x + 1 < t$. Then $(x + 1)^2 < t^2 < x$. By Exercise 12(b) of Section 7.1, we get that $x + 1 < x$, a contradiction. Therefore, $t \le x + 1$ for all $t \in S$. Hence $x + 1$ is an upper bound for S.

Now we can apply the Least Upper Bound Axiom to S to conclude that S has a least upper bound. Call it y. Clearly $y > 0$. We will prove that $y^2 = x$ by showing that if $y^2 < x$ or if $y^2 > x$, we get a contradiction.

Assume that $y^2 < x$. Then $\dfrac{x - y^2}{2y + 1} > 0$. By Corollary 7.2.4, there is a positive integer n such that $\dfrac{1}{n} < \dfrac{x - y^2}{2y + 1}$. Hence we have

$$\left(y + \frac{1}{n}\right)^2 = y^2 + \frac{2y}{n} + \frac{1}{n^2}$$
$$= y^2 + \frac{1}{n}\left(2y + \frac{1}{n}\right)$$
$$\le y^2 + \frac{1}{n}(2y + 1)$$
$$< y^2 + (x - y^2)$$
$$= x.$$

Thus $y + \dfrac{1}{n} \in S$, a contradiction to the fact that y is an upper bound for S.

Now suppose that $y^2 > x$. Then $\dfrac{y^2 - x}{2y} > 0$. By Corollary 7.2.4, there is a positive integer m such that $\dfrac{1}{m} < \dfrac{y^2 - x}{2y}$. Therefore,

$$\left(y - \frac{1}{m}\right)^2 = y^2 - \frac{2y}{m} + \frac{1}{m^2}$$
$$> y^2 - \frac{2y}{m}$$
$$> y^2 - (y^2 - x)$$
$$= x.$$

Now if $w \in S$, then $w^2 < x < \left(y - \dfrac{1}{m}\right)^2$ from which it follows that $w < y - 1/m$. (See Exercise 12(b) of Section 7.1.) Hence $y - 1/m$ is an upper bound for S, contradicting the fact that y is the *least* upper bound of S. We can now conclude that $y^2 = x$. ●

If x is a positive real number, then there is only *one* positive real number y such that $y^2 = x$. (See Exercise 14.) We use the usual symbol $\sqrt{x}$ for y.

● HISTORICAL COMMENTS: THE LOGICAL FOUNDATION OF **R**

The Intermediate Value Theorem, which we stated in Chapter 3, is an important theorem of calculus, and it can be made obvious by drawing a picture: if f is a continuous function on the closed interval $[a, b]$, and $f(a) < 0 < f(b)$, then there is a $c \in (a, b)$ such that $f(c) = 0$; that is, the graph of f must cross the x-axis between a and b.

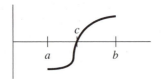

We did not prove the theorem in Chapter 3, in part because it involves the nature of **R**: the real numbers must be represented by a continuous line. Mathematicians in the 19th century realized that to rigorously prove the theorem (and many others from calculus), a firm, logical basis for **R** was needed, and several attempts were made to give one.

At the time, the rational numbers were thought to be completely under-stood, or at least natural enough to need no formal definition. So the problem was how to define the irrational numbers in such a way as to catch the continuity of the correspondence between **R** and the line. One approach was taken by **J. W. R. Dedekind** (1831–1916), who, remarkably, took his cue from ideas that can be found in Euclid's Elements.

Dedekind's insight was that each point divides the line into two sets: those to the left and those to the right. He then starts with the rational numbers, and identifies real numbers a with ordered pairs of sets of rationals: those less than or equal to a and those greater. For example, 0 is identified with (A_0, B_0), where $A_0 = \{r \in \mathbf{Q} \mid r \le 0\}$ and $B_0 = \{r \in \mathbf{Q} \mid r > 0\}$. More importantly, $\sqrt{2}$, an irrational number, is identified with $(A_{\sqrt{2}}, B_{\sqrt{2}})$, where $A_{\sqrt{2}} = \{r \in \mathbf{Q} \mid r \le 0 \text{ or } r^2 \le 2\}$ and $B_{\sqrt{2}} = \{r \in \mathbf{Q} \mid r > 0 \text{ and } r^2 > 2\}$.

Thus **R** becomes a set of ordered pairs of sets of rational numbers. (The ordered pairs have become known as *Dedekind cuts*.) On this complicated set, Dedekind defined operations $(+, \cdot, -, \div)$ and order $(<, >, =)$, and then proved many of the expected properties, such as the transitivity of $>$. He expressed the continuity of the line by proving that if **R** is divided into two nonempty sets A and B with every element of A less than every element of B, then there is exactly one real number $a = (A, B)$. Among other things, the Intermediate Value Theorem becomes provable in this logical system.

Needless to say, some mathematicians did not find this approach entirely satisfactory. However, it soon became clear that any successful design would have to be quite sophisticated. For example, **Georg Cantor** (1845–1918) defined an irrational number as a sequence of rational numbers $r_1, r_2, r_3, \ldots$ that eventually cluster: that is, for any $\varepsilon > 0$, $|r_n - r_m| < \varepsilon$ for n and m sufficiently large. (Notice that the number is not the "limit" of the sequence;

Cantor was *defining* the number, so he identified it with the sequence itself.) Cantor then faced the same difficulties as Dedekind: he defined order and operations, and proved all the expected theorems. His real numbers were also infinite sets, in his case sequences of rationals.

One final remark: it is fascinating to realize that the historical development of the logical basis for the real numbers came in "reverse." The simplest reals are the integers, followed by the rationals, and finally the irrationals. Yet the first commonly accepted axiom system for $\mathbf{Z}$ was given by **Guiseppe Peano** (1858–1932), *after* Dedekind's definition of irrationals!

Exercises 7.2

1. Find the least upper bound and greatest lower bound (if they exist) for the following sets. (No proofs necessary.)
   (a) $S = [-1, \infty)$
   (b) $S = \{x \in \mathbf{Q} \mid x^2 < 5\}$
   (c) $S = \{x \in \mathbf{Q} \mid x^3 < 4\}$
   (d) $S = \left\{ 1 - \dfrac{1}{n} \,\middle|\, n \in \mathbf{Z}^+ \right\}$
   (e) $S = \left\{ 1 + \dfrac{(-1)^n}{n} \,\middle|\, n \in \mathbf{Z}^+ \right\}$
   (f) $S = \mathbf{Z}$

2. Let S be a nonempty subset of $\mathbf{R}$ that is bounded below. Prove that S has a greatest lower bound.

3. Prove that the least upper bound of a nonempty subset S of $\mathbf{R}$, if it exists, is unique.

4. Let $a, b \in \mathbf{R}$ such that $a < b$. Let $S = (a, b)$.
   (a) Prove that lub $S = b$.
   (b) Prove that glb $S = a$.

5. Let $S \subseteq \mathbf{R}, S \neq \varnothing$. Suppose that there exist an upper bound x for S such that $x \in S$. Prove that $x =$ lub S.

6. Let $S \subseteq \mathbf{R}, S \neq \varnothing$. Suppose that there exists a lower bound y for S such that $y \in S$. Prove that $y =$ glb S.

7. Let $a, b \in \mathbf{R}$ such that $a < b$. Let $S = [a, b]$.
   (a) Prove that lub $S = b$.
   (b) Prove that glb $S = a$.

8. Let S and T be nonempty bounded subsets of $\mathbf{R}$ such that $S \subseteq T$.
   (a) Prove that lub $S \leq$ lub T.
   (b) Prove that glb $S \geq$ glb T.

9. Let S be a nonempty set of real numbers. Complete the following statements:

(a) S is not bounded above if . . . .
(b) S is not bounded below if . . . .
(c) S is not bounded if . . . .

10. Let S be a nonempty set of real numbers that is bounded above. Let $y = \operatorname{lub} S$. Prove that for every positive real number ε, there is a real number $z \in S$ such that $z > y - \varepsilon$.

11. Let S be a nonempty set of real numbers that is bounded below. Let $y = \operatorname{glb} S$. Prove that for every positive real number ε, there is a real number $z \in S$ such that $z < y + \varepsilon$.

12. Without using the Least Upper Bound Axiom, prove that $\mathbf{Q}$ satisfies the Archimedean Principle; that is, if $x, y \in \mathbf{Q}$, then there exists a positive integer n such that $nx > y$.

13. (a) Write the negation of the Least Upper Bound Axiom.
(b) Write the negation of the Archimedean Principle.
(c) Write the negation of Theorem 7.2.6.

14. Theorem 7.2.7 shows the existence, for each positive real number x, of a positive real number y such that $y^2 = x$. Prove that y is unique: if there exists a real number z such that $z^2 = x$, then $z = y$ or $z = -y$.

15. Prove that if $x, y \in \mathbf{R}$ such that $0 < x < y$, then $\sqrt{x} < \sqrt{y}$.

16. Let $x, y \in \mathbf{R}$ and suppose that $x < y + \varepsilon$ for all positive real numbers ε. Prove that $x \le y$.

17. Prove that between any two real numbers there are infinitely many rational numbers.

18. Prove that between any two real numbers there is an irrational number.

Discussion and Discovery Exercises

You may remember that the rational numbers can be characterized as those real numbers whose decimal expansion contains a sequence of digits that repeats infinitely. For example,

$$\frac{1}{3} = 0.3\overline{3}$$

$$\frac{22}{7} = 3.142857\overline{142857}.$$

(Here, the bar means that the sequence will repeat forever.)

Conversely, the irrational numbers are those real numbers with no such repeating pattern. For example,

$$\sqrt{2} = 1.414213562\ldots$$
$$\pi = 3.141592653\ldots.$$

D1. Does this description make you think that it might be harder to prove that a number is irrational than rational? Explain.

D2. Use this description to argue that between every two real numbers there are both a rational number and an irrational number. Criticize the logical validity of your argument.

7.3   THE COMPLEX NUMBERS

In Chapter 8 we will be studying solutions of polynomial equations. Finding such solutions can be difficult and some equations have no solutions in a given field. The polynomial equation $x^2 + 1 = 0$, for example, has no solutions in the field of real numbers. To find a solution we need to find a set containing the reals that also contains an element whose square is -1. Such a set is the complex numbers and remarkably, as we will see in Chapter 8, the Fundamental Theorem of Algebra tells us that this set is "rich" enough to contain the solutions of all polynomial equations with real coefficients.

We begin by defining the complex numbers and proving that they form a field.

Definition 7.3.1 The set **C** of **complex numbers** is the set of all ordered pairs (a, b) of real numbers together with two binary operations of addition and multiplication defined as follows:

$$(a, b) + (c, d) = (a + c, b + d)$$
$$(a, b) \cdot (c, d) = (ac - bd, ad + bc)$$

The definition of multiplication seems unusual but, as we will see, it gives us an element whose square is -1.

These two binary operations make **C** a field. Of course, all of the axioms must be verified, but these are routine and some are left as exercises. (See Exercise 1.) The additive identity is $(0, 0)$ and the additive inverse of (a, b) is $(-a, -b)$. The multiplicative identity is $(1, 0)$ and multiplication is easily seen to be commutative. It is easy to show that if $a \neq 0$ or $b \neq 0$, then the multiplicative inverse of (a, b) is

$$\left(\frac{a}{a^2 + b^2}, \frac{-b}{a^2 + b^2} \right).$$

So we have the following result.

Theorem 7.3.2   **C** is a field with respect to the binary operations defined in Definition 7.3.1.

   Some comments are in order before we proceed. First, by the nature of our definition, two complex numbers (a, b) and (c, d) are equal if and only if $a = c$ and $b = d$. Another important fact to note is that **C** is not an ordered field since -1 is a square in **C**. (See Exercise 16 of Section 7.1.) Finally, it is common to identify the real number a with the point $(a, 0)$ in **C**. Because these points form a subset of **C** that is closed under addition and multiplication (check!), we can consider the field **R** as a *subfield* of **C**; that is, a subset of **C** which is itself a field with respect to the binary operations of addition and multiplication in **C**.

   Now suppose we consider the point $(0, 1)$, which we denote i. Then $i^2 = (0, 1)(0, 1) = (-1, 0)$, and since we identify $(-1, 0)$ with the real number -1, we have the important property that $i^2 = -1$. As discussed previously, this fact distinguishes the complex numbers as a set containing the real numbers and also containing a solution to the equation $x^2 + 1 = 0$.

   If $(a, b) \in$ **C**, $(a, b) = (a, 0) + (0, b) = (a, 0) + (b, 0)(0, 1) = a + bi$. From now on, we will use the expression $a + bi$ for the complex number (a, b). This notation makes computations with complex numbers easier to perform. Addition and multiplication with complex numbers can be done in the usual way that we treat algebraic expressions of real numbers, with the added rule that $i^2 = -1$. That is,

$$(a + bi)(c + di) = ac + bci + adi + bdi^2$$
$$= (ac - bd) + (ad + bc)i.$$

   If $z = a + bi \in$ **C**, then a is called the **real part** of z, denoted by Re z, and b is called the **imaginary part** of z, denoted by Im z. So two complex numbers are equal if and only if their real and imaginary parts are equal.

   We must not lose sight of the fact that a complex number can be represented as a point in a plane, which we call the **complex plane**. We will see later in this section that important properties of complex numbers can be derived by writing them in polar coordinates.

Conjugation and Absolute Value

Definition 7.3.3 If $z = a + bi \in$ **C**, then $a - bi$ is called the **complex conjugate** of z and is denoted by $\bar{z}$. The **absolute value** of z is $\sqrt{a^2 + b^2}$ and is denoted by $|z|$.

   Geometrically, the absolute value of $z = a + bi$ is the length of the line segment from the origin to the point (a, b).

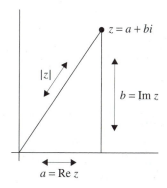

The following lemma lists some important properties of the complex conjugate. The proof is left as an exercise.

Lemma 7.3.4   Let $z, w \in \mathbf{C}$. Then

1. $\overline{z + w} = \overline{z} + \overline{w}$.
2. $\overline{zw} = \overline{z}\,\overline{w}$.
3. $z\overline{z} = |z|^2 \in \mathbf{R}$, and therefore is a nonnegative real number.
4. $z + \overline{z} = 2\,\mathrm{Re}\,z \in \mathbf{R}$.
5. $z^{-1} = \dfrac{\overline{z}}{|z|^2}$ if $z \neq 0$.

Part 5 of Lemma 7.3.4 gives a straightforward way to compute the multiplicative inverse of a complex number.

EXAMPLE 1 $(6 - 7i)^{-1} = \dfrac{6 + 7i}{6^2 + 7^2} = \dfrac{6}{85} + \dfrac{7}{85}i.$

Solutions of Equations

As we have seen, the polynomial equation $x^2 + 1 = 0$ has a solution in the complex numbers. In fact, it has two solutions: i and $-i$. As we noted at the beginning, the field $\mathbf{C}$ of complex numbers contains the solutions of every polynomial equation $p(x) = 0$ with real coefficients. The fact that such a polynomial equation has a solution in $\mathbf{C}$ is the content of the Fundamental Theorem of Algebra, which we will discuss in more detail in Chapter 8. Along the way, however, we will see special cases of this theorem.

The first and easiest special case is the equation $ax + b = 0$ where a, $b \in \mathbf{R}$, $a \neq 0$. The unique solution is $x = -b/a$.

The general quadratic equation $ax^2 + bx + c = 0$ with $a, b, c \in \mathbf{R}$

and $a \neq 0$ can be solved using the well-known quadratic formula, giving the solution:

$$x = \frac{-b \pm \sqrt{b^2 - 4ac}}{2a}.$$

If $b^2 - 4ac = 0$, there is one real solution, and if $b^2 - 4ac > 0$, there are two real solutions. If, however, $b^2 - 4ac < 0$, the quadratic formula gives two nonreal solutions in $\mathbf{C}$.

The general cubic equation $ax^3 + bx^2 + cx + d = 0$ also has a general formula for its solution, but it is considerably more complicated than the quadratic formula and will not be given here. We will discuss the cubic formula and some special cubic equations in Chapter 8, however.

Notice that the quadratic formula implies that if the complex number z is a solution of $ax^2 + bx + c = 0$, then its complex conjugate is also a solution. A more general result is the following:

Theorem 7.3.5   Let $p(x) = a_n x^n + a_{n-1} x^{n-1} + \ldots + a_1 x + a_0$ be a polynomial of degree n, $n \geq 1$, with real coefficients. If $z \in \mathbf{C}$ is a solution of $p(x) = 0$, then its complex conjugate is also a solution.

PROOF: First note that if r is a real number, then $\bar{r} = r$ and that for any z in $\mathbf{C}$, $\overline{z^k} = \bar{z}^k$ for all positive integers k. (See Exercise 4.) Thus if $p(z) = 0$, then

$$\begin{aligned}
p(\bar{z}) &= a_n \bar{z}^n + a_{n-1} \bar{z}^{n-1} + \ldots + a_1 \bar{z} + a_0 \\
&= \overline{a_n} \, \overline{z}^n + \overline{a_{n-1}} \, \overline{z}^{n-1} + \ldots + \overline{a_1} \, \overline{z} + \overline{a_0} \\
&= \overline{a_n z^n} + \overline{a_{n-1} z^{n-1}} + \ldots + \overline{a_1 z} + \overline{a_0} \\
&= \overline{a_n z^n + a_{n-1} z^{n-1} + \ldots + a_1 z + a_0} \\
&= \overline{a_n z^n + a_{n-1} z^{n-1} + \ldots + a_1 z + a_0} \\
&= \overline{p(z)} \\
&= \bar{0} \\
&= 0. \quad \bullet
\end{aligned}$$

EXAMPLE 2   Let $p(x) = x^3 + x^2 - x + 2$. By trial and error, one can see that $p(-2) = 0$ and that $p(x) = (x + 2)(x^2 - x + 1)$. Using the quadratic formula, we see that the solutions of $x^2 - x + 1 = 0$ are $1/2 + \sqrt{3}/2 \, i$ and $1/2 - \sqrt{3}/2 \, i$. These two solutions together with -2 give all of the solutions of the equation $x^3 + x^2 - x + 2 = 0$. This fact is not obvious but will follow from results in Chapter 8.

In Chapter 8, we will also discuss ways of finding *integer* solutions of polynomial equations.

Polar Form

Let $z = x + yi$ be a nonzero complex number. As a point in the complex plane, z can be represented in polar coordinates. Let $r = |z|$ and θ be the angle that the line segment from $(0, 0)$ to (x, y) makes with the positive x-axis. Then $x = r \cos \theta$, $y = r \sin \theta$, and $z = r(\cos \theta + i \sin \theta)$. This is called the **polar form** of z, and θ is called the **argument** of z, denoted by arg z.

Notice that there is a subtlety here. There are many different choices for θ: any of the values θ, $\theta \pm 2\pi$, $\theta \pm 4\pi$, ... could also be used. We will call any of them arg z.

Another inconvenience is that in the case $z = 0$, we can write $z = |z|(\cos \theta + i \sin \theta)$, for any angle θ. For this reason, we do not define arg 0.

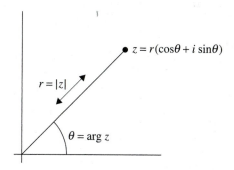

A convention: in writing a complex number in polar form, it is customary to write the i before sin θ, whereas in standard form, we write $a + bi$ with i written last.

EXAMPLE 3   We will write $z = -2 + 2\sqrt{3}i$ in polar form. First we see that $|z| = 4$, so that $z = 4(-1/2 + \sqrt{3}/2\, i)$. Hence $\cos \theta = -1/2$ and $\sin \theta = \sqrt{3}/2$, so $\theta = \frac{2\pi}{3}$. We thus have $z = 4(\cos \frac{2\pi}{3} + i \sin \frac{2\pi}{3})$.

The following result shows how to compute the product of complex numbers when expressed in polar form.

Theorem 7.3.6   Let $z, w \in \mathbf{C}$, where $z = r(\cos \theta + i \sin \theta)$ and $w = s(\cos \phi + i \sin \phi)$. Then $zw = rs(\cos(\theta + \phi) + i \sin(\theta + \phi))$.

PROOF: Using the definition of multiplication in $\mathbf{C}$ and some well-known trigonometric formulas, we have

$$zw = rs \cos \theta \cos \phi + irs \cos \theta \sin \phi + irs \sin \theta \cos \phi + i^2 rs \sin \theta \sin \phi$$
$$= rs(\cos \theta \cos \phi - \sin \theta \sin \phi) + irs(\cos \theta \sin \phi + \sin \theta \cos \phi)$$
$$= rs(\cos(\theta + \phi) + i \sin(\theta + \phi)). \quad \bullet$$

The following corollary, known as DeMoivre's Theorem, gives a convenient way of computing positive integer powers of a complex number.

Corollary 7.3.7   **DeMoivre's Theorem**   Let $z = r(\cos \theta + i \sin \theta) \in \mathbf{C}$. Then for any $n \in \mathbf{Z}^+$, $z^n = r^n(\cos n\theta + i \sin n\theta)$.

PROOF: The proof is by induction on n. The result obviously holds for $n = 1$. Suppose it is true for a positive integer k. Then $z^k = r^k(\cos k\theta + i \sin k\theta)$, and so

$$z^{k+1} = z^k z$$
$$= r^k r(\cos(k\theta + \theta) + i \sin(k\theta + \theta))$$
$$= r^{k+1}(\cos(k+1)\theta + i \sin(k+1)\theta).$$

Therefore, the formula holds for $k + 1$, and so for all positive integers n by induction. ●

EXAMPLE 4   We compute z^5 for $z = -2 + 2\sqrt{3}i$. As we saw in Example 3, in polar form,

$$z = 4 \left(\cos \frac{2\pi}{3} + i \sin \frac{2\pi}{3} \right).$$

So by DeMoivre's Theorem,

$$z^5 = 4^5 \left(\cos \frac{10\pi}{3} + i \sin \frac{10\pi}{3} \right)$$

$$= 2^{10} \left(\cos \frac{4\pi}{3} + i \sin \frac{4\pi}{3} \right)$$

$$= 2^{10} \left(-\frac{1}{2} - \frac{\sqrt{3}}{2} i \right)$$

$$= -2^9(1 + \sqrt{3}i).$$

Complex Roots

An important application of DeMoivre's Theorem is the computation of nth roots of complex numbers.

Definition 7.3.8 Let $z \in \mathbf{C}$ and $n \in \mathbf{Z}^+$. An **nth root** of z is a complex number w such that $w^n = z$.

Theorem 7.3.9   Let $0 \neq z = r(\cos \theta + i \sin \theta) \in \mathbf{C}, n \in \mathbf{Z}^+$. Then z has exactly n nth roots, given by

$$w_k = r^{1/n} \left(\cos \left(\frac{\theta + 2\pi k}{n} \right) + i \sin \left(\frac{\theta + 2\pi k}{n} \right) \right),$$

for $k = 0, 1, 2, \ldots, n - 1$.

PROOF: By DeMoivre's Theorem, it follows easily that $w_k^n = z$ for all k, where $0 \le k \le n - 1$. It is also easy to see that the w_ks are distinct. So z has at least n nth roots.

Let $w = s(\cos \phi + i \sin \phi)$ be an nth root of z. Then

$$w^n = s^n(\cos n\phi + i \sin n\phi)$$
$$= z$$
$$= r(\cos \theta + i \sin \theta)$$

so $s^n = r$ and therefore $s = r^{1/n}$.

Also, $\cos n\phi = \cos \theta$ and $\sin n\phi = \sin \theta$, so $n\phi = \theta + 2t\pi$ for some $t \in$ **Z**. Then $\phi = (\theta + 2t\pi)/n$, and by the Division Algorithm (Theorem 5.3.1), there exist integers q and k such that $t = nq + k$ where $0 \le k < n$. It follows that $\phi = (\theta + 2k\pi)/n + 2q\pi$, and so $w = w_k$. We can now conclude that the w_ks give all of the nth roots of z. ●

EXAMPLE 5   We compute the third (cube) roots of $z = -2 + 2\sqrt{3}i$. We know from Example 3 that $z = 4(\cos \frac{2\pi}{3} + i \sin \frac{2\pi}{3})$. So the third roots of z are given by

$$w_k = 4^{1/3} \left(\cos \left(\frac{2\pi}{9} + \frac{2\pi k}{3} \right) + i \sin \frac{2\pi}{9} + \frac{2\pi k}{3} \right)$$

for $k = 0, 1, 2$. We get:

$$w_0 = 4^{1/3} \left(\cos \frac{2\pi}{9} + i \sin \frac{2\pi}{9} \right),$$

$$w_1 = 4^{1/3} \left(\cos \frac{8\pi}{9} + i \sin \frac{8\pi}{9} \right),$$

$$w_2 = 4^{1/3} \left(\cos \frac{14\pi}{9} + i \sin \frac{14\pi}{9} \right).$$

An important special case of the last theorem is the computation of the nth roots of $z = 1$, called the nth **roots of unity**: w is an nth root of unity if $w^n = 1$.

Theorem 7.3.10   For each $n \in$ **Z**$^+$, there are exactly n nth roots of unity, given by

$$\omega_k = \cos \frac{2\pi k}{n} + i \sin \frac{2\pi k}{n}$$

for $0 \le k \le n - 1$.

PROOF: We leave it as an exercise. ●

EXAMPLE 6   The third or cube roots of 1 are

$$\omega_0 = 1,$$
$$\omega_1 = \cos\frac{2\pi}{3} + i\sin\frac{2\pi}{3} = -\frac{1}{2} + \frac{\sqrt{3}}{2}i,$$
$$\omega_2 = \cos\frac{4\pi}{3} + i\sin\frac{4\pi}{3} = -\frac{1}{2} - \frac{\sqrt{3}}{2}i.$$

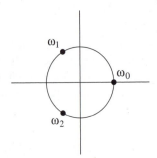

EXAMPLE 7   The fourth roots of 1 are $1, i, -1, -i$.

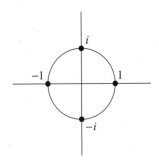

Corollary 7.3.11   Let $n \in \mathbf{Z}^+$. Let $\omega = \cos\frac{2\pi}{n} + i\sin\frac{2\pi}{n}$. Then the nth roots of unity are $1, \omega, \omega^2, \omega^3, \ldots, \omega^{n-1}$.

PROOF: We leave it as an exercise. ●

The complex number ω of Corollary 7.3.11 is called a **primitive** nth root of unity, since every other nth root of unity is a power of ω.

● HISTORICAL COMMENTS: EXTENDING THE COMPLEX NUMBERS

Our definition of the complex numbers, as a set of ordered pairs of real numbers, was given by the Irish mathematician **W. R. Hamilton** (1805–1865). Of course, mathematicians had long been using them, taking i as $\sqrt{-1}$, an "imaginary" number. But Hamilton's definition took the mystery away by giving a logical basis for **C**. Again, it is surprising that this milestone preceded any logical foundation for **R**, just as Dedekind's work predated Peano's. (See the Historical Comments in Section 7.2.)

In true mathematical spirit, Hamilton wondered if it were possible to continue extending the number system. Specifically, he searched for a way to define multiplication on triples (a, b, c) (or sums $a + bi + cj$ for some j) of real numbers that would result in a field. It turned out to be a harder proposition than he imagined. He spent 15 years on the problem, to the point where his daughters would tease him at the dinner table: "Daddy, can you multiply triples?" There is no record of his reaction. As the story goes, he was walking with his wife one day across the Brougham Bridge in Dublin. Apparently his mind was elsewhere because he had a flash of insight: don't multiply triples, but rather *quadruples*. Then he stopped and carved a formula into the bridge:

$$i^2 = j^2 = k^2 = ijk = -1.$$

Hamilton had discovered what he came to call the *quaternions* **H**. To be specific, $\mathbf{H} = \{a + bi + cj + dk \mid a, b, c, d \in \mathbf{R}\}$, where $i^2 = j^2 = k^2 = -1$, and interestingly, multiplication is *not* commutative: $ij = -ji = k$, $ik = -ki = -j$, and $jk = -kj = i$. Hamilton was able to prove that, except for commutativity, all the axioms of a field hold for **H**. He had thus extended **C**, at the price of only one of the axioms.

Hamilton was so entranced by his discovery that he spent the rest of his life studying this new set of numbers and espousing their use. Although he may have understandably overestimated their importance, they have proven extraordinarily useful in many areas, notably in modern physics with its conception of space time as a four-dimensional universe.

The reason for Hamilton's struggle became clear only later, when **F. G. Frobenius** (1849–1917) proved that the problem was not a lack of creativity on Hamilton's part. Frobenius showed that the only way to extend the multiplication on **R** to n-tuples $(a_1, a_2, \ldots, a_n)$ of reals and preserve the field axioms is through Hamilton's construction of **C**, and the only way to do so without commutativity is by his construction of **H**. This is an amazing, far-reaching result, and is a testament to Hamilton's achievement. Not only did he find the only two ways to naturally extend the real numbers in the manner of the complex numbers, but by removing the "natural" axiom of commutativity, encouraged the study of abstract, algebraic objects.

Exercises 7.3

1. (a) Verify that addition in **C** is associative and commutative.
   (b) Verify that multiplication in **C** is associative and commutative.
   (c) Verify the distributive laws in **C**.
   (d) Prove that every nonzero element in **C** has a multiplicative inverse.

2. Let $z = -4 + 2i$ and $w = 6 - 5i$. Compute the following and express your answer in the form $a + bi$.
   (a) $2z + 5w$ (b) zw (c) z^2 (d) w^3
   (e) z^{-1} (f) w^{-1} (g) $\dfrac{z}{w}$ (h) $(z + w)^2$

3. Prove Lemma 7.3.4.

4. Let $z \in \mathbf{C}$. Prove that $z = \bar{z}$ if and only if $z \in \mathbf{R}$.

5. Let $z_1, z_2 \in \mathbf{C}$. Prove that $|z_1 + z_2| \leq |z_1| + |z_2|$ and $|z_1 z_2| = |z_1||z_2|$.

6. Prove that if $z \in \mathbf{C}$ and k is any positive integer, then $\bar{z}^k = \overline{z^k}$.

7. Express the following complex numbers in polar form:
   (a) $-3 + 3i$ (b) $6 - 2\sqrt{3}i$ (c) -5 (d) $-2i$
   (e) $-3 - 3i$ (f) $6 + 2\sqrt{3}i$ (g) $5 + i$ (h) $6 - 2i$

8. Let $0 \neq z = r(\cos \theta + i \sin \theta)$, $w = s(\cos \phi + i \sin \phi)$.

   (a) Prove that $z^{-1} = \dfrac{1}{r}(\cos(-\theta) + i \sin(-\theta))$.

   (b) Let $n \in \mathbf{Z}^+$. Prove that $z^{-n} = \dfrac{1}{r^n}(\cos(-n\theta) + i \sin(-n\theta))$.

   (c) Prove that $\dfrac{w}{z} = \dfrac{s}{r}(\cos(\phi - \theta) + i \sin(\phi - \theta))$.

9. Let $z = 3\sqrt{3} - 9i$ and $w = -2 + 2i$. Compute:
   (a) z^3 (b) z^8 (c) w^6 (d) w^{19}
   (e) z^{-3} (f) z^{128} (g) $z^{12}w^{12}$ (h) w^{-100}

10. Let z and w be as in Exercise 9.
    (a) Compute the 4th roots of z.
    (b) Compute the 3rd roots of w.

11. Prove Theorem 7.3.10.

12. Find the 5th, 6th, and 8th roots of unity.

13. Let $n \in \mathbf{Z}^+$.
    (a) Prove that the sum of the nth roots of unity is zero.
       Hint: Factor the polynomial $x^n - 1$.
    (b) Determine the product of the nth roots of unity.

14. Suppose that $u \in \mathbf{C}$ is an nth root of unity and $w \in \mathbf{C}$ is an nth root of z, $0 \neq z \in \mathbf{C}$. Prove that the nth roots of z are $w, wu, wu^2, \ldots, wu^{n-1}$.

15. Prove Corollary 7.3.11.

Discussion and Discovery Exercises

D1. The complex number ω is a **primitive** nth root of unity if every other nth root of unity is a power of ω.
    (a) Which of the third roots of unity are primitive? Which of the fourth? Fifth? Sixth? Seventh?

(b) Based on your answers to part (a), speculate on the general case: when is an nth root of unity primitive? Can you prove that your answer is correct?

D2. The exponential function e^x is defined on $\mathbf{R}$, but can be extended to a function on $\mathbf{C}$ by taking $e^z = e^{x+iy} = e^x(\cos y + i \sin y)$. (There are, of course, many other ways to extend to $\mathbf{C}$.)

    (a) Verify that this particular extension satisfies the rules of exponentiation:

$$e^z e^w = e^{z+w},$$

$$e^{-z} = \frac{1}{e^z}.$$

    (b) Give a simple expression for $e^{i\theta}$, for $\theta \in \mathbf{R}$. (This is known as *Euler's Formula*.)

    (c) Show that if $z \neq 0$, then $z = re^{i\theta}$, where $r = |z|$ and $\theta = \arg z$.

    (d) In part (a), you were not asked to verify the expected rule $(e^z)^w = e^{zw}$. Why not?

POLYNOMIALS

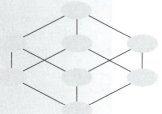

8.1   POLYNOMIALS

The Algebra of Polynomials

In this chapter we begin a systematic study of polynomials. Our main goal is to develop some techniques for solving equations of the form $f(x) = 0$ where $f(x)$ is a polynomial. The coefficients of such polynomials will usually be real numbers. Sometimes we will restrict our attention to polynomials with just rational coefficients or even integer coefficients. We will also consider polynomials whose coefficients are complex numbers.

In fact, we will first look at the set of polynomials whose coefficients are in some unspecified field. The definition of a field was given in Section 7.1.

> **Definition 8.1.1** Let F be a field. A **polynomial over F in the variable x** is an expression of the form
>
> $$f(x) = a_0 + a_1 x + a_2 x^2 + \ldots + a_n x^n$$
>
> where $a_0, a_1, a_2, \ldots, a_n \in F$ and $n \in \mathbf{Z}$, $n \geq 0$.
>
> The elements $a_0, a_1, a_2, \ldots, a_n$ are called the **coefficients** of $f(x)$. If $a_n \neq 0$, so that n is the highest power of x that appears, then a_n is called the **leading coefficient** of $f(x)$; a_0 is called the **constant term** of $f(x)$; and n is called the **degree** of $f(x)$, denoted $\deg f(x)$.
>
> If $f(x) = a_0 \neq 0$, then $f(x)$ is called a **nonzero constant polynomial** and has degree 0. If $f(x) = 0$, $f(x)$ is called the **zero polynomial**. The zero polynomial is not assigned a degree.
>
> We let $F[x]$ denote the set of all polynomials with coefficients in F.

. What makes $F[x]$ interesting to study as a set is that it has many properties similar to the set $\mathbf{Z}$ of integers. First $F[x]$ has two binary operations; namely, ordinary addition and multiplication of polynomials.

The formal definitions of addition and multiplication in $F[x]$ are as follows. Let $f(x) = a_0 + a_1x + a_2x^2 + \ldots + a_nx^n$ and $g(x) = b_0 + b_1x + b_2x^2 + \ldots + b_mx^m$ be in $F[x]$. If $n \neq m$, say $n > m$, then we can also write $g(x) = b_0 + b_1x + b_2x^2 + \ldots + b_nx^n$ where $b_i = 0$ for all $i > m$. Notice that the degree of $g(x)$ is still m, but writing $g(x)$ this way makes the notation needed to define addition in $F[x]$ easier.

We define

$$f(x) + g(x) = (a_0 + b_0) + (a_1 + b_1)x + (a_2 + b_2)x^2 + \ldots + (a_n + b_n)x^n$$

and

$$f(x)g(x) = a_0b_0 + (a_0b_1 + a_1b_0)x + (a_0b_2 + a_1b_1 + a_2b_0)x^2 + \ldots + a_nb_mx^{n+m}.$$

These binary operations are really just defined in the usual way we define addition and multiplication of real-valued polynomial functions. The coefficient c_i of x^i in the product $f(x)g(x)$ can be expressed by the following formula:

$$c_i = a_0b_i + a_1b_{i-1} + a_2b_{i-2} + \ldots + a_ib_0$$

where it is understood that $a_i = 0$ if $i > n$ and $b_j = 0$ if $j > m$.

It is easy to see that addition and multiplication in $F[x]$ are associative and commutative. For example, if $f(x)$, $g(x)$, $h(x) \in F[x]$, where

$$f(x) = a_0 + a_1x + a_2x^2 + \ldots + a_nx^n,$$
$$g(x) = b_0 + b_1x + b_2x^2 + \ldots + b_nx^n,$$

and

$$h(x) = c_0 + c_1x + c_2x^2 + \ldots + c_nx^n,$$

then

$$
\begin{aligned}
(f(x) + g(x)) + h(x) &= ((a_0 + b_0) + (a_1 + b_1)x + (a_2 + b_2)x^2 + \ldots + (a_n + b_n)x^n) \\
&\quad + (c_0 + c_1x + c_2x^2 + \ldots + c_nx^n) \\
&= ((a_0 + b_0) + c_0) + ((a_1 + b_1) + c_1)x \\
&\quad + ((a_2 + b_2) + c_2)x^2 + \ldots + ((a_n + b_n) + c_n)x^n \\
&= (a_0 + (b_0 + c_0)) + (a_1 + (b_1 + c_1))x \\
&\quad + (a_2 + (b_2 + c_2))x^2 + \ldots + (a_n + (b_n + c_n))x^n \\
&= a_0 + a_1x + a_2x^2 + \ldots + a_nx^n + (b_0 + c_0) + (b_1 + c_1)x \\
&\quad + \ldots + (b_n + c_n)x^n \\
&= f(x) + (g(x) + h(x)).
\end{aligned}
$$

Notice that we needed the fact that addition in F is associative.

Similar proofs show that addition is commutative and multiplication is associative and commutative. (See the exercises.)

The zero polynomial is the additive identity of $F[x]$, and the additive inverse of $a_0 + a_1x + a_2x^2 + \ldots + a_nx^n$ is $-(a_0) + (-a_1)x + (-a_2)x^2 + \ldots + (-a_n)x^n$.

Since F has an identity 1, the constant polynomial $f(x) = 1$ is the multiplicative identity of $F[x]$. We leave as an exercise the fact that distributive laws hold in $F[x]$.

EXAMPLE 1   If $f(x) = 5x^3 + 7x^2 - 9x + 2$ and $g(x) = 4x^5 - x^3 + 8$ in $\mathbf{R}[x]$, then

$$f(x) + g(x) = 4x^5 + 4x^3 + 7x^2 - 9x + 10$$

and

$$f(x)g(x) = 20x^8 + 28x^7 - 41x^6 + x^5 + 9x^4 + 38x^3 + 56x^2 - 72x + 16.$$

In the next example, we consider polynomials whose coefficients are in the field $\mathbf{Z}_5$ of integers mod 5. (See Section 5.5 for the definition of $\mathbf{Z}_n$.)

EXAMPLE 2   If $f(x) = [3]x^2 + [2]x + [4]$ and $g(x) = [1]x^3 + [4]x^2 + [1]$ in $\mathbf{Z}_5[x]$,

then

$$f(x) + g(x) = [1]x^3 + [2]x^2 + [2]x$$

and

$$f(x)g(x) = [3]x^5 + [4]x^4 + [2]x^3 + [4]x^2 + [2]x + [4].$$

The following theorem gives us information about the degree of the sum and product of two polynomials.

Theorem 8.1.2   Let F be a field and let $f(x), g(x) \in F[x]$, where $f(x) \neq 0$, $g(x) \neq 0$, and $f(x) + g(x) \neq 0$. Then

1. $\deg(f(x) + g(x)) \leq \max\{\deg f(x), \deg g(x)\}$. Moreover, if $\deg f(x) \neq \deg g(x)$, then equality holds.
2. $\deg(f(x)g(x)) = \deg f(x) + \deg g(x)$.

PROOF: Let $f(x) = a_0 + a_1x + a_2x^2 + \ldots + a_nx^n$ and $g(x) = b_0 + b_1x + b_2x^2 + \ldots + b_mx^m$, where $a_n \neq 0$ and $b_m \neq 0$.

1. If $n > m$, then by the definition of addition in $F[x]$, x^n is the highest power of x that appears in $f(x) + g(x)$, so $\deg(f(x) + g(x)) = n$. Similarly, if $m > n$, then $\deg(f(x) + g(x)) = m$. If $n = m$, then the highest power of x that appears in $f(x) + g(x)$ is x^n unless $a_n = -b_n$, in which case the highest power of x that appears is less than n.
2. We leave this as an exercise. ●

As pointed out previously, many properties of the set $F[x]$ are similar to properties of the set of integers $\mathbf{Z}$. Recall that the only elements of $\mathbf{Z}$

that have multiplicative inverses are 1 and -1. Similarly, the only elements of $F[x]$ that have multiplicative inverses are the nonzero constant polynomials. This fact is a consequence of Theorem 8.1.2 and is left as an exercise. In particular, it follows that $F[x]$ is not a field.

The Division Algorithm

Another point of similarity is that there is a Division Algorithm for Polynomials similar to the Division Algorithm for Integers, as stated in Theorem 5.3.1. Also, the notion of an **irreducible** polynomial, to be defined in Section 8.2, mirrors the concept of a prime number. We will first prove the Division Algorithm for Polynomials and then derive some consequences having to do with finding zeros of polynomials.

Throughout the rest of this chapter, F will denote a field.

Theorem 8.1.3   **Division Algorithm for Polynomials** Let $f(x)$, $g(x) \in F[x]$, $g(x) \neq 0$. Then there exist unique polynomials $q(x)$ and $r(x)$ in $F[x]$ such that $f(x) = g(x)q(x) + r(x)$, where $r(x) = 0$ or $0 \le \deg r(x) < \deg g(x)$.

PROOF: We first prove the existence of $q(x)$ and $r(x)$. Let $S = \{h(x) \in F[x] \mid h(x) = f(x) - g(x)q(x), \text{ for some } q(x) \in F[x]\}$; that is, S is the set of all polynomials in $F[x]$ that can be expressed as $f(x)$ minus a multiple of $g(x)$.

Clearly, $S \neq \varnothing$. If the zero polynomial is in S, then $f(x) = g(x)q(x)$ for some $q(x) \in F[x]$. The theorem then follows with $r(x) = 0$.

Suppose the zero polynomial is not in S. Then let $r(x)$ be a polynomial in S having the smallest degree of all polynomials in S. Such a polynomial exists by the Well-Ordering Principle. We have $f(x) = g(x)q(x) + r(x)$ for some $q(x) \in F[x]$. So it remains to prove that $\deg r(x) < \deg g(x)$.

Suppose that $\deg r(x) \ge \deg g(x)$. Let

$$g(x) = b_0 + b_1 x + b_2 x^2 + \ldots + b_m x^m$$

where $b_m \neq 0$, and

$$r(x) = c_0 + c_1 x + c_2 x^2 + \ldots + c_t x^t$$

where $c_t \neq 0$. Let

$$r_1(x) = r(x) - c_t b_m^{-1} x^{t-m} g(x).$$

(Note that $t \ge m$.) Then

$$r_1(x) = f(x) - g(x)(q(x) + c_t b_m^{-1} x^{t-m}),$$

so $r_1(x) \in S$. But $\deg r_1(x) < t = \deg r(x)$, a contradiction to the fact that $r(x)$ has minimal degree in S. Therefore we must have $\deg r(x) < \deg g(x)$. This completes the proof that $q(x)$ and $r(x)$ exist. We leave the proof of uniqueness as an exercise. ●

If $f(x) = a_0 + a_1x + a_2x^2 + \ldots + a_nx^n \in F[x]$ and $c \in F$, then we write $f(c)$ to denote the element $a_0 + a_1c + a_2c^2 + \ldots + a_nc^n \in F$. We say that we are *substituting* c for x in $f(x)$ or *evaluating* $f(x)$ at c. Of course, $f(c)$ is in F, not $F[x]$.

If $f(x) = g(x) + h(x)$ where $g(x)$ and $h(x)$ are in $F[x]$, then by the definition of addition in $F[x]$, it follows that $f(c) = g(c) + h(c)$. Similarly, if $f(x) = g(x)h(x)$, then the definition of multiplication in $F[x]$ implies that $f(c) = g(c)h(c)$.

Corollary 8.1.4 Let $f(x) \in F[x]$ and $c \in F$. Then there exists $q(x)$ in $F[x]$ such that $f(x) = (x - c)q(x) + f(c)$.

PROOF: Applying the Division Algorithm for Polynomials with $g(x) = x - c$, there exist $q(x)$ and $r(x)$ in $F[x]$ such that $f(x) = (x - c)q(x) + r(x)$ where $r(x) = 0$ or deg $r(x) < \deg(x - c) = 1$. Hence $r(x)$ must be a constant polynomial, say $r(x) = d$. Substituting c for x we get $d = r(c) = f(c)$. The corollary then follows. ●

EXAMPLE 3 Let $f(x) = x^3 + 7x^2 - 2x + 1 \in \mathbf{R}[x]$. Then if $f(x)$ is divided by $x + 2$ the remainder is $f(-2) = 25$.

Zeros of Polynomials

> **Definition 8.1.5** Let $f(x) \in F[x]$ and $c \in F$. We say that c is a **zero** of $f(x)$ if $f(c) = 0$. We also say that c is a **root** or a **solution** of the equation $f(x) = 0$.

The following result, known as the Factor Theorem, is an important result relating zeros of polynomials to factoring a polynomial.

Corollary 8.1.6 **Factor Theorem** Let $f(x) \in F[x]$ and $c \in F$. Then c is a zero of $f(x)$ if and only if there is a $q(x)$ in $F[x]$ such that $f(x) = (x - c)q(x)$.

PROOF: The proof is immediate from Corollary 8.1.4. ●

EXAMPLE 4 Suppose we wish to find the zeros of the polynomial $f(x) = x^3 - 2x + 1 \in \mathbf{R}[x]$. A quick inspection yields the fact that 1 is a zero of $f(x)$ and thus $x - 1$ is a factor of $f(x)$. Dividing $x - 1$ into $f(x)$ we get $f(x) = (x - 1)(x^2 + x - 1)$. Next, using the quadratic formula, we see that the zeros of $x^2 + x - 1$ are $(-1 \pm \sqrt{5})/2$. But any zero of a factor of $f(x)$ is also a zero of $f(x)$ itself. Thus 1, $(-1 + \sqrt{5})/2$, and $(-1 - \sqrt{5})/2$ are zeros of $f(x)$.

EXAMPLE 5   Let $f(x) = x^2 + 1 \in \mathbf{R}[x]$. Clearly $f(x)$ has no zeros in $\mathbf{R}$. But since $\mathbf{R} \subseteq \mathbf{C}$, we can consider $f(x) \in \mathbf{C}[x]$. As we have seen, $f(x)$ has two zeros in $\mathbf{C}$, i and $-i$, and $f(x)$ factors into $(x - i)(x + i)$.

These examples leave some unanswered questions. First, is it possible that we have not found all of the zeros of these polynomials? For example, could these polynomials be factored in another way? The answer to both of these questions is no. That we have found all of the zeros of these polynomials follows from the next corollary: a polynomial of degree n has at most n zeros. The fact that factorization of polynomials is essentially unique will be proved in the next section.

Another question involves how we found the zero $x = 1$ in Example 4. In that case a little bit of trial and error calculation yielded the answer fairly quickly. But what if the coefficients or the degree of the polynomial had been larger? Simple trial and error computation would have been impractical, and besides, there may be no integer zeros at all. We will explore some of these questions in Section 8.3.

Corollary 8.1.7   Let $f(x) \in F[x]$, $\deg f(x) = n$. Then $f(x)$ has at most n zeros in F.

PROOF: The proof is by induction on n. Let $P(n)$ be the statement "Any polynomial $f(x)$ in $F[x]$ of degree n has at most n zeros in F." If $n = 0$, then $f(x)$ is a nonzero constant function and thus has no zeros in F. Therefore, $P(0)$ is true. If $n = 1$, then $f(x) = a_1 x + a_0$, where a_1 and a_0 are in F and $a_1 \neq 0$. It is easy to see that $-a_1^{-1}a_0$ is the unique zero of $f(x)$ and so $P(1)$ is true.

Now suppose $P(n)$ is true. Then any polynomial in $F[x]$ of degree n has at most n zeros in F. We now wish to prove that $P(n + 1)$ is true; that is, we want to show that any polynomial in $F[x]$ of degree $n + 1$ has at most $n + 1$ zeros in F.

Let $f(x)$ have degree $n + 1$ in $F[x]$. If $f(x)$ has no zeros in F, there is nothing to prove.

Suppose then that $f(x)$ has at least one zero c in F. Then by Corollary 8.1.6, $f(x) = (x - c)g(x)$ for some $g(x)$ in $F[x]$. Since $\deg f(x) = n + 1$ and $\deg(x - c) = 1$, $\deg g(x) = n$. Therefore, by the induction hypothesis, $g(x)$ has at most n zeros in F. If d is a zero of $g(x)$, it is clear that d is a zero of $f(x)$. Conversely, if d is a zero of $f(x)$, $d \neq c$, then $0 = f(d) = (d - c)g(d)$ implies that d is a zero of $g(x)$.

Therefore, the zeros of $f(x)$ are c together with any zeros of $g(x)$. It now follows that $f(x)$ has at most $n + 1$ zeros in F. The induction proof is now complete. ●

EXAMPLE 6   Let p be an odd prime number and $\mathbf{Z}_p$ the set of congruence classes mod p. Recall that, by Theorem 7.1.9, $\mathbf{Z}_p$ is a field. Let $[a] \in \mathbf{Z}_p$. Consider the polynomial $f(x) = x^2 - [a] \in \mathbf{Z}_p[x]$. By Corollary 8.1.7, $f(x)$ has at most two zeros in $\mathbf{Z}_p$. If $[b]$ is a zero of $f(x)$, then it is clear that $[-b]$ is also a zero. Therefore, if $(a, p) = 1$, $f(x)$ has either no zeros or two zeros in $\mathbf{Z}_p$. Both cases can occur. For example, if $p = 7$, then $x^2 - [3] \in \mathbf{Z}_7[x]$ has no zeros in $\mathbf{Z}_7$ and $x^2 - [2]$ has two zeros.

We can express these results in the language of congruences. If $[b] \in \mathbf{Z}_p$ is a zero of $x^2 - [a]$, then $[b]^2 = [a]$ or $b^2 \equiv a(\mathrm{mod}\ p)$. In other words, $x^2 - [a]$ has a zero in $\mathbf{Z}_p$ if and only if there exists b in $\mathbf{Z}$ such that $b^2 \equiv a(\mathrm{mod}\ p)$. It follows then that for a given odd prime p and a given a in $\mathbf{Z}$ such that $(a, p) = 1$, the congruence $x^2 \equiv a(\mathrm{mod}\ p)$ has either two incongruent solutions or no solutions. The two examples cited in the previous paragraph imply that there is no integer b such that $b^2 \equiv 3(\mathrm{mod}\ 7)$, and there exist 2 integers b, not congruent to each other mod 7, such that $b^2 \equiv 2(\mathrm{mod}\ 7)$. In fact, 3 and 4 are two such solutions.

EXAMPLE 7   Let $f(x) = x^3 - 2x + 1 \in \mathbf{R}[x]$. In Example 4 we found three zeros of $f(x)$. It now follows from Corollary 8.1.7 that these are the only zeros of $f(x)$.

HISTORICAL COMMENTS: SOLUTIONS OF POLYNOMIAL EQUATIONS BY RADICALS

The quadratic formula has been known for centuries; even the Babylonians of 2000 B.C. solved quadratic equations. For a long time negative solutions were disregarded and full acceptance of complex numbers as solutions did not gain hold until the 19th century.

The beauty of the quadratic formula is that it provides explicit solutions of a quadratic equation in terms of the coefficients of the quadratic polynomial. We say that the quadratic formula gives a *solution by radicals* for the equation $ax^2 + bc + c = 0$ because the solution involves just the algebraic operations of addition, subtraction, multiplication, division, and taking a square root; explicitly, the square root of $b^2 - 4ac$.

A similar formula for solving a cubic equation $ax^3 + bx^2 + cx + d = 0$ was discovered by **Niccolo Tartaglia** (1499?–1557) in the 16th century. Tartaglia did not want his formula to be known. It was quite common in those days for mathematicians to keep their method of solving a problem a secret so that they could challenge other mathematicians to solve it. In 1539, however, Tartaglia revealed his method to the mathematician **Girolamo Cardano** (1501–1576) on the condition that Cardano would not divulge it to anyone. Cardano came to the conclusion that Tartaglia's method of solving a cubic was the same as that of **Scipione dal Ferro** (1465–1526), a professor of mathematics at the University of Bologna. So Cardano published his own version of the method and gave credit to both Tartaglia and dal Ferro for having come up with the same solution. Needless to say, Tartaglia was not pleased and a dispute ensued as to who came up with the formula first, Tartaglia or dal Ferro.

The cubic formula is similar to the quadratic formula in the sense that it gives a formula for the roots in terms of the coefficients, using only algebraic operations of addition, subtraction, multiplication, division, and taking square roots and cube roots. However, the formula is much more complicated.

Specifically, the equation $x^3 + bx^2 + cx + d = 0$ can be solved as shown next. (We can always make the leading coefficient 1 by dividing through by it.) Let

$$\Delta = b^2c^2 - 4c^3 - 4b^3d - 27d^2 + 18bcd,$$

$$\gamma = -b^3 + \frac{9}{2}bc - \frac{27}{2}d + \frac{3}{2}\sqrt{-3\Delta},$$

$$\delta = -b^3 + \frac{9}{2}bc - \frac{27}{2}d - \frac{3}{2}\sqrt{-3\Delta}.$$

Let $\omega = -\frac{1}{2} + \frac{\sqrt{3}}{2}i$, the cubic third root of unity. Then the zeros, α_1, α_2, α_3 of the cubic equation $x^3 + bx^2 + cx + d = 0$ are given by

$$\alpha_1 = \frac{1}{3}[-b + \sqrt[3]{\gamma} + \sqrt[3]{\delta}],$$

$$\alpha_2 = \frac{1}{3}[-b + \omega^2\sqrt[3]{\gamma} + \omega\sqrt[3]{\delta}],$$

$$\alpha_3 = \frac{1}{3}[-b + \omega\sqrt[3]{\gamma} + \omega^2\sqrt[3]{\delta}].$$

Since there are three cube roots of a complex number (see Section 7.3), we have to specify which cube roots of γ and δ are chosen. In fact we can choose any cube root of γ and then the cube root of δ is chosen so that the equation $\sqrt[3]{\gamma}\sqrt[3]{\delta} = b^2 - 3c$ is satisfied.

For example, if we solve the equation $x^3 - 3x + 2 = 0$, we get: $\Delta = 0$, $\gamma = -27$, $\delta = -27$, thus giving the solutions $\alpha_1 = -2$, $\alpha_2 = \alpha_3 = 1$.

These solutions of course could have been obtained by noting that 1 is a solution, then factoring $x - 1$ and applying the quadratic formula to the remaining factor. In fact, if there is an easily obtained integer solution, then this latter method is preferable to applying the cubic formula.

Consider the equation $x^3 + 2x^2 - x - 2 = 0$. Using the cubic formula just given, we get the solution $\alpha_1 = \frac{1}{3}\{-2 + \sqrt[3]{10 + 9\sqrt{3}i} + \sqrt[3]{10 - 9\sqrt{3}i}\}$. Now it can be shown that this expression for α_1 is in fact equal to 1! The other two solutions are -1 and -2. Clearly, the cubic formula is not always the most efficient method for solving a cubic equation.

Shortly after the cubic equation was solved, a solution by radicals of the quartic equation $ax^4 + bx^3 + cx^2 + dx + e = 0$ was discovered by **Lodovico Ferrari** (1522–1565), a pupil of Cardano. This formula is even more complicated than the cubic and will not be discussed here.

The next step of course was to find a solution by radicals for the quintic equation $ax^5 + bx^4 + cx^3 + dx^2 + ex + f = 0$. But no such formula was found despite repeated efforts. By the 19th century, mathematicians such as Lagrange and Gauss became convinced that no such formula existed for the quintic equation and indeed for any polynomial equation of degree greater than four. Their intuition proved to be correct and a proof that the general polynomial equation of degree greater than four was not solvable by radicals was given by **Niels Henrik Abel** (1802–1829) in the early 19th century.

Exercises 8.1

1. Let F be a field.
   (a) Prove that addition in $F[x]$ is commutative.
   (b) Prove that multiplication in $F[x]$ is commutative.
   (c) Prove that multiplication in $F[x]$ is associative.
   (d) Prove that the distributive laws hold in $F[x]$.

2. Theorem 8.1.2 says that if $\deg f(x) \neq \deg g(x)$, then $\deg(f(x) + g(x)) = \max\{\deg f(x), \deg g(x)\}$. Write the converse of this statement. Prove or disprove.

3. Let F be a field.
   (a) Let $f(x), g(x) \in F[x]$. Prove that if $f(x)g(x) = 0$, then $f(x) = 0$ or $g(x) = 0$.
   (b) Let $f(x), g(x), h(x) \in F[x]$. Suppose that $f(x)g(x) = f(x)h(x)$ and that $f(x) \neq 0$. Prove that $g(x) = h(x)$.
   (*Note:* This exercise shows that a cancellation law holds in $F[x]$.)

4. Prove part 2 of Theorem 8.1.2.

5. Let F be a field. Let $f(x) \in F[x]$. Prove that $f(x)$ has a multiplicative inverse in $F[x]$ if and only if $f(x)$ is a nonzero constant polynomial; that is, there exists $c \in F$, $c \neq 0$, such that $f(x) = c$.

6. Let $f(x) = 8x^3 - 6x^2 + 5$ and $g(x) = x^4 + x^3 - 10x + 2$ in $\mathbf{R}[x]$. Compute:
   (a) $f(x) + g(x)$ (b) $f(x)g(x)$ (c) $f(x)^2$

7. Let $f(x) = [4]x^3 - [6]x^2 + [2]x + [3]$ and $g(x) = [2]x^2 + [4]x - [5]$ in $\mathbf{Z}_7[x]$.
   (a) Compute:
       (i) $f(x) + g(x)$ (ii) $f(x)g(x)$ (iii) $f(x)^2$
   (b) Prove that the additive inverse of $f(x)$ is the polynomial $[3]x^3 - [1]x^2 + [5]x + [4]$.

8. Prove that $q(x)$ and $r(x)$ of the Division Algorithm for Polynomials are unique.

9. For $f(x)$ and $g(x)$ given next, find $q(x)$ and $r(x)$ that satisfy the conditions of the Division Algorithm:
   (a) $f(x) = x^5 + 6x^3 - x^2 + 14x + 1$, $g(x) = x^2 + 5$ in $\mathbf{Q}[x]$.
   (b) $f(x) = [3]x^3 + [4]$, $g(x) = [3]x^2 + [4]x - [1]$ in $\mathbf{Z}_5[x]$.

10. Use Corollary 8.1.4 to find the remainder if $x^5 - 7x^3 + 2x - 6$ is divided by $x + 4$.

11. Find all the real zeros of the following polynomials:
    (a) $f(x) = x^3 + 2x^2 + 2x + 1$
    (b) $f(x) = x^4 + 2x^3 + 3x^2 + 4x + 2$
    (c) $f(x) = x^5 + 2x^4 + x + 2$

12. Find all the complex zeros of the polynomials in Exercise 11.

13. Let $f(x) = [2]x^3 + [4]x^2 - [1]x + [2] \in \mathbf{Z}_7[x]$. Find all the zeros of $f(x)$ in $\mathbf{Z}_7$.

14. Solve the following congruences, if possible:
    (a) $x^2 \equiv 5 (\text{mod } 11)$
    (b) $x^2 \equiv 8 (\text{mod } 13)$

15. For each of the following primes p, find all integers a between 1 and $p - 1$ for which the congruence $x^2 \equiv a(\text{mod } p)$ is solvable, and then find those a for which it is not solvable.
    (a) $p = 7$
    (b) $p = 11$
    (c) $p = 13$

16. Write the negation of Corollary 8.1.7.

17. Prove that the equation $y^2 = 7x + 3$ has no integer solutions.

18. If $f(x) = a_0 + a_1x + a_2x^2 + \ldots + a_nx^n \in F[x]$, we define its *derivative* $Df(x)$ by the formula $Df(x) = a_1 + 2a_2x + 3a_3x^2 + \ldots + na_nx^{n-1}$, if $n \geq 1$ and $Df(x) = 0$ if $n = 0$ or if $f(x)$ is the zero polynomial, based on the usual formula for the derivative of a real-valued polynomial. (But note that the coefficients are in an arbitrary field F and not necessarily the real numbers, so we make no mention of limits.)
    (a) Prove that if $f(x)$ has degree $n \geq 1$, then deg $Df(x) \leq n - 1$.
    (b) Prove that if $f(x) \in \mathbf{R}[x]$ and deg $f(x) = n \geq 1$, then deg $Df(x) = n - 1$.
    (c) Give an example of the field F and a polynomial $f(x) \in F[x]$ of degree n for some $n \geq 1$ such that deg $Df(x) < n - 1$.
    (d) Let $f(x), g(x) \in F[x]$. Prove that
        (i) $D(f(x) + g(x)) = Df(x) + Dg(x)$
        (ii) $D(f(x)g(x)) = Df(x)g(x) + f(x)Dg(x)$

19. Let a be a nonzero real number and let $f(x) = x^3 - a$.
    (a) Prove that $f(x)$ has only one real zero, namely, $\alpha = \sqrt[3]{a}$.
    (b) Find the other two (complex) zeros of $f(x)$.

20. Use the cubic formula to find the zeros of the polynomial $x^3 + 3x + 2 = 0$.

21. Verify that $\frac{1}{3}\{-2 + \sqrt[3]{10 + 9\sqrt{3i}} + \sqrt[3]{10 - 9\sqrt{3i}}\} = 1$.

22. Find all of the zeros of $f(x) = x^4 + 3x^2 - 4$.

Discussion and Discovery Exercises

For the first two problems, review the techniques you learned in first-year calculus for graphing polynomials, in particular quadratic and cubic polynomials.

D1. Let $f(x) = ax^2 + bx + c \in \mathbf{R}[x]$, $a \neq 0$.
  (a) Give an example of a quadratic polynomial with no real zeros, then one with exactly one real zero, then one with exactly two real zeros.
  (b) Prove that if $b^2 < 4ac$, then $f(x)$ has no real zeros.
  (c) Prove that if $b^2 = 4ac$, then $f(x)$ has exactly one real zero.
  (d) Prove that if $b^2 > 4ac$, then $f(x)$ has exactly two real zeros.

D2. Let $f(x) = ax^3 + bx^2 + cx + d \in \mathbf{R}[x]$, $a \neq 0$.
  (a) Prove that $f(x)$ has one, two, or three real zeros.
  (b) Give an example of a cubic polynomial with exactly one real zero, then one with exactly two real zeros, then one with exactly three real zeros.
  (c) Prove that if $b^2 \leq 3ac$, then $f(x)$ has exactly one real zero.
  (d) Suppose that $b^2 > 3ac$. Show that no conclusion beyond the one stated in part (a) can be drawn about the number of real zeros that $f(x)$ may have.

D3. Can you deduce any general conjectures from your answers to Exercise 15? Test your conjecture using some other primes.

D4. Generalize Exercise 19 to the polynomial $f(x) = x^n - a$.

8.2 UNIQUE FACTORIZATION

Irreducible Polynomials

In this section we consider the question of how polynomials factor. In the previous section, we saw that a zero of a polynomial yields a factor of that polynomial. But not all polynomials have a zero in a given field. This does not mean, however, that such a polynomial does not factor. For example, the polynomial $x^4 + x^3 + 2x^2 + x + 1$ factors into $(x^2 + 1)(x^2 + x + 1)$. Neither of these polynomials, $x^2 + 1$ or $x^2 + x + 1$, have zeros in $\mathbf{R}$ and therefore they cannot be factored any more in $\mathbf{R}[x]$. (They can, however, be factored in $\mathbf{C}[x]$.)

Polynomials that cannot be factored are called **irreducible**. A formal definition is given next. Irreducible polynomials have properties similar to the prime numbers. Just as the Unique Factorization Theorem for the Integers says that every integer is the unique product of prime numbers, we show in this section that every polynomial in $F[x]$ can be written as a product of irreducible polynomials in an essentially unique way.

It would be wrong to take unique factorization for granted. Not every set has this property. For example, consider the set $S = \{a + b\sqrt{-5} \mid a, b \in \mathbf{Z}\}$. Like the integers, S has two binary operations of addition and multiplication. In S we can write $9 = (3)(3)$ and $9 = (2 + \sqrt{-5})(2 - \sqrt{-5})$. It is not obvious, but in fact none of the elements 3, $2 + \sqrt{-5}$, or $2 - \sqrt{-5}$ can be factored in S. So 9 can be factored in two different ways in S. Thus S does not have the unique factorization property.

Definition 8.2.1 Let F be a field and let $f(x) \in F[x]$ such that deg $f(x) \geq 1$. Then $f(x)$ is said to be **irreducible over** F if $f(x)$ cannot be written as a product of two polynomials in $F[x]$ having smaller degree than $f(x)$.

The condition that the factors have smaller degree than $f(x)$ is to avoid trivial factorizations obtained by factoring out a constant. For example, $x^2 + x + 1$ is irreducible over $\mathbf{R}$, but we could write $x^2 + x + 1 = \frac{1}{2}(2x^2 + 2x + 2)$. We do not consider this a proper factorization.

It is not sufficient just to say that a polynomial is irreducible. It is necessary to say that it is irreducible over a particular field. For example, $x^2 - 2$ is irreducible over $\mathbf{Q}$ but not over $\mathbf{R}$ and $x^2 + 1$ is irreducible over $\mathbf{Q}$ and over $\mathbf{R}$ but not over $\mathbf{C}$.

In general, it is a difficult problem to determine whether a polynomial is irreducible or not. Any polynomial of degree 1 is clearly irreducible over any field. For polynomials of degree 2 or 3, we have the following result.

Lemma 8.2.2   Let $f(x) \in F[x]$ with deg $f(x) = 2$ or 3. Then $f(x)$ is irreducible over F if and only if $f(x)$ has no zeros in F.

PROOF: We leave it as an exercise. ●

The lemma is not true in general for polynomials of degree 4 or greater. The polynomial $x^4 + 2x^2 + 1 = (x^2 + 1)^2$ is an example of a polynomial with no zeros in $\mathbf{R}$ but not irreducible over $\mathbf{R}$ either. The implication the other way is always true for polynomials of degree ≥ 2, however; that is, if $f(x)$ is an irreducible polynomial in $F[x]$ and deg $f(x) \geq 2$, then $f(x)$ has no zeros in F. The proof is left as an exercise.

In order to prove the Unique Factorization Theorem for Polynomials, we need to introduce the concepts of divisibility and greatest common divisor of polynomials. The definitions and properties are similar to the analogous ones for integers that we discussed in Chapter 5.

Definition 8.2.3 A polynomial $f(x)$ is called a **monic** polynomial if its leading coefficient is 1.

Definition 8.2.4 Let $f(x)$, $g(x) \in F[x]$. We say that $f(x)$ is an **associate** of $g(x)$ if there is an element c in F, $c \neq 0$, such that $f(x) = cg(x)$.

If $f(x)$ is an associate of $g(x)$, then it is easy to see that $g(x)$ is an associate of $f(x)$ since $g(x) = c^{-1}f(x)$. Consequently, we can simply say that $f(x)$ and $g(x)$ are **associate to each other** or simply **associate**.

It is easy to see that every polynomial in $F[x]$ is associate to a monic polynomial and that if $f(x) \in F[x]$ is irreducible over F, then any associate of $f(x)$ in $F[x]$ is also irreducible over F. We leave the details as exercises.

Definition 8.2.5 Let $f(x)$, $g(x) \in F[x]$. We say that $f(x)$ **divides** $g(x)$ if there exists $h(x) \in F[x]$ such that $f(x)h(x) = g(x)$. We write $f(x) \mid g(x)$. We also say $f(x)$ is a **divisor** or a **factor** of $g(x)$.

Lemma 8.2.6   Let $p(x) \in F[x]$ such that deg $p(x) \geq 1$. If $p(x)$ is irreducible over F, then the only divisors of $p(x)$ are the nonzero constant polynomials and the associates of $p(x)$.

PROOF: We leave it as an exercise.

Note that this lemma is analogous to the result that the only divisors in **Z** of a prime number p are ± 1 and $\pm p$. Recall that ± 1 are the only elements of **Z** with multiplicative inverses and that the nonzero constant polynomials are the only elements of $F[x]$ with multiplicative inverses.

Greatest Common Divisors

In the proof of unique factorization of integers, we used the fact that if a prime number divides a product, then it must divide one of the factors of that product. A similar result holds for irreducible polynomials. To prove it, we must introduce the notion of a greatest common divisor of two polynomials.

Definition 8.2.7   Let $f(x)$ and $g(x)$ be nonzero polynomials in $F[x]$. Then a **greatest common divisor** of $f(x)$ and $g(x)$ is a monic polynomial $d(x)$ such that

1. $d(x)$ divides both $f(x)$ and $g(x)$,
2. if $h(x) \in F[x]$ divides both $f(x)$ and $g(x)$, then $h(x)$ divides $d(x)$.

Just as in the case with integers, we must prove that a greatest common divisor of two polynomials always exists. Note that the proof of the following theorem is similar to the existence theorem for greatest common divisors of integers (Theorem 5.3.5).

Theorem 8.2.8   Let $f(x)$ and $g(x)$ be nonzero polynomials in $F[x]$. Then a greatest common divisor of $f(x)$ and $g(x)$ exists and is unique. Moreover, if $g(x)$ is the greatest common divisor of $f(x)$ and $g(x)$, then there exist polynomials $s(x)$ and $t(x)$ such that $d(x) = f(x)s(x) + g(x)t(x)$.

PROOF: Let $S = \{f(x)s(x) + g(x)t(x) \mid s(x), t(x) \in F[x]\}$; S is the set of all polynomials in $F[x]$ that can be written in the form $f(x)s(x) + g(x)t(x)$ for some polynomials $s(x)$ and $t(x)$ in $F[x]$. Since $f(x) = f(x)1 + g(x)0$ and $g(x) = f(x)0 + g(x)1$, both $f(x)$ and $g(x)$ are in S and therefore S is a nonempty set.

By the Well-Ordering Principle, there is a polynomial $d_0(x)$ in S of smallest degree. Any constant times $d_0(x)$ is still in S and has the same degree as $d_0(x)$. Multiplying $d_0(x)$ by the reciprocal of its leading coefficient gives a monic polynomial $d(x)$ of smallest degree in S. We claim that $d(x)$ satisfies the definition of a greatest common divisor of $f(x)$ and $g(x)$.

Now since $d(x) \in S$, there exist polynomials $s(x)$ and $t(x)$ in $F[x]$ such that $d(x) = f(x)s(x) + g(x)t(x)$.

We will show that $d(x)$ divides $f(x)$. By the Division Algorithm, there are polynomials $q(x)$ and $r(x)$ such that $f(x) = d(x)q(x) + r(x)$ where $r(x) = 0$ or $\deg r(x) < \deg d(x)$. We then have

$$\begin{aligned}
r(x) &= f(x) - d(x)q(x) \\
&= f(x) - (f(x)s(x) + g(x)t(x))q(x) \\
&= f(x)(1 - s(x)q(x)) - g(x)t(x)q(x).
\end{aligned}$$

This implies that $r(x)$ is in S. It follows that $r(x) = 0$, since $d(x)$ has minimal degree in S. Hence $f(x) = d(x)q(x)$ and $d(x)$ divides $f(x)$.

A similar proof shows that $d(x)$ divides $g(x)$.

Let $h(x) \in F[x]$ such that $h(x)$ divides both $f(x)$ and $g(x)$. It follows from Exercise 8 that $h(x)$ divides $f(x)s(x) + g(x)t(x) = d(x)$. Hence $d(x)$ is a greatest common divisor of $f(x)$ and $g(x)$.

The proof of uniqueness is left as an exercise. ●

We will write g.c.d. to abbreviate greatest common divisor.

Now we come to the problem of actually *computing* the g.c.d. of two polynomials. Recall that in Section 5.3, we had an algorithm for computing the g.c.d. of two integers by repeated application of the Division Algorithm for Integers. A similar process works by polynomials. The key result that makes this process work is the fact that if $f(x) = g(x)h(x) + r(x)$, then the g.c.d. of $f(x)$ and $g(x)$ is equal to the g.c.d. of $g(x)$ and $r(x)$. (See Exercise 13.)

EXAMPLE I   Let $f(x) = x^6 + x^5 - 4x^4 + 2x^3 - 11x^2 + x - 6$ and $g(x) = x^4 - 4x^3 - 20x^2 - 4x - 21$. Applying the Division Algorithm to $f(x)$ and $g(x)$, we get

$$f(x) = g(x)(x^2 + 5x + 36) + 250(x^3 + 3x^2 + x + 3).$$

By the previous remarks, then, the g.c.d. of $f(x)$ and $g(x)$ is equal to the g.c.d. of $g(x)$ and $250(x^3 + 3x^2 + x + 3)$ and this in turn is equal to the g.c.d. of $g(x)$ and $x^3 + 3x^2 + x + 3$.

Applying the Division Algorithm again, this time to $g(x)$ and $x^3 + 3x^2 + x + 3$, we obtain the equation $g(x) = (x^3 + 3x^2 + x + 3)(x - 7)$. From this equation we can conclude that the g.c.d. of $g(x)$ and $x^3 + 3x^2 + x + 3$ is $x^3 + 3x^2 + x + 3$.

Thus the g.c.d. of $f(x)$ and $g(x)$ is $x^3 + 3x^2 + x + 3$.

Another method for finding the g.c.d. of two polynomials is to factor them. In this example, we can factor $f(x)$ and $g(x)$ as follows: $f(x) = (x - 2)(x + 3)(x^2 + 1)^2$ and $g(x) = (x + 3)(x - 7)(x^2 + 1)$.

Now it is easy to see that the g.c.d. of $f(x)$ and $g(x)$ is $(x + 3)(x^2 + 1) = x^3 + 3x^2 + x + 3$.

Note that we did not show how the factorization of $f(x)$ and $g(x)$ was obtained. Factoring a polynomial, even one of relatively low degree, is in general a difficult problem. We discuss some methods for doing so later in this section and in the next section.

Definition 8.2.9   Two polynomials $f(x)$ and $g(x)$ in $F[x]$ are called **relatively prime** if the g.c.d. of $f(x)$ and $g(x)$ is 1.

Corollary 8.2.10   Let $f(x), g(x) \in F[x]$. Then $f(x)$ and $g(x)$ are relatively prime if and only if there exist polynomials $s(x)$ and $t(x)$ such that $f(x)s(x) + g(x)t(x) = 1$.

PROOF: If $f(x)$ and $g(x)$ are relatively prime, then the result follows immediately from Theorem 8.2.8. Conversely, suppose that there exist polynomials $s(x)$ and $t(x)$ such that $f(x)s(x) + g(x)t(x) = 1$. Then 1 is an element of the set S defined in the proof of Theorem 8.2.8. But it was shown in the proof of Theorem 8.2.8 that the g.c.d. of $f(x)$ and $g(x)$ is the monic polynomial of smallest degree in S. Since 1 is monic and deg $1 = 0$, no monic polynomial in S can have smaller degree than the polynomial 1. Hence 1 is the g.c.d. of $f(x)$ and $g(x)$. ●

Corollary 8.2.11   Let $f(x)$, $g(x)$, $h(x) \in F[x]$. Suppose that $f(x)$ divides $g(x)h(x)$ and $f(x)$ and $g(x)$ are relatively prime. Then $f(x)$ divides $h(x)$.

PROOF: The proof is similar to the proof of Theorem 5.3.9 and is left as an exercise. ●

Corollary 8.2.12   Let $p(x)$ be an irreducible polynomial in $F[x]$. If $p(x)$ divides $f(x)g(x)$ where $f(x)$, $g(x) \in F[x]$, then $p(x)$ divides $f(x)$ or $p(x)$ divides $g(x)$.

PROOF: The proof is similar to the proof of Proposition 5.4.4 and is left as an exercise. ●

Corollary 8.2.13   Let $p(x)$ be an irreducible polynomial in $F[x]$. If $p(x)$ divides $f_1(x) \cdot f_2(x) \cdot \ldots \cdot f_m(x)$, where each $f_i(x) \in F[x]$, then $p(x)$ divides $f_i(x)$ for some $i = 1, 2, \ldots, m$.

PROOF: We leave it as an exercise. ●

The Unique Factorization Theorem

We now come to the main result of this section, the Unique Factorization Theorem for Polynomials. The reader should refer to the proof of Theorem 5.4.7, the Unique Factorization Theorem for Integers, since the proofs of these two theorems are very similar.

Theorem 8.2.14   **Unique Factorization Theorem for Polynomials** Let $f(x) \in F[x]$, deg $f(x) \geq 1$. Then $f(x)$ is irreducible or is a product of irreducible polynomials in $F[x]$. Moreover this product is unique in the following sense: if

$$
\begin{aligned}
f(x) &= p_1(x) \cdot p_2(x) \cdot \ldots \cdot p_m(x) \\
&= q_1(x) \cdot q_2(x) \cdot \ldots \cdot q_n(x)
\end{aligned}
$$

where each $p_i(x)$ and $q_j(x)$ is irreducible in $F[x]$, then $m = n$, and after renumbering, if necessary, $p_i(x)$ is associate to $q_i(x)$ for each $i = 1, 2, \ldots, m$.

PROOF: The proof that $f(x)$ is a product of irreducible polynomials proceeds by induction on deg $f(x)$. Moreover, we use the Second Principle of Induction (Theorem 5.2.3). If deg $f(x) = 1$, then $f(x)$ is already irreducible and so there is nothing to prove.

Suppose the theorem is true for all polynomials of degree less than n. Assume deg $f(x) = n$. If $f(x)$ is irreducible, there is nothing to prove. If not, we can write $f(x) = g(x)h(x)$ where deg $g(x) < n$ and deg $h(x) < n$. By the induction hypothesis, both $g(x)$ and $h(x)$ are products of irreducible polynomials. Hence $f(x)$ is a product of irreducibles.

We now prove uniqueness. Suppose

$$
\begin{aligned}
f(x) &= p_1(x) \cdot p_2(x) \cdot \ldots \cdot p_m(x) \\
&= q_1(x) \cdot q_2(x) \cdot \ldots \cdot q_n(x)
\end{aligned}
$$

where $p_i(x)$ and $q_j(x)$ are irreducible in $F[x]$. We can assume, without loss of generality, that $m \leq n$. Since $p_1(x)$ divides $f(x)$, $p_1(x)$ divides $q_1(x) \cdot q_2(x) \cdot \ldots \cdot q_n(x)$. Therefore by Corollary 8.2.13, $p_1(x)$ divides $q_j(x)$ for some $j = 1, 2, \ldots, n$. Renumber the $q_j(x)$s so that $q_j(x) = q_1(x)$. It follows that $p_1(x)$ and $q_1(x)$ are associate. (See Exercise 19.) So $q_1(x) = c_1 p_1(x)$ for some constant c_1.

We thus get the equation

$$p_1(x) \cdot p_2(x) \cdot \ldots \cdot p_m(x) = c_1 p_1(x) \cdot q_2(x) \cdot \ldots \cdot q_n(x).$$

By Exercise 3(b) of Section 8.1, we can cancel $p_1(x)$ to get

$$p_2(x) \cdot \ldots \cdot p_m(x) = c_1 \cdot q_2(x) \cdot \ldots \cdot q_n(x),$$

where $c_1 \in F$. Proceeding in the same way and renumbering if necessary, after m steps, we get that $p_i(x)$ and $q_i(x)$ are associate for $i = 1, 2, \ldots, m$. If $m < n$, we would get

$$1 = c_1 \cdot c_2 \cdot \ldots \cdot c_m \cdot q_{m+1}(x) \cdot \ldots \cdot q_n(x),$$

where $c_1, c_2, \ldots, c_m \in F$, a contradiction. Therefore $m = n$ and the proof is complete. ●

EXAMPLE 2 Let $f(x) = x^4 - 2x^3 - x^2 + 4x - 2 \in \mathbf{Q}[x]$. Since 1 is a zero of this polynomial, $x - 1$ is a factor of $f(x)$.

Dividing by $x - 1$, we get $f(x) = (x - 1)(x^3 - x^2 - 2x + 2)$. We see that 1 is also a zero of $x^3 - x^2 - 2x + 2$. Dividing by $x - 1$ again, we get $f(x) = (x - 1)^2(x^2 - 2)$.

Now $x^2 - 2$ has no zeros in $\mathbf{Q}$ (why not?), so it is irreducible over $\mathbf{Q}$. Therefore, $(x - 1)^2(x^2 - 2)$ is the factorization of $f(x)$ into a product of irreducible elements of $\mathbf{Q}[x]$.

Notice that $f(x) \in \mathbf{R}[x]$ since $\mathbf{Q} \subseteq \mathbf{R}$. In $\mathbf{R}[x]$, $x^2 - 2$ factors into $(x - \sqrt{2})(x + \sqrt{2})$. Therefore, $f(x) = (x - 1)^2(x - \sqrt{2})(x + \sqrt{2})$ is the factorization of $f(x)$ into a product of irreducible elements of $\mathbf{R}[x]$. This is also a factorization of $f(x)$ into a product of irreducible elements of $\mathbf{C}[x]$.

EXAMPLE 3 Let $f(x) = 2x^3 - 8x^3 + 8x - 6 \in \mathbf{Q}[x]$. Since 3 is a zero of $f(x)$, we can write $f(x) = 2(x - 3)(x^2 - x + 1)$. Using the quadratic formula, we see that $x^2 - x + 1$ is irreducible over $\mathbf{Q}$ and over $\mathbf{R}$. Therefore this is the correct factorization of $f(x)$ into a product of irreducible elements of $\mathbf{Q}[x]$ and $\mathbf{R}[x]$.

In $\mathbf{C}$, $x^2 - x + 1$ has two zeros: $\frac{1}{2} + \frac{\sqrt{3}}{2} i$ and $\frac{1}{2} - \frac{\sqrt{3}}{2} i$. So in $\mathbf{C}[x]$, the factorization of $f(x)$ into a product of irreducibles is $f(x) = 2(x - 3)(x - \frac{1}{2} - \frac{\sqrt{3}}{2} i)(x - \frac{1}{2} + \frac{\sqrt{3}}{2} i)$.

EXAMPLE 4 In the previous examples, finding a zero of the polynomial was the key step in computing the factorization. However, a polynomial with no zeros may still have a factorization. For example, let $f(x) = x^4 + 2$. Clearly $f(x)$ has no zeros in $\mathbf{R}$. So if $f(x)$ does factor over $\mathbf{R}$, it must factor into a product of two polynomials of degree 2. We write

$$x^4 + 2 = (x^2 + ax + b)(x^2 + cx + d),$$

where a, b, c, d are in $\mathbf{R}$. Multiplying out these factors, we get

$$x^4 + 2 = x^4 + (a + c)x^3 + (ac + b + d)x^2 + (ad + bc)x + bd.$$

This leads to the equations

$a + c = 0,$
$ac + b + d = 0,$
$ad + bc = 0,$
$bd = 2.$

Thus, $c = -a$ and $d = 2/b$. Substituting into $ad + bc = 0$ gives $2a/b = ab$ or $2a = ab^2$. You can easily check that $a = 0$ is not a possibility. So dividing by a gives $b^2 = 2$ or $b = \pm\sqrt{2}$. Hence $d = 2/b = \pm\sqrt{2}$. Substituting these values into $ac + b + d = 0$, we get $a^2 = \pm 2\sqrt{2}$. Since $a^2 = -2\sqrt{2}$ is impossible, we get the following solutions:

$$a = \sqrt[4]{8}; \qquad b = \sqrt{2}; \qquad c = -\sqrt[4]{8}; \qquad d = \sqrt{2}.$$

Therefore,

$$x^4 + 2 = (x^2 + \sqrt[4]{8}x + \sqrt{2})(x^2 - \sqrt[4]{8}x + \sqrt{2}).$$

So we have established that $x^4 + 2$ is irreducible over $\mathbf{Q}$ but not over $\mathbf{R}$. The two quadratic factors are irreducible over $\mathbf{R}$: if they were not, then $x^4 + 2$ would have linear factors and therefore would have a real zero, which it does not.

Over $\mathbf{C}$, these two quadratic factors are not irreducible. To see this, we show that $x^4 + 2$ has 4 zeros in $\mathbf{C}$. A complex number z is a zero of $x^4 + 2$ if and only if z is a fourth root of -2. Using Theorem 7.3.9, we get the solutions:

$$z_1 = \sqrt[4]{2}\left(\frac{1}{\sqrt{2}} + \frac{1}{\sqrt{2}}i\right),$$

$$z_2 = \sqrt[4]{2}\left(-\frac{1}{\sqrt{2}} + \frac{1}{\sqrt{2}}i\right),$$

$$z_3 = \sqrt[4]{2}\left(-\frac{1}{\sqrt{2}} - \frac{1}{\sqrt{2}}i\right),$$

$$z_4 = \sqrt[4]{2}\left(\frac{1}{\sqrt{2}} - \frac{1}{\sqrt{2}}i\right).$$

So, in $\mathbf{C}[x]$, we have

$$x^4 + 2 = (x - z_1)(x - z_2)(x - z_3)(x - z_4).$$

The Unique Factorization Theorem for Polynomials can be re-stated another way. Let $f(x) = p_1(x) \cdot p_2(x) \cdot \ldots \cdot p_m(x)$, where each $p_i(x)$ is irreducible in $F[x]$. Let a_i be the leading coefficient of $p_i(x)$. Then $p_i(x) = a_i q_i(x)$, where $q_i(x)$ is a monic irreducible polynomial in $F[x]$ associate to $p_i(x)$. Let $a = a_1 \cdot a_2 \cdot \ldots \cdot a_m$. Then $f(x) = a \cdot q_1(x) \cdot q_2(x) \cdot \ldots \cdot q_m(x)$.

If we group together those $q_i(x)$s that are associate to each other and use the fact that if two polynomials are associate then one is a constant times the others, then we can write $f(x)$ in the form

$$f(x) = c \cdot P_1(x)^{m_1} \cdot P_2(x)^{m_2} \cdot \ldots \cdot P_r(x)^{m_r},$$

where $c \in F$ and each $P_i(x)$ is a monic irreducible polynomial in $F[x]$. Note that each $P_i(x)$ equals some $q_j(x)$ and the exponent m_i is the number of times that $q_j(x)$ or one of its associates appears in the product $q_1(x) \cdot q_2(x) \cdot \ldots \cdot q_m(x)$. Therefore, no $P_i(x)$ is associate to $P_j(x)$ if $i \neq j$. So we have the following.

Corollary 8.2.15   Let $f(x) \in F[x]$, deg $f(x) \geq 1$. Then there exist a unique element c in F, unique monic irreducible polynomials $P_1(x), P_2(x), \ldots, P_r(x)$ in $F[x]$, and unique integers $m_1, m_2, \ldots, m_r$ such that

$$f(x) = c \cdot P_1(x)^{m_1} \cdot P_2(x)^{m_2} \cdot \ldots \cdot P_r(x)^{m_r}.$$

Moreover, $P_i(x)$ is not associate to $P_j(x)$ if $i \neq j$, for all $i, j = 1, 2, \ldots, r$.

We conclude this section with a generalization of Corollary 8.1.6.

Definition 8.2.16   Let $f(x) \in F[x]$ and let c be a zero of $f(x)$. We say c is a **zero of $f(x)$ of multiplicity m** if $(x - c)^m$ is a factor of $f(x)$ but $(x - c)^{m+1}$ is not. A zero of multiplicity 1 is called a **simple** zero; a zero of multiplicity greater than 1 is called a **multiple** zero.

EXAMPLE 5   Let $f(x) = (x + 3)^4(x - 7)^3(x - 9)$. Then -3 is a zero of $f(x)$ of multiplicity 4, 7 is a zero of multiplicity 3, and 9 is a zero of multiplicity 1.

Theorem 8.2.17   Let $f(x) \in F[x]$ and let $c_1, c_2, \ldots, c_t$ be the distinct zeros of $f(x)$ with multiplicities $m_1, m_2, \ldots, m_t$ respectively. Then

$$f(x) = (x - c_1)^{m_1}(x - c_2)^{m_2} \ldots (x - c_t)^{m_t} g(x),$$

where $g(x) \in F[x]$ and $g(x)$ has no zeros in F.

PROOF: By Corollary 8.2.15, we can write

$$f(x) = c \cdot P_1(x)^{n_1} \cdot P_2(x)^{n_2} \cdot \ldots \cdot P_r(x)^{n_r},$$

where each $P_i(x)$ is a monic irreducible polynomial in $F[x]$ and no two of them are associate.

By the definition of multiplicity, $(x - c_i)^{m_i}$ is a factor of $f(x)$ and $(x - c_i)^{m_i+1}$ is not, for each $i = 1, 2, \ldots, t$. Since $x - c_i$ is monic irreducible, by uniqueness of factorization (see Corollary 8.2.15), each $x - c_i$ equals some $P_j(x)$ of degree 1 and $m_i = n_j$.

Conversely, if some $P_k(x)$ has degree 1, then $P_k(x) = x + a_k$ for some a_k in F. It follows that $-a_k$ is a zero of $f(x)$ and so $P_k(x) = x - c_i$ for some i. Again by unique factorization, we have $n_k = m_i$.

Consequently, exactly t of the polynomials $P_j(x)$ have degree 1. Renumber the $P_j(x)$s so that the first t of them have degree 1 and so that $P_j(x) = x - c_j$ and $n_j = m_j$ for $j = 1, 2, \ldots, t$.

Let $g(x) = c \cdot P_{t+1}(x) \cdot P_{t+2}(x) \cdot \ldots \cdot P_r(x)$ if $r > t$ or $g(x) = c$ if $r = t$. Then $f(x) = (x - c_1)^{m_1}(x - c_2)^{m_2} \ldots (x - c_t)^{m_t} g(x)$. Since $\deg P_j(x) \geq 2$ for $j = t + 1, \ldots, r$, $g(x)$ can have no linear factors and therefore $g(x)$ has no zeros in F. ●

Corollary 8.2.18   Let $f(x) \in F[x]$ and let $c_1, c_2, \ldots, c_t$ be the distinct zeros of $f(x)$ with multiplicities $m_1, m_2, \ldots, m_t$ respectively. Then $m_1 + m_2 + \ldots + m_t \leq \deg f(x)$.

PROOF: We leave it as an exercise. ●

●   HISTORICAL COMMENTS: THE FUNDAMENTAL THEOREM
    OF ALGEBRA

In the historical comments at the end of Section 8.1, we saw that much effort was spent trying to find extensions of the quadratic formula to higher degree polynomials, with the result that appropriate formulas for third- and fourth-degree polynomials were discovered in the 16th century and the impossibility of a formula for polynomials of degree higher than four was proved in the 19th century. During this same stretch of time, the 16th to the 19th century, a related problem was being discussed: whether or not every real polynomial of degree n has n zeros (counting multiplicity) and where those zeros actually exist. It was known of course that not every real polynomial has zeros in **R** and that some have zeros in **C**.

Albert Girard (1595–1632) was the first mathematician to assert that every real polynomial of degree n has n zeros. He did not claim that all of the zeros had to be in **C** but allowed for the possibility that there may be a bigger field than **C** in which all of the zeros exist. Euler took Girard's assertion a step further by claiming that every real polynomial has all of its zeros in **C**. He was able to prove his assertion for all polynomials of degree ≤ 6.

Attempts to prove the theorem for all polynomials were made by Euler, **Jean le Rond D'Alembert** (1717–1783), **Joseph Louis Lagrange** (1736–1813), and **Pierre Simon Laplace** (1749–1827). All of these attempts fell short, mainly because, in trying to prove that a real polynomial has n zeros, the mathematicians assumed that such a polynomial has at least one zero in **C**. This flaw was pointed out by **Carl Friedrich Gauss** (1777–1855), one of the greatest mathematicians of all time. Gauss gave the first proof of what has come to be known as the Fundamental Theorem of Algebra: that every polynomial with real coefficients (or even complex coefficients) has at least one zero in **C**. From the Fundamental Theorem, it is then fairly straightforward to show that a real polynomial of degree n has n zeros in **C**. We will discuss these matters in the next section.

Gauss's first proof of the Fundamental Theorem of Algebra was contained in his doctoral thesis in 1799. Even this proof contained some statements that Gauss did not prove rigorously and only years later, with the development of topology, were the gaps in the proof filled. However, Gauss subsequently gave three additional proofs of the Fundamental Theorem of Algebra and these proofs were completely rigorous.

Exercises 8.2

1. Prove that $4x^2 - 5x + 2$ is irreducible over **Q** and **R** but not over **C**.

2. Factor the following polynomials into irreducibles over **Q**, **R**, and **C**.
   (a) $x^2 - 5$
   (b) $2x^3 - 11x^2 + 16x - 7$
   (c) $x^5 - 3x^4 + 2x^3 + 7x^2 - 9x - 10$

3. Let $f(x)$ be an irreducible polynomial in $F[x]$. Prove that if deg $f(x) \geq 2$, then $f(x)$ has no zeros in F.

4. Prove Lemma 8.2.2.

5. Prove that every polynomial in $F[x]$ is associate to a monic polynomial.

6. Let $f(x)$ be an irreducible polynomial in $F[x]$. Prove that any associate of $f(x)$ is irreducible in $F[x]$.

7. Let $f(x), g(x) \in F[x]$. Suppose that $f(x)$ divides $g(x)$.
   (a) Prove that any associate of $f(x)$ divides $g(x)$.
   (b) Prove that $f(x)$ divides any associate of $g(x)$.

8. Let $f(x), g(x), h(x) \in F[x]$. Suppose that $f(x)$ divides $g(x)$ and $f(x)$ divides $h(x)$. Prove that $f(x)$ divides $s(x)g(x) + t(x)h(x)$ for all $s(x), t(x)$ in $F[x]$.

9. Let $f(x), g(x) \in F[x]$.
   (a) Suppose that $f(x)$ divides $g(x)$ and $g(x)$ divides $f(x)$. Prove that $f(x)$ and $g(x)$ are associate.
   (b) State the converse of part (a). Prove or disprove.
   (c) State the contrapositive of part (a).

10. Write $f(x) \sim g(x)$ if $f(x)$ and $g(x)$ are associate. Prove that this defines an equivalence relation on $F[x]$.

11. (a) Prove Lemma 8.2.6.
    (b) State the converse of Lemma 8.2.6. Prove or disprove.

12. Prove the uniqueness part of Theorem 8.2.8.

13. Prove that if $f(x) = g(x)h(x) + r(x)$, then the g.c.d. of $f(x)$ and $g(x)$ is equal to the g.c.d. of $g(x)$ and $r(x)$.

14. Find the g.c.d. of $f(x)$ and $g(x)$:
    (a) $f(x) = 3x^4 - 3x^2$ and $g(x) = x^2 - 2x - 8$
    (b) $f(x) = x^6 - 5x^5 + 10x^4 - 14x^3 + 16x^2 - 8x$ and $g(x) = x^4 + x^3 - 4x^2 + 2x - 12$

15. Prove Corollary 8.2.11.

16. Let $f(x), g(x) \in F[x]$.
    (a) Prove that if $f(x)$ and $g(x)$ have a common zero in F, then they are not relatively prime.
    (b) State the converse of part (a). Prove or disprove.

17. Prove Corollary 8.2.12.

18. Prove Corollary 8.2.13.

19. Suppose that $p(x)$ and $q(x)$ are irreducible in $F[x]$ and $p(x)$ divides $q(x)$. Prove that $p(x)$ and $q(x)$ are associate.

20. Find all irreducible polynomials of degree 2 and degree 3 in $\mathbf{Z}_2[x]$.

21. Find all irreducible polynomials of degree 2 and degree 3 in $\mathbf{Z}_3[x]$.

22. Prove Corollary 8.2.18.

23. Let $\leq$ be the relation on $F[x]$ defined by $f(x) \leq g(x)$ if $\deg f(x) \leq \deg g(x)$. Determine whether this relation is reflexive, symmetric, transitive, or antisymmetric. Give proofs for each of your assertions.

24. Let $f(x) \in F[x]$ and let $Df(x)$ denote its derivative as defined in Exercise 18 of Section 8.1.
    (a) Let $c \in F$ be a zero of $f(x)$. Prove that c is a simple zero of $f(x)$ if and only if c is not a zero of $Df(x)$.
    (b) Suppose that $f(x)$ and $Df(x)$ are relatively prime. Prove that all of the zeros of $f(x)$ are simple.
    (c) State the converse of part (b). Prove that this converse is false in general but is true if F is the field $\mathbf{C}$ of complex numbers.

25. Let $a \in \mathbf{C}$, $a \neq 0$ and let $f(x) = x^n - a$. Prove that all the zeros of $f(x)$ are simple.

26. Let p be a prime number and let $a \in \mathbf{Z}_p$. Let $f(x) = x^p - a \in \mathbf{Z}_p[x]$. Prove that all of the zeros of $f(x)$ are multiple zeros.

Discussion and Discovery Exercises

D1. Compare and contrast the proofs of Theorem 5.3.5 and Theorem 8.2.8. Describe the similarities in the proofs as well as their differences.

D2. Repeat Exercise D1 for the proofs of Theorem 5.4.7 and Theorem 8.2.14.

D3. Explain why, in the algorithm for computing the g.c.d. of two polynomials, the process will eventually terminate.

8.3. POLYNOMIALS OVER **C**, **R**, AND **Q**

The Fundamental Theorem of Algebra

In this section we apply some of the results of this chapter to polynomials over **R** and **C**. First, using the Fundamental Theorem of Algebra, we see how polynomials over the complex numbers factor into irreducibles and then we do the same thing for polynomials over the real numbers. In particular, we will see that the only irreducible polynomials in $C[x]$ are the polynomials of degree 1 and the only irreducible polynomials in $R[x]$ are the polynomials of degree 1 and the polynomials of degree 2 with no real zeros. Then we consider polynomials in $Q[x]$. Unlike $R[x]$ and $C[x]$, there are irreducible polynomials in $Q[x]$ of every degree. For example, one can show, although it is not obvious, that if p is a prime number, then the polynomial $x^n + p$ is irreducible over **Q** for every positive integer n. This is a consequence of the Eisenstein criterion, which we prove in this section.

The key result needed for this section is the *Fundamental Theorem of Algebra*. As we saw in the historical comments at the end of Section 8.2, this is a very deep result and was first proven by Gauss. Since then, many different proofs have been devised, but they all require advanced techniques. You may see a proof in a later course (abstract algebra, complex analysis, or one of several other possibilities); here we will simply state this important theorem.

Theorem 8.3.1   **Fundamental Theorem of Algebra** Let $f(x) \in C[x]$, deg $f(x) \geq 1$. Then $f(x)$ has at least one zero in **C**.

Note that this theorem applies to polynomials with real coefficients since $R \subseteq C$. For example, the polynomial $f(x) = 3x^4 + 4x^3 + 2$ has no real zeros. (This can be seen easily by graphing.) But the Fundamental Theorem of Algebra tells us that $f(x)$ has at least one zero in **C**. In fact, it will have four zeros in **C** if we count multiplicity. More generally, we have the following:

Theorem 8.3.2   Let $f(x) \in C[x]$, deg $f(x) = n \geq 1$. Then $f(x)$ has exactly n zeros in **C**, counting multiplicity. Specifically, if $c_1, c_2, \ldots, c_t$ are the distinct zeros of $f(x)$ in **C**, with multiplicities $m_1, m_2, \ldots, m_t$ respectively, then $f(x) = a(x - c_1)^{m_1}(x - c_2)^{m_2} \ldots (x - c_t)^{m_t}$ where $a \in C$; that is, $f(x)$ factors into all linear factors in $C[x]$ and $m_1 + m + \ldots + m_t = n$.

PROOF: By the Fundamental Theorem of Algebra, $f(x)$ has at least one zero in $\mathbf{C}$. By Theorem 8.2.17, we can write $f(x) = (x - c_1)^{m_1}(x - c_2)^{m_2} \ldots (x - c_t)^{m_t}$ where $g(x)$ has no zeros in $\mathbf{C}$.

Since, again by the Fundamental Theorem of Algebra, every polynomial of degree ≥ 1 in $\mathbf{C}[x]$ has at least one zero in $\mathbf{C}$, we must have $\deg g(x) = 0$, so $g(x) = a$ for some a in $\mathbf{C}$. The theorem is now proved.

Corollary 8.3.3   No polynomial in $\mathbf{C}[x]$ of degree at least 2 is irreducible.

PROOF: We leave it as an exercise.

EXAMPLE I   Consider the polynomial $f(x) = x^n - 1$ where n is a fixed positive integer. In Section 7.3 we learned that the zeros of $f(x)$ are the nth roots of unity $\omega_0, \omega_1, \omega_2, \ldots, \omega_{n-1}$, where $\omega_k = \cos 2\pi k/n + i \sin 2\pi k/n$ for $0 \leq k \leq n - 1$. So we can write $x^n - 1 = (x - \omega_0)(x - \omega_1) \ldots (x - \omega_{n-1})$.

Real Polynomials

We now consider polynomials over the field of real numbers $\mathbf{R}$. We will show that the irreducible polynomials in $\mathbf{R}[x]$ are the polynomials of degree 1 and the polynomials of degree 2 that have no real zeros.

Definition 8.3.4 Let $f(x) = ax^2 + bx + c \in \mathbf{R}[x]$. Then the real number $b^2 - 4ac$ is called the **discriminant** of $f(x)$.

Lemma 8.3.5   A polynomial of degree 2 in $\mathbf{R}[x]$ is irreducible over $\mathbf{R}$ if and only if its discriminant is less than 0.

PROOF: This proof, which follows easily from Lemma 8.2.2 and the quadratic formula, is left as an exercise.

Before we prove our main result about polynomials in $\mathbf{R}[x]$, recall that we proved in Theorem 7.3.5 that if $z \in \mathbf{C}$ is a zero of a polynomial $f(x)$ in $\mathbf{R}[x]$, then its complex conjugate $\bar{z}$ is also a zero. It is also true that their multiplicities as zeros of $f(x)$ are equal.

Lemma 8.3.6   Let $f(x) \in \mathbf{R}[x]$, $\deg f(x) \geq 1$. Let z be a complex nonreal zero of $f(x)$. Then z and $\bar{z}$ have the same multiplicities as zeros of $f(x)$.

PROOF: Let m and n be the multiplicities of z and $\bar{z}$, respectively. Assume that $m > n$. Then $f(x) = (x - z)^m(x - \bar{z})^n g(x)$, where $g(x) \in \mathbf{R}[x]$ and $g(z) \neq 0$ and $g(\bar{z}) \neq 0$.

Let $h(x) = (x - z)(x - \bar{z})$. Then $h(x) = x^2 - (z + \bar{z})x + z\bar{z}$, which is a polynomial in $\mathbf{R}[x]$, since both $z + \bar{z}$ and $z\bar{z}$ are real numbers.

Then $f(x) = h(x)^n(x - z)^{m-n}g(x)$.

The polynomial $(x - z)^{m-n}g(x)$ must have real coefficients since it is equal to $f(x)/h(x)^n$, which is a quotient of two polynomials with real coefficients. But $(x - z)^{m-n}g(x)$ has z as a zero and not $\bar{z}$, contradicting Theorem 7.3.5. Therefore, the assumption that $m > n$ must be false. A similar proof shows that $n > m$ is false also. Thus $m = n$. ●

Theorem 8.3.7 Let $f(x) \in \mathbf{R}[x]$, $\deg f(x) \geq 1$. Then $f(x)$ can be factored in $\mathbf{R}[x]$ into a product of linear polynomials and quadratic polynomials of negative discriminant.

PROOF: Let $r_1, r_2, \ldots, r_t$ be the distinct real zeros of $f(x)$, if any, with multiplicities $m_1, m_2, \ldots, m_t$ respectively. The complex, nonreal zeros of $f(x)$ occur in conjugate pairs. Let $z_1, \bar{z}_1, z_2, \bar{z}_2, \ldots, z_k, \bar{z}_k$ be the distinct complex, nonreal zeros of $f(x)$, if any, with multiplicities $n_1, n_2, \ldots, n_k$, respectively. (Note: Here we use Lemma 8.3.6.)

By Theorem 8.3.2, we can write:

$$f(x) = a(x - r_1)^{m_1}(x - r_2)^{m_2} \ldots (x - r_t)^{m_t}(x - z_1)^{n_1}(x - \bar{z}_1)^{n_1} \\ \ldots (x - z_k)^{n_k}(x - \bar{z}_k)^{n_k},$$

where $a \in \mathbf{R}$ is the leading coefficient of $f(x)$.

Let $g_i(x) = (x - z_i)(x - \bar{z}_i)$ for $i = 1, 2, \ldots, k$.

Then $g_i(x) = x^2 - (z_i + \bar{z}_i)x + z_i\bar{z}_i$, which is a polynomial in $\mathbf{R}[x]$, since both $z_i + \bar{z}_i$ and $z_i\bar{z}_i$ are real numbers. Also, $g_i(x)$ has no real zeros so it must be irreducible and have negative discriminant by Lemma 8.3.5. ●

Corollary 8.3.8 No polynomial in $\mathbf{R}[x]$ of degree at least 3 is irreducible.

PROOF: We leave it as an exercise. ●

Factoring polynomials of degree 3 or better over the reals is in general a difficult problem. Even finding zeros of such polynomials is difficult. The following result from first-year calculus, a special case of *Rolle's Theorem,* is helpful.

Lemma 8.3.9 Let $f(x) \in \mathbf{R}[x]$. Between any two distinct real zeros of $f(x)$ there is a point where its derivative $Df(x)$ is zero. Therefore if $f(x)$ has m distinct real zeros, then $Df(x)$ has at least $m - 1$ real zeros.

Another result of calculus, a consequence of the *Intermediate Value Theorem,* says the following:

Lemma 8.3.10 Let $f(x) \in \mathbf{R}[x]$. If $f(r_1) > 0$ and $f(r_2) < 0$ for some r_1 and r_2 in $\mathbf{R}$, then $f(x)$ has at least one zero between r_1 and r_2.

EXAMPLE 2 Let $f(x) = x^3 + 4x - 9$. Then $Df(x) = 3x^2 + 4$, which is never 0. It follows from Lemma 8.3.9 that $f(x)$ has at most one real zero. Since $f(1) = -4$ and $f(2) = 7$, then by Lemma 8.3.10, $f(x)$ has a zero between 1 and 2.

Polynomials over Q

In general, determining whether or not a polynomial is irreducible over the field of rational numbers $\mathbf{Q}$ is a difficult problem. One way to factor a polynomial is to find zeros, although, as we have seen, a polynomial with no zeros in a particular field may still be factorable over that field. In what immediately follows we will concern ourselves with ways of finding rational zeros of polynomials in $\mathbf{Q}[x]$. Then we will consider some tests for irreducibility.

Let $g(x) \in \mathbf{Q}[x]$. If we multiply $g(x)$ by the least common denominator of its coefficients, we get a polynomial with integer coefficients. More specifically, there is an integer s such that $f(x) = sg(x)$ is a polynomial with integer coefficients. Clearly, any zero of $g(x)$ is a zero of $f(x)$ and vice versa. Therefore, in developing criteria for finding zeros of polynomials in $\mathbf{Q}[x]$, it suffices to consider polynomials with integer coefficients. We will let $\mathbf{Z}[x]$ denote the set of such polynomials. Note that $\mathbf{Z}[x]$ is closed under addition and multiplication of polynomials.

Theorem 8.3.11 Let $f(x) = a_0 + a_1x + a_2x^2 + \ldots + a_nx^n \in \mathbf{Z}[x]$. Let $\frac{r}{s}$ be a zero of $f(x)$ in $\mathbf{Q}$, where $(r, s) = 1$. Then r divides a_0 and s divides a_n.

PROOF: Substituting $\frac{r}{s}$ for x and multiplying through by s^n gives the equations

$$a_nr^n + a_{n-1}r^{n-1}s + \ldots + a_1rs^{n-1} + a_0s^n = 0,$$
$$a_0s^n = -r(a_nr^{n-1} + a_{n-1}r^{n-2}s + \ldots + a_1s^{n-1}).$$

This last equation implies that r divides a_0s^n and since $(r, s) = 1$, we must have that r divides a_0. (Why?)

We can also write

$$a_nr^n = -s(a_{n-1}r^{n-1} + \ldots + a_1rs^{n-2} + a_0s^{n-1}).$$

Thus s divides a_nr^n and since $(r, s) = 1$, s divides a_n. ●

Corollary 8.3.12 Let $f(x)$ be a monic polynomial in $\mathbf{Z}[x]$. Then any rational zero of $f(x)$ must be an integer that divides the constant term of $f(x)$.

PROOF: We leave it as an exercise. ●

EXAMPLE 3 Let $f(x) = 30x^3 - 17x^2 - 3x + 2$. By Theorem 8.3.11, if $\frac{r}{s}$ is a rational zero of $f(x)$, then $r \mid 2$ and $s \mid 30$. The possible values of r and s are: $r = \pm 1$ or

± 2 and $s = \pm 1, \pm 2, \pm 3, \pm 5, \pm 6, \pm 10, \pm 15,$ or ± 30. Plugging in values, we get that the zeros of $f(x)$ are $\frac{1}{2}, -\frac{1}{3}$, and $\frac{2}{5}$. We can write

$$f(x) = 30\left(x - \frac{1}{2}\right)\left(x + \frac{1}{3}\right)\left(x - \frac{2}{5}\right)$$
$$= (2x - 1)(3x + 1)(5x - 2).$$

EXAMPLE 4 Let $f(x) = 12x^3 - 3x + 2$. Any rational zero $\frac{r}{s}$ of $f(x)$ must satisfy $r = \pm 1$, ± 2 and $s = \pm 1, \pm 2, \pm 3, \pm 4, \pm 6, \pm 12$. After trying all possible values, we see that $f(x)$ has no rational zeros.

We can then ask how many real zeros $f(x)$ has. Simply graphing $f(x)$ using methods of first-year calculus reveals that $f(x)$ has only one real zero. Therefore, $f(x)$ has one irrational real zero and two nonreal complex zeros. We can also conclude that $f(x)$ is irreducible over **Q** but not over **R**.

EXAMPLE 5 We will use Corollary 8.3.12 to prove that $\sqrt{2}$ is irrational. (Recall that we proved this another way in Section 5.4.) Since $\sqrt{2}$ is a zero of the polynomial $x^2 - 2$ in **Z**$[x]$, Corollary 8.3.12 tells us that if $\sqrt{2}$ is rational it must be an integer that divides 2. But clearly ± 1 and ± 2 are not zeros of $x^2 - 2$.

We now give some criteria for a polynomial $f(x)$ in **Z**$[x]$ to be irreducible over **Q**. We first state without proof a result due to Gauss.

Theorem 8.3.13 Let $f(x) \in$ **Z**$[x]$. If $f(x)$ is irreducible over **Z**, then $f(x)$ is irreducible over **Q**.

Theorem 8.3.13 says that if a polynomial with integer coefficients cannot be factored into two polynomials of smaller degree with integer coefficients, then it cannot be factored into two polynomials of smaller degree with rational coefficients.

Theorem 8.3.14 **Eisenstein Irreducibility Criterion** Let $f(x) = a_0 + a_1x + a_2x^2 + \ldots + a_nx^n \in$ **Z**$[x]$. Suppose there exists a prime p such that p divides a_j for all $j = 0, 1, 2, \ldots, n - 1$, p does not divide a_n, and p^2 does not divide a_0. Then $f(x)$ is irreducible over **Q**.

PROOF: The proof is by contradiction. Suppose $f(x)$ factors into two polynomials $g(x)$ and $h(x)$ in **Z**$[x]$ of smaller degree than $f(x)$.

Let $g(x) = b_0 + b_1x + b_2x^2 + \ldots + b_mx^m$ and $h(x) = c_0 + c_1x + c_2x^2 + \ldots + c_rx^r$. Then $a_0 = b_0c_0$, and since p divides a_0, p must divide b_0 or c_0. Also, since p^2 does not divide a_0, p cannot divide both b_0 and c_0. For the sake of argument, we assume p divides c_0 but not b_0.

Now $a_n = b_mc_r$, so p does not divide b_m and p does not divide c_r. Let k be the smallest integer such that p does not divide c_k. Then $0 < k \leq r < n$.

Consider the coefficient $a_k = b_0c_k + b_1c_{k-1} + \ldots + b_kc_0$. Since p divides the integers $a_k, c_0, c_1, \ldots, c_{k-1}$, it follows that p divides b_0c_k. But this is a contradiction since p does not divide b_0 and p does not divide c_k.

Therefore $f(x)$ must be irreducible over $\mathbf{Z}$, and by the previous theorem it must also be irreducible over $\mathbf{Q}$. ●

EXAMPLE 6   Let $f(x) = 7x^6 - 9x^4 + 6x^2 + 15$. Applying the Eisenstein Criterion with $p = 3$, we get that $f(x)$ is irreducible over $\mathbf{Q}$.

EXAMPLE 7   Let $f(x) = x^n + p$, where n is some positive integer and p is a prime. By the Eisenstein Criterion, $f(x)$ is irreducible over $\mathbf{Q}$. Thus we see that $\mathbf{Q}[x]$ contains irreducible polynomials of arbitrarily high degree.

The method of determining irreducibility that we consider next is called *reducing mod p*.

Definition 8.3.15 Let $f(x) = a_0 + a_1x + a_2x^2 + \ldots + a_nx^n \in \mathbf{Z}[x]$, deg $f(x) \geq 1$. Let p be a prime number. Then the polynomial

$$f_p(x) = [a_0] + [a_1]x + [a_2]x^2 + \ldots + [a_n]x^n \in \mathbf{Z}_p[x]$$

is called the **reduction** of $f(x)$ mod p. As usual, the notation $[a]$ means the congruence class of a mod p.

Lemma 8.3.16   Let $f(x), g(x), h(x) \in \mathbf{Z}[x]$.

1. deg $f_p(x) \leq$ deg $f(x)$ and deg $f_p(x) =$ deg $f(x)$ if and only if p does not divide a_n.
2. If $f(x) = g(x)h(x)$, then $f_p(x) = g_p(x)h_p(x)$.

PROOF: We leave it as an exercise. ●

Theorem 8.3.17   Let $f(x) \in \mathbf{Z}[x]$, deg $f(x) \geq 1$. Let p be a prime and let $f_p(x)$ be the reduction of $f(x)$ mod p. If $f_p(x)$ is irreducible over $\mathbf{Z}_p$ and deg $f_p(x) =$ deg $f(x)$, then $f(x)$ is irreducible over $\mathbf{Q}$.

PROOF: It suffices to prove that $f(x)$ is irreducible over $\mathbf{Z}$. Suppose, on the contrary, that $f(x) = g(x)h(x)$ where $g(x)$ and $h(x)$ are polynomials in $\mathbf{Z}[x]$ both of degree less than the degree of $f(x)$.

Then $f_p(x) = g_p(x)h_p(x)$ and

deg $g_p(x) \leq$ deg $g(x) <$ deg $f(x) =$ deg $f_p(x)$,

deg $h_p(x) \leq$ deg $h(x) <$ deg $f(x) =$ deg $f_p(x)$.

This implies that $f_p(x)$ is not irreducible over $\mathbf{Z}_p$, a contradiction. ●

EXAMPLE 8   Let $f(x) = x^4 + 6x^3 - 5x + 3$. We try reduction mod 2 and get $f_2(x) = x^4 + x + [1]$. If $f_2(x)$ has a linear factor, then it must have a zero in $\mathbf{Z}_2$. But $f_2([0]) = [1]$ and $f_2([1]) = [1]$.

Thus if $f_2(x)$ factors, it must have two irreducible quadratic factors. But the only irreducible polynomial of degree 2 in $\mathbf{Z}_2[x]$ is $x^2 + x + [1]$. (Check. See Exercise 20 of Section 8.2.) A calculation shows that $f_2(x) \neq (x^2 + x + [1])(x^2 + x + [1])$.

Therefore $f_2(x)$ is irreducible over $\mathbf{Z}_2$ and by the previous theorem $f(x)$ is irreducible over $\mathbf{Q}$.

EXAMPLE 9   Let $f(x) = x^3 + 17x + 36$. One way to check irreducibility over $\mathbf{Q}$ is by determining whether $f(x)$ has any zeros in $\mathbf{Q}$. As we have seen, it suffices to look at the divisors of 36. Since there are many of these, we will try reduction mod p instead.

$f_2(x) = x^3 + x$ is not irreducible over $\mathbf{Z}_2$.

$f_3(x) = x^3 + [2]x$ is not irreducible over $\mathbf{Z}_3$.

$f_5(x) = x^3 + [2]x + [1]$ has no zeros in $\mathbf{Z}_5$ and therefore is irreducible over $\mathbf{Z}_5$.

Thus $f(x)$ is irreducible over $\mathbf{Q}$.

Note: To apply the reduction method, we just have to show that $f_p(x)$ is irreducible for some p that does not divide the leading coefficient of $f(x)$. However, it is possible to have a polynomial $f(x)$ in $\mathbf{Z}[x]$ that is irreducible over $\mathbf{Q}$ but for which $f_p(x)$ is not irreducible over $\mathbf{Z}_p$ for any prime p. An example is $x^4 + 1$. The proof that this is not irreducible over $\mathbf{Z}_p$ for any prime p is nontrivial.

HISTORICAL COMMENTS: GALOIS'S THEORY

In Section 8.1 we noted that Abel proved that the general polynomial equation of degree greater than 4 is not solvable by radicals. But the story does not end there. Some polynomial equations of degree greater than 4 *can* be solved by radicals. For example, Gauss was able to prove that the equation $x^n - 1 = 0$, the so-called binomial equation, is solvable by radicals. (We saw in Section 7.3 that the roots of this equation are the nth roots of unity.)

What Gauss did not do was give a criterion for determining whether or not *any* given polynomial equation is solvable by radicals. This more difficult problem was solved by **Evariste Galois** (1811–1832).

Galois was a young genius whose work on the theory of polynomial equations was either ignored or misunderstood by the leading mathematicians of his time. Two of his papers were submitted to Cauchy, who lost them, and another to Fourier, who died soon after, with the result that this paper was also lost. Another of Galois's papers was sent to Poisson, who returned it because he thought it unintelligible. Unfortunately much of Galois's time was spent in prison because of radical political activities, and in May 1832 he was challenged to a duel. The night before the duel was to take place, Galois wrote up as best he could the results of his work. The next day he was killed in the duel. Years later, his work was recognized for its significance and was published in leading mathematical journals. An entire branch of abstract algebra developed from Galois's work and today is known as Galois Theory. Its consequences are still being felt in modern mathematical research.

Exercises 8.3

1. Prove Corollary 8.3.3.

2. Prove Lemma 8.3.5.

3. Prove that every polynomial in $\mathbf{R}[x]$ of odd degree has at least one real zero.
   (Hint: The Intermediate Value Theorem will help.)

4. Determine the number of real zeros of the following polynomials. Find the integer closest to each zero.
   (a) $f(x) = 2x^3 + 3x^2 - 12x + 4$
   (b) $f(x) = x^3 - 9x^2 + 27x - 2$
   (c) $f(x) = x^5 - 20x + 28$

5. Let $f(x) = x^4 - 2x^3 + 4x^2 - 3x + 2$.
   (a) Factor $f(x)$ into irreducibles over $\mathbf{R}$.
      (Hint: The irreducible factors have integer coefficients.)
   (b) Prove that $f(x)$ has no real zeros.
   (c) Find all of the complex zeros of $f(x)$.

6. Let ω be the primitive eighth root of unity $\cos\dfrac{2\pi}{8} + i\sin\dfrac{2\pi}{8}$.
   (a) Find a fourth-degree polynomial with integer coefficients of which ω is a zero.
   (b) Find a second-degree polynomial with real coefficients of which ω is a zero.
      (Hint: Factor the polynomial $x^8 - 1$.)

7. Prove Corollary 8.3.8.

8. Let $f(x) \in \mathbf{R}[x]$. Lemma 8.3.9 says that if $f(x)$ has m distinct real zeros, then its derivative $Df(x)$ has at least $m - 1$ real zeros. Is the converse of this statement true? Prove your answer.

9. Let $f(x) \in \mathbf{R}[x]$. Prove that if $Df(x)$ has n real zeros, then $f(x)$ has at most $n + 1$ real zeros.

10. Prove Lemma 8.3.10.

11. Let $g(x) \in \mathbf{Q}[x]$.
    (a) Prove that there exists a nonzero integer s such that the polynomial $f(x) = sg(x)$ is a polynomial with integer coefficients.

(b) Prove that c is a zero of $g(x)$ if and only if c is a zero of $f(x)$.

12. Find the rational zeros, if any, of the following polynomials and then find their irreducible factors over **Q**, **R**, and **C** respectively.
    (a) $x^3 - x^2 - 3x + 2$
    (b) $x^4 - 6x^3 + 7x^2 - 12x + 10$
    (c) $12x^3 - 8x^2 - 3x + 2$

13. Use the method of Example 5 to prove the following.
    (a) Prove that $\sqrt{3}$ is irrational.
    (b) Prove that $\sqrt[3]{2}$ is irrational.
    (c) Prove that $\sqrt[n]{2}$ is irrational for every $n \in \mathbf{Z}, n \geq 2$.
    (d) Prove that if p is a prime number, then $\sqrt[n]{p}$ is irrational for every $n \in \mathbf{Z}, n \geq 2$.

14. Let $f(x) = x^3 - 3x + 5$. Show that $f(x)$ has one irrational real zero and two complex nonreal zeros.

15. Prove Corollary 8.3.12.

16. State the converse of Theorem 8.3.13. Prove or disprove.

17. Prove Lemma 8.3.16.

18. Prove that the following polynomials are irreducible over **Q**.
    (a) $f(x) = 4x^5 - 7x^3 + 21x^2 + 28$
    (b) $f(x) = x^4 + 7x^3 - 9x^2 + 3x - 1$
    (c) $f(x) = x^3 + 4x^2 - 9x + 96$

19. Let $f(x) = x^4 + 1$.
    (a) Prove that $f(x)$ is irreducible over **Q**.
    (b) Prove that $f_p(x)$ is not irreducible over $\mathbf{Z}_p$ for $p = 2, 3, 5,$ and 7.
    (c) Factor $f(x)$ in $\mathbf{R}[x]$.
    (d) Find all the zeros of $f(x)$ in **C**.
    (e) Use part (c) and the method of partial fractions to compute the indefinite integral $\int \dfrac{1}{x^4 + 1}\, dx$.

Discussion and Discovery Exercises

D1. Look up the proof of Lemma 8.3.9 in a calculus text and write out the proof.

D2. One consequence of Galois's theory of equations is the following theorem: Let $f(x) \in \mathbf{Q}[x]$ be irreducible over **Q** and have degree n where n is a prime number greater than 3. Suppose that $f(x)$ has exactly two nonreal zeros in **C**. Then the equation $f(x) = 0$ is not solvable by radicals.

   Use this theorem to show that the polynomial equation $x^5 - 20x + 5 = 0$ is not solvable by radicals.

D3. In this problem we will see how to express the roots of the equation $x^5 - 1 = 0$ as radicals. The roots of this equation are the fifth roots of unity: $1, \zeta, \zeta^2, \zeta^3, \zeta^4$, where $\zeta = \cos\dfrac{2\pi}{5} + i\sin\dfrac{2\pi}{5}$.

   (a) Prove that ζ^4 is the complex conjugate of ζ.
   (b) Let $\alpha = \zeta + \zeta^4$. Prove that α is a zero of the polynomial $x^2 + x - 1$.
   (c) Prove that $\cos\dfrac{2\pi}{5} = \dfrac{-1 + \sqrt{5}}{4}$.
   (d) Prove that $x^5 - 1 = 0$ is solvable by radicals.
   (e) Use your answer to (c) to construct a regular pentagon inside a circle of radius 1 using only a compass and straightedge.

D4. A complex number α is called an *algebraic number* if α is a zero of a polynomial $f(x)$ in $\mathbf{Q}[x]$, where $f(x)$ is not the zero polynomial. For instance, $\sqrt{2}$ and i are complex numbers since they are zeros of the polynomials $x^2 - 2$ and $x^2 + 1$, respectively. On the other hand, π is not an algebraic number, although this fact is not easy to prove.

   Let α be an algebraic number. Let $S_\alpha = \{f(x) \in \mathbf{Q}[x] \mid f(\alpha) = 0\}$; that is, S_α is the set of all polynomials with rational coefficients that have α as a zero. (Note that the zero polynomial is an element of S_α.)

   Let $P_\alpha(x)$ be the monic polynomial of smallest degree in S_α. Then $P_\alpha(x)$ is called the *minimal polynomial* of α.

   The polynomial $x^2 - 2$ is the minimal polynomial of $\sqrt{2}$ since no polynomial of smaller degree with rational coefficients has $\sqrt{2}$ as a zero. $\sqrt{2}$ is a zero of the polynomial $x - \sqrt{2}$ but this polynomial does not have rational coefficients. Similarly, the minimal polynomial of i is $x^2 + 1$.

   (a) Explain why the minimal polynomial of the algebraic number α has to exist.
   (b) Prove that the elements of S_α are precisely the polynomials that have $P_\alpha(x)$ as a factor; that is, if $f(x) \in S_\alpha$, then $P_\alpha(x)$ divides $f(x)$. *(Hint: Use the Division Algorithm.)*
   (c) Prove that $P_\alpha(x)$ is irreducible.
   (d) Prove that $P_\alpha(x)$ is the only monic polynomial of smallest degree in S_α.
   (e) Prove that if $g(x)$ is a monic irreducible polynomial in $\mathbf{Q}[x]$ and $g(\alpha) = 0$, then $g(x) = P_\alpha(x)$.

D5. Using the notation of the previous problem, find $P_\alpha(x)$ if
   (a) $\alpha = \sqrt{3}$.
   (b) $\alpha = \sqrt[3]{5}$.
   (c) $\alpha = \sqrt[4]{2}$.
   (d) $\alpha = \sqrt{2} + \sqrt{3}$.

ANSWERS AND HINTS TO SELECTED EXERCISES

1. (a) statement
   (d) neither
   (g) statement
   (j) statement

2. (a) This statement has a universal quantifier: for all rectangles T, the area of T is its length times its width.
   (d) This statement has a universal quantifier: for all integers n and m, if n is even and m is odd, then $n + m$ is even.
   (g) This statement has a universal quantifier: for all real-valued functions f, if f is differentiable on $[a, b]$, then f is continuous on $[a, b]$.
   (i) This statement has a universal quantifier and an existential quantifier: for every positive real number x, there is a real number y such that $y = \sqrt{x}$.

3. (a) There is a rectangle T such that the area of T is not its length times its width.
   (d) There is an even integer n and an odd integer m such that $n + m$ is odd.
   (g) There is a real-valued function f that is differentiable on $[a, b]$, but is not continuous on $[a, b]$.
   (i) There is a positive real number x that has no square root. Or: there is a positive real number x such that for all real numbers y, $y \neq \sqrt{x}$.

5. (a) For every real number x, $x^2 + x + 1 \neq 0$.
   (d) There is a polynomial function f that is not continuous at some point.

7. (a) For all integers n and m, if n and m are even, then $n + m$ is divisible by 4.

295

9. (a) A real-valued function f is *not increasing* on $[a, b]$ if there exist x_1, $x_2 \in [a, b]$ such that $x_1 < x_2$ and $f(x_1) \geq f(x_2)$.

Exercises 1.2

2. (a)

P	Q	$(\neg P) \vee Q$
T	T	T
T	F	F
F	T	T
F	F	T

(c)

P	Q	$\neg[(\neg P) \wedge Q]$
T	T	T
T	F	T
F	T	F
F	F	T

4. (a) Either August is sometimes not a hot month or September is never cool.

(c) All cars are either uncomfortable or expensive.

5. (a) There exist real numbers x and y such that $xy = 0$ but $x \neq 0$ and $y \neq 0$.

(d) For every rational number r, either $r \leq 1$ or $r \geq 2$.

7. (a) For all integers n and m, if nm is even, then n is even and m is even.

10. (a)

P	Q	$P \wedge Q$	$Q \wedge P$
T	T	T	T
T	F	F	F
F	T	F	F
F	F	F	F

11. (a) S and T are equivalent statements. We will show that S and T are either both true or both false. Suppose that S is a true statement. Then if a is an assigned value of the variable x, $P(a) \wedge Q(a)$ is true. Hence $P(a)$ is true and $Q(a)$ is true. It follows that $\forall x, P(x)$ is a true statement and $\forall x, Q(x)$ is a true statement. Therefore, $(\forall x, P(x)) \wedge (\forall x, Q(x))$ is a true statement.

Now suppose that T is true. Then $\forall x, P(x)$ is true and $\forall x, Q(x)$ is true. Let a be an assigned value of the variable x. Then $P(a)$ is true and $Q(a)$ is true. Thus $P(a) \wedge Q(a)$ is true. It follows that S is a true statement.

12. (a)

P	Q	R	$(P \wedge \neg Q) \vee \neg(R)$
T	T	T	F
T	T	F	T
T	F	T	T
T	F	F	T
F	T	T	F
F	T	F	T
F	F	T	F
F	F	F	T

15. (a) By Exercise 13(c), $P \vee [(Q \wedge R) \wedge S)]$ is equivalent to $(P \vee (Q \wedge R)) \wedge (P \vee S)$, which in turn is equivalent to $((P \vee Q) \wedge (P \vee R)) \wedge (P \vee S)$. By Exercise 13(b), this last statement form is equivalent to $(P \vee Q) \wedge (P \vee R) \wedge (P \vee S)$.

16. (a) If P is true, then either $P \wedge Q$ or $P \wedge \neg Q$ is true depending on whether or not Q is true or false. Hence, if P is true, the statement form $(P \wedge Q) \vee (P \wedge \neg Q) \vee (\neg P \wedge Q) \vee (\neg P \wedge \neg Q)$ is true. On the other hand, if P is false, then either $\neg P \wedge Q$ or $\neg P \wedge \neg Q$ is true, again depending on the truth or falsity of Q. Thus if P is false, $(P \wedge Q) \vee (P \wedge \neg Q) \vee (\neg P \wedge Q) \vee (\neg P \wedge \neg Q)$ is true. It follows that $(P \wedge Q) \vee (P \wedge \neg Q) \vee (\neg P \wedge Q) \vee (\neg P \wedge \neg Q)$ is a tautology.

Exercises 1.3

1. (a) P: For every polygon T, if T is a hexagon, then T has six sides.

       $\neg P$:   There is a hexagon T such that T does not have six sides.

  (d) P: For every real number x, if x is positive, then there is a real number y such that $y = \sqrt{x}$.

       $\neg P$:   There is a real number x such that x is positive and for all real numbers y, $y \neq \sqrt{x}$.

3. (a)

P	Q	$\neg(P \Rightarrow Q)$	$P \wedge \neg Q$
T	T	F	F
T	F	T	T
F	T	F	F
F	F	F	F

5. (a) For every integer n, if n is odd, then n^3 is odd.

  (c) The original statement is true. Let n be an odd integer. Then we can write $n = 2t + 1$ for some $t \in \mathbf{Z}$. Then $n^3 = (2t + 1)^3 = 8t^3 + 12t^2 + 6t + 1 = 2(4t^3 + 6t^2 + 3t) + 1$, so that n^3 is odd.

7. (a) For all integers x, y, and z, if x, y, and z are consecutive integers, then $x^2 + y^2 + z^2$ is even.

(c) The original statement is false since $2^2 + 3^2 + 4^2 = 29$, and 29 is odd.

11. Suppose that $[\forall x(P(x) \Rightarrow Q(x))] \wedge [\forall x(Q(x) \Rightarrow R(x))]$ is a true statement. Then the statements $\forall x(P(x) \Rightarrow Q(x))$ and $\forall x(Q(x) \Rightarrow R(x))$ are both true statements. Let a be an assigned value of the variable x. Then the statements $P(a) \Rightarrow Q(a)$ and $Q(a) \Rightarrow R(a)$ are both true. By Exercise 10, the statement $P(a) \Rightarrow R(a)$ is true. It now follows that the statement $\forall x(P(x) \Rightarrow R(x))$ is true.

12. We compare truth tables.

P	Q	R	$P \Rightarrow (Q \vee R)$	$(P \wedge \neg Q) \Rightarrow R$
T	T	T	T	T
T	T	F	T	T
T	F	T	T	T
T	F	F	F	F
F	T	T	T	T
F	T	F	T	T
F	F	T	T	T
F	F	F	T	T

17. (a) The statement "P is a sufficient condition for Q" is the statement "If n is a multiple of 4, then n^2 is a multiple of 4." This is a true statement. If n is a multiple of 4, then we can write $n = 4t$ for some integer t. Then $n^2 = (4t)^2 = 4(4t^2)$, proving that n^2 is a multiple of 4.

(d) The statement "Q is a necessary condition for P" is logically equivalent to the statement "P is a sufficient condition for Q" and therefore by part (a) it must be a true statement.

Exercises 1.4

1.

P	Q	$\neg Q \Rightarrow \neg P$	$P \Rightarrow Q$
T	T	T	T
T	F	F	F
F	T	T	T
F	F	T	T

4. (a) *Contrapositive*: If the food is not spoiled, then the power is on.
   Converse: If the food is spoiled, then the power went off.

6. (a) Because n and m are even, we can write $n = 2t$ and $m = 2s$ for some $t, s \in \mathbf{Z}$. Then $n + m = 2(t + s)$. Therefore, $n + m$ is even.

8. Jonathan cannot conclude anything.

12. (a) The statement $\forall n, m \in \mathbf{Z}, P(n, m) \Rightarrow Q(n, m)$ is true. Let $n, m \in \mathbf{Z}$ and suppose n and m are odd. Then there exist integers s and t such that $n = 2s + 1$ and $m = 2t + 1$. Thus $nm = (2s + 1)(2t + 1) = 4st + 2t + 2s + 1 = 2(2st + t + s) + 1$. Since $2st + t + s$ is an integer, it follows that nm is odd.

  (b) The statement $\forall n, m \in \mathbf{Z}, Q(n, m) \Rightarrow P(n, m)$ is true. Prove it by proving its contrapositive.

15. (a) This statement is false. The negation is the statement "There exists an integer n such that n is even but $\sqrt{n}$ is not an even integer." Since the integer 2 is even and $\sqrt{2}$ is not an even integer because it is not an integer, the negation is true.

22. (a) We first suppose that P and Q are logically equivalent. We must prove that the statement $P \Leftrightarrow Q$ is always true. If P is true, then Q is true, since P and Q are logically equivalent. Hence from its truth table $P \Leftrightarrow Q$ is true. On the other hand, if P is false, then Q is false and therefore $P \Leftrightarrow Q$ is again true by its truth table. It now follows that $P \Leftrightarrow Q$ is a tautology.

  Now suppose that $P \Leftrightarrow Q$ is a tautology. Then $P \Leftrightarrow Q$ is never false. Now $P \Leftrightarrow Q$ is false precisely when either P is true and Q is false or P is false and Q is true. Thus P and Q are either both true or both false and hence are logically equivalent.

Exercises 2.1

1. (a) $\{15, 16, 17, 18, 19, 20, 21\}$
   (d) $\{\pi, 4\pi, 9\pi\}$
   (g) $\{2, 5, 13, 17, 29, 37, 41, 53, 61, 73\}$

2. (a) $\{11, 12, 13, 14, 15, \ldots\}$
   (d) $\{3, 7, 11, 15, 19, 23, \ldots\}$
   (g) $\{0, \pi, -\pi, 2\pi, -2\pi, 3\pi, -3\pi, \ldots\}$

4. (a) $\{x \in \mathbf{Z} \mid |x| \le 6\}$
   (d) $\{x \in \mathbf{Z} \mid x = 4t \text{ for some } t \in \mathbf{Z}\}$
   (g) $\left\{ n\dfrac{\pi}{2} \,\middle|\, n \in \mathbf{Z} \right\}$ or $\left\{ x \in \mathbf{R} \,\middle|\, x = n\dfrac{\pi}{2} \text{ for some } n \in \mathbf{Z} \right\}$

5. (a) $\{x \in \mathbf{Z} \mid x = 5t \text{ for some } t \in \mathbf{Z}\}$

7. (a) $[-1, 5)$ (d) $(-\infty, -1)$ (g) $\varnothing$

8. (a) $\{x \in \mathbf{R} \mid -4 < x < 4\}$
   (e) $\{x \in \mathbf{R} \mid x > 16 \text{ or } x < -2\}$

9. (a) For all integers n, if n is a multiple of 9, then n is a multiple of 3.

(d) *Negation*: There is an integer n such that n is a multiple of 9 and n is not a multiple of 3.

(f) *Converse*: For all integers n, if n is a multiple of 3, then n is a multiple of 9. The converse is false because 3 is a multiple of 3 and not a multiple of 9.

10. (a) Let $x \in A$. Then $x = ns$ for some $s \in \mathbf{Z}$. Since n is a multiple of m, there exists $t \in \mathbf{Z}$ such that $n = tm$. Then $x = ns = tms = m(ts) \in B$. Thus $A \subseteq B$.

17. (a) Let A be the set of all hexagons and B the set of all polygons with six sides. U is the set of all curves in the plane.

(d) Let $A = \{x \in \mathbf{Z} \mid x = 6t$ for some integer $t\}$ and $B = \{x \in \mathbf{Z} \mid x$ is even$\}$. $U = \mathbf{Z}$.

18. (a) If an integer n is not a multiple of 3, then n is not a multiple of 9.

(c) If a polygon does not have six sides, then it is not a hexagon.

(f) For all integers x, if x is odd, then $x \neq 6t$ for all integers t.

19. (a) $\{x \in \mathbf{Z} \mid x \leq 5\}$

(d) $\{x \in \mathbf{R} \mid x \leq -2$ or $x > 4\}$

20. (a) $A - B = \{1, 2, 3, 4, 5, 8, 10\}$

$B - A = \{11, 12\}$

(e) $A - B = \{x \in \mathbf{R} \mid x \geq 10\}$

$B - A = \{x \in \mathbf{R} \mid x \leq 4\}$

21. Only (b) and (d) are equivalent to $A \subseteq B$.

Exercises 2.2

---

1. (a) $A \cup B = \{0, 1, 2, 3, 4, 5, 8\}$

(d) $A - B = \{2, 3, 5\}$

(g) $A \cap (B \cup C) = \{1, 2, 4, 5\}$

(j) $(A \cup B) \cap (A \cup C) = \{1, 2, 3, 4, 5\}$

2. (a) $\overline{A} = \{0, 6, 7, 8, \ldots, 20\}$

(d) $\overline{A} \cup \overline{B} = \{0, 2, 3, 5, 6, \ldots, 20\}$

4. (a) $(0, 4]$

(d) $[2, 4]$

5. (a) $(-\infty, 1) \cup (4, \infty)$

(d) $(0, 1)$

7. (a) *Hint*: Let $A = (a, b)$ and $B = (c, d)$ where $a \leq c$. Show that $c < b$ and then prove that $A \cup B = (a, b)$ or (a, d).

9. (a) $A \cup B = \{n \in \mathbf{Z} \mid n$ is even or n is a multiple of 5$\}$

(d) $B - A = \{n \in \mathbf{Z} \mid n$ is even but n is not a multiple of 10$\}$

13. (a) Let $x \in A \cap B$. Then $x \in A$ and $x \in B$. Since $x \in A$, $x \leq 10$ and

since $x \in B$, $x > 5$. So, $5 < x \leq 10$. It follows that $x \in (5, 10]$. Therefore, $A \cap B \subseteq (5, 10]$.

Conversely, let $x \in (5, 10]$. Then $5 < x \leq 10$, implying that $x \leq 10$ and $x > 5$. Hence $x \in A$ and $x \in B$ or in other words, $x \in A \cap B$. Thus $(5, 10] \subseteq A \cap B$. It now follows that $A \cap B = (5, 10]$.

14. (a) Let $x \in A \cap (A \cup B)$. Then $x \in A$ and $x \in A \cup B$. In particular, $x \in A$, so $A \cap (A \cup B) \subseteq A$. Conversely, let $x \in A$. Then $x \in A \cup B$ by part 5 of Proposition 2.2.2. Since x is also in A, we have that $x \in A \cap (A \cup B)$. Therefore, $A \subseteq A \cap (A \cup B)$. It follows that $A \cap (A \cup B) = A$.

16. (a) We want to prove that if A is a subset of C and B is a subset of C, then $A \cup B$ is a subset of C. Let $x \in A \cup B$. Then $x \in A$ or $x \in B$. If $x \in A$, then $x \in C$ since A is a subset of C. If $x \in B$, then $x \in C$ since B is also a subset of C. In either case, x is an element of C. Therefore, $A \cup B \subseteq C$.

22. (a) $\begin{aligned}(A \cup B) \cap \overline{A} &= \overline{A} \cap (A \cup B)\\ &= (\overline{A} \cap A) \cup (\overline{A} \cap B)\\ &= \varnothing \cup (\overline{A} \cap B)\\ &= \overline{A} \cap B\\ &= B - A \text{ by Example 5.}\end{aligned}$

24. (a) First suppose that $A = A - B$. We want to prove that $A \cap B = \varnothing$. Suppose, on the contrary, that $A \cap B \neq \varnothing$. Then there exists $x \in A \cap B$. Since $x \in A \cap B$, $x \in A$ and $x \in B$. But if $x \in A$, then $x \in A - B$ since $A = A - B$. This implies that $x \notin B$. Now we have the contradiction that $x \in B$ and $x \notin B$. Therefore, we can conclude that $A \cap B = \varnothing$.

Conversely, suppose that $A \cap B = \varnothing$. We want to show that $A = A - B$. Any element of $A - B$ is an element of A, by the definition of $A - B$. So $A - B \subseteq A$. Now let $x \in A$. Then $x \notin B$, otherwise x would be in $A \cap B$, contradicting the fact that $A \cap B = \varnothing$. Thus $x \in A - B$. This proves that $A \subseteq A - B$. Therefore, $A = A - B$.

27. $A \times B = \{(-1, 7), (-1, 12), (-1, 19), (-1, 21), (0, 7), (0, 12), (0, 19), (0, 21)\}$.

28. *Hint:* Assume that $r + x$ is rational and derive a contradiction.

Exercises 2.3

1. (a) $\mathbf{P}(A) = \{\varnothing, \{4\}, \{7\}, \{10\}, \{4, 7\}, \{4, 10\}, \{7, 10\}, \{4, 7, 10\}\}$

3. $\mathbf{P}(\mathbf{P}(A)) = \{ \varnothing, \{\varnothing\}, \{\{0\}\}, \{\{1\}\}, \{\{0, 1\}\}, \{\varnothing, \{0\}\}, \{\varnothing, \{1\}\}, \{\varnothing, \{0, 1\}\}, \{\{0\}, \{1\}\}, \{\{0\}, \{0, 1\}\}, \{\{1\}, \{0, 1\}\}, \{\varnothing, \{0\}, \{1\}\}, \{\varnothing, \{0\}, \{0, 1\}\}, \{\varnothing, \{1\}, \{0, 1\}\}, \{\{0\}, \{1\}, \{0, 1\}\}, \{\varnothing, \{0\}, \{1\}, \{0, 1\}\} \}$

5. (a) false   (d) true   (g) true

7. $\displaystyle\bigcup_{i=1}^{n} A_i = \{1, 2, 3, \dots, n, n + 1\}$

11. $\displaystyle\bigcup_{i=1}^{\infty} A_i = [1, \infty), \bigcap_{i=1}^{\infty} A_i = \varnothing$

13. $\displaystyle\bigcup_{i=1}^{\infty} A_i = [1, 2], \bigcap_{i=1}^{\infty} A_i = \{1\}$

20. One partition of **Z** is {**E**, **O**}, where **E** and **O** are the sets of even and odd integers, respectively. A partition of **R** is {{$x \in \mathbf{R} \mid x > 1$}, {$x \in \mathbf{R} \mid x < 1$}, {1}}.

27. (a) 1/13

28. (a) 3/5 is not in C.

Exercises 3.1

---

3. (a) Im $f = \mathbf{O}$, the set of odd integers
     $f(X) = \{4t + 1 \mid t \in \mathbf{Z}\}$
   (d) Im $f = [-1, 1]$
     $f([0, \frac{\pi}{2}]) = [0, 1]$
   (g) Im $f = \mathbf{Z}$
     $f(\mathbf{E} \times \mathbf{Z}) = \mathbf{E}$

4. (a) not a function
   (c) a function. Im $f = \{x \in \mathbf{R} \mid x \geq 0\}$
   (e) a function. Im $f = \mathbf{R}$

5. (a) By definition, Im $f \subseteq \mathbf{R}$, so we only have to show that $\mathbf{R} \subseteq$ Im f. Let $y \in \mathbf{R}$. Then $f(\frac{y-5}{6}) = 6(\frac{y-5}{6}) + 5 = y$. Thus $y \in$ Im f.

8. (a) We will prove that $f([0, 1]) = [0, 1]$. If $0 \leq x \leq 1$, then multiplying these inequalities by x gives $0 \leq x^2 \leq x \leq 1$, so that $f(x) \in [0, 1]$. Hence $f([0, 1]) \subseteq [0, 1]$. Conversely, suppose that $y \in [0, 1]$. Then $0 \leq y \leq 1$. Let $x = \sqrt{y}$. Then $y = f(x)$ and $x \geq 0$. If $x > 1$, then $x^2 = xx > 1x > 1$. But $y = x^2 \leq 1$. Therefore, $x \in [0, 1]$ and we have shown that $y \in f([0, 1])$. Hence $[0, 1] \subseteq f([0, 1])$. Therefore $f([0, 1]) = [0, 1]$.

10. (a) Let $n \in \mathbf{Z}$. If n is even, then $f(n) = n + 2$ is even. If n is odd, then $f(n) = 2n$ is even also. Therefore Im $f \subseteq \mathbf{E}$. Now let $y \in \mathbf{E}$. Then $y - 2$ is even and $f(y - 2) = y - 2 + 2 = y$, proving that $y \in$ Im f. Thus $\mathbf{E} \subseteq$ Im f. We can conclude that Im $f = \mathbf{E}$.

12. (a) Im $f = \mathbf{R}$
    (d) $f([-1, 2]) = [-6, 14]$

17. (a) $f^{-1}(B) = A$
    (d) $f^{-1}(\{b\}) = \varnothing$

18. (a) $f^{-1}(W_1) = \mathbf{O}$ $f^{-1}(W_2) = \{1\}$ $f^{-1}(W_3) = \{0\}$
    (d) $f^{-1}(W_1) = \varnothing$ $f^{-1}(W_2) = \mathbf{R}$ $f^{-1}(W_3) = \{0\}$
    (g) $f^{-1}(W_1) = \{0\}$ $f^{-1}(W_2) = \varnothing$

19. (a) $f^{-1}(\{11, 12, 13, 14, 15\}) = \{5, 22, 24, 26, 28, 30\}$

21. (a) By the definition of inverse image, $f^{-1}(f(X)) = \{a \in A \mid f(a) \in f(X)\}$. So if $a \in X$, then $f(a) \in f(X)$, proving that $a \in f^{-1}(f(X))$. Hence $X \subseteq f^{-1}(f(X))$.

Exercises 3.2

1. (a) f is surjective. By definition, Im $f \subseteq \mathbf{R}$, so we only have to show that $\mathbf{R} \subseteq$ Im f. Let $y \in \mathbf{R}$. Then $f(\frac{y-1}{2}) = 2(\frac{y-1}{2}) + 1 = y$. Thus $y \in$ Im f.
   (d) f is surjective. Let $y \in \mathbf{R}^*$. Let $x = \frac{1}{y}$. Then $f(x) = \frac{1}{x} = \frac{1}{1/y} = y$. So $y \in$ Im f.
   (g) f is surjective. Let $y \in \mathbf{R}^+$. Let $x = \ln y$. Note that x exists since $y > 0$. Then $f(x) = e^{\ln y} = y$. Therefore, $y \in$ Im f.
   (j) f is not surjective. For example, $(0, 1)$ is not in Im f.

2. (a) f is not surjective because $f(n)$ is odd for all n in $\mathbf{Z}$.
   (d) f is surjective. To see this, let $x \in \mathbf{Z}$. Let $n = 2x + 1$. Then n is odd so $f(n) = f(2x + 1) = \frac{2x+1-1}{2} = x$. So x is in the image of f.

12. (a) f is injective. Suppose that x and $y \in \mathbf{R}$ and $f(x) = f(y)$. Then $2x + 1 = 2y + 1$. It follows that $2x = 2y$ and therefore $x = y$.
    (d) f is injective. Suppose that $x, y \in \mathbf{R}^*$ and $f(x) = f(y)$. Then $\frac{1}{x} = \frac{1}{y}$ and thus $x = y$.
    (g) f is not injective since $f(1, -1) = f(2, -2) = 0$.

13. (b) f is injective. To prove this, suppose that $f(n) = f(m)$ where $n, m \in \mathbf{Z}$. If n, m are both even, then we get immediately that $n = m$. If n and m are both odd, then we get $2n - 1 = 2m - 1$, which implies that $n = m$. If n is even and m is odd, we get $n = 2m - 1$. But this cannot happen since n is even and $2m - 1$ is odd. Therefore, this case cannot occur. For the cases that can occur, we have shown that whenever $f(n) = f(m)$, then $n = m$. So f is injective.
    (d) f is not injective since $f(2) = f(5) = 2$.

24. (a) $f: \mathbf{R} \to \mathbf{R}$ defined by $f(x) = x$.
    $f: \mathbf{R} \to \mathbf{R}$ defined by $f(x) = x^3$.

$$f: \mathbf{R} \to \mathbf{R} \text{ defined by } f(x) = \begin{cases} x^2, & \text{if } x \geq 0 \\ -x^2, & \text{if } x < 0. \end{cases}$$

Exercises 3.3

1. (a) $fg(x) = \sin^2 x + 1$
   $gf(x) = \sin(x^2 + 1)$

2. (a) Let $n \in \mathbf{Z}$. If n is even, $(fg)(n) = f(g(n)) = f(n) = 2n$. If n is odd, $(fg)(n) = f(g(n)) = f(2n - 1) = 2(2n - 1) = 4n - 2$.
   For any $n \in \mathbf{Z}$, $(gf)(n) = g(f(n)) = g(2n) = 2n$.

3. (a) $g, h \in \mathbf{F}(\mathbf{R})$, $g(x) = x^8$, $h(x) = 3x^2 - 11$
   (d) $g: \mathbf{R} \to \mathbf{R}$, $g(x) = x - 1$; $h: \{0, 1\} \to \mathbf{R}$, $h(0) = 2$, $h(1) = 1$

7. (a) *Hint*: Since your objective is to show that f is injective, you should start the proof by letting $a, b, \in A$ and assume that $f(a) = f(b)$. Then prove that $a = b$. To do that you will have to use the fact that gf is injective in some way.

10. (a) Not invertible. f is not injective.
    (d) Invertible. $f^{-1}(x) = \frac{x-3}{7}$
    (g) Not invertible. f is not injective and not surjective.
    (j) Invertible. $f^{-1}(x) = \ln x$

11. (a) Let $x, y \in \mathbf{R}$, $x \neq 1$, and $y \neq 1$, and suppose that $f(x) = f(y)$. Then $\frac{x+1}{x-1} = \frac{y+1}{y-1}$, $(x - 1)(y + 1) = (x + 1)(y - 1)$, $xy - y + x - 1 = xy + y - x - 1$, $-y + x = y - x$, $2x = 2y$, $x = y$.

13. (a) The six permutations are defined by
    $f_1(a) = a, f_1(b) = b, f_1(c) = c$ $f_4(a) = b, f_4(b) = c, f_4(c) = a$
    $f_2(a) = a, f_2(b) = c, f_2(c) = b$ $f_5(a) = c, f_5(b) = a, f_5(c) = b$
    $f_3(a) = b, f_3(b) = a, f_3(c) = c$ $f_6(a) = c, f_6(b) = b, f_6(c) = a$

Exercises 4.1

1. (a) Let $a, b \in \mathbf{R} - \{0\}$. Then $a \div b = \frac{a}{b}$ is a real number since $b \neq 0$ and $\frac{a}{b} \neq 0$ since $a \neq 0$. Therefore, $a \div b \in \mathbf{R} - \{0\}$, proving that division is a binary operation on $\mathbf{R} - \{0\}$.

2. (a) i. Let $n, m, p \in \mathbf{Z}$. Then
   $(n * m) * p = (n + m + 2) * p = n + m + 2 + p + 2 = n + m + p + 4$
   $n * (m * p) = n * (m + p + 2) = n + m + p + 2 + 2 = n + m + p + 4$.

   Therefore, $*$ is associative.
   ii. Let n and $m \in \mathbf{Z}$. Then $n * m = n + m + 2 = m + n + 2 = m * n$. Therefore, $*$ is commutative.

(d) i. Let $n, m, p \in \mathbf{Z}$. Then

$$n * (m * p) = \max(n, \max(m, p)) = \max(n, m, p)$$
$$= \max(\max(n, m), p) = (n * m) * p.$$

Therefore, $*$ is associative.

ii. Let n and $m \in \mathbf{Z}$. Then $n * m = \max(n, m) = \max(m, n) = m * n$. Therefore, $*$ is commutative.

(g) i. $(2 * 3) * 4 = 2^6 * 4 = 64 * 4 = 2^{256}$, $2 * (3 * 4) = 2 * (2^{12}) = 2 * 4096 = 2^{8192}$. Therefore, $*$ is not associative.

ii. If $x, y \in \mathbf{R}$, then $x * y = 2^{xy} = 2^{yx} = y * x$. Therefore, $*$ is commutative.

4. (a) Let $f, g \in \mathbf{F}(\mathbf{R})$. $(f + g)(x) = f(x) + g(x) = g(x) + f(x) = (g + f)(x)$ for all x in $\mathbf{R}$. Therefore, $f + g = g + f$ and so addition on $\mathbf{F}(\mathbf{R})$ is commutative.

7. (a) Let $n \in \mathbf{Z}$. Then $n * (-2) = (-2) * n = n + (-2) + 2 = n$. Therefore -2 is the identity element.

(d) If $n \in \mathbf{Z}^+$, then $n * 1 = 1 * n = \max(n, 1) = n$. Therefore, 1 is the identity w.r.t. $*$.

(g) Suppose there exists an identity element e w.r.t. $*$. Then $e * (-1) = 2^{-e} = -1$. But this cannot happen since every power of 2 is positive. Thus there is no identity.

10. (a) Let $n \in \mathbf{Z}^+$. Then $n * (-n - 4) = (-n - 4) * n = n + (-n - 4) + 2 = -2$. Thus $-n - 4$ is the inverse of n w.r.t. $*$.

(d) Let $n \in \mathbf{Z}^+$. Suppose that n has an inverse x w.r.t. $*$. Then $n * x = \max(n, x) = 1$. Thus $n = x = 1$. Therefore, only 1 has an inverse w.r.t. $*$.

11. *Hint:* Consider the matrix equation $\begin{pmatrix} a & b \\ c & d \end{pmatrix}\begin{pmatrix} x & y \\ z & w \end{pmatrix} = \begin{pmatrix} 1 & 0 \\ 0 & 1 \end{pmatrix}$. Solve for $x, y, z,$ and w in terms of $a, b, c,$ and d. You will see that a solution exists if and only if $ad - bc \neq 0$. Check that your answer gives a solution to $\begin{pmatrix} x & y \\ z & w \end{pmatrix}\begin{pmatrix} a & b \\ c & d \end{pmatrix} = \begin{pmatrix} 1 & 0 \\ 0 & 1 \end{pmatrix}$ as well.

16. (a) Let $a, b, \in \mathbf{E}$. Then $a = 2x$ and $b = 2y$ for some $x, y \in \mathbf{Z}$. Then $ab = (2x)(2y) = 2(2xy) \in \mathbf{E}$. Therefore, $\mathbf{E}$ is closed under multiplication.

19. (a) Let $a, b, \in \mathbf{E}$. Then $a = 2x$ and $b = 2y$ for some $x, y \in \mathbf{Z}$. Then $a * b = (2x) * (2y) = 2x + 4y = 2(x + 2y) \in \mathbf{E}$. Therefore $\mathbf{E}$ is closed under $*$.

22. (a) $\{f \in \mathbf{F}(A) \mid f \text{ is injective}\}$ is closed under composition by part 2 of Proposition 3.3.4.

23. (a) Let $S = \{f \in \mathbf{F}(\mathbf{R}) \mid f(0) = 0\}$. Let $f, g \in S$. Then $(f + g)(0) = f(0) + g(0) = 0 + 0 = 0$. Thus $f + g \in S$, so S is closed under addition.

(d) Let $S = \{f \in \mathbf{F}(\mathbf{R}) \mid f(-x) = f(x) \text{ for all } x \in \mathbf{R}\}$. Let $f, g \in S$. Then $(f + g)(-x) = f(-x) + g(-x) = f(x) + g(x) = (f + g)(x)$ for all $x \in \mathbf{R}$. Hence $f + g \in S$. Therefore S is closed under addition.

25. (a) Let $S = \{f \in \mathbf{F}(\mathbf{R}) \mid f(0) = 0\}$. Let $f, g \in S$. Then $fg(0) = f(g(0)) = f(0) = 0$. Therefore, $fg \in S$, proving that S is closed under composition.

(d) Let $S = \{f \in \mathbf{F}(\mathbf{R}) \mid f(-x) = f(x) \text{ for all } x \in \mathbf{R}\}$. Let $f, g \in S$. Then $fg(-x) = f(g(-x)) = f(g(x)) = fg(x)$. Thus $fg \in S$, so S is closed under composition.

26. (a) Let $S = \left\{ \begin{pmatrix} a & b \\ c & d \end{pmatrix} \in M_2(\mathbf{R}) \ \middle| \ b = 0 \right\}$. Let $\begin{pmatrix} a & b \\ c & d \end{pmatrix}$ and $\begin{pmatrix} x & y \\ z & w \end{pmatrix} \in S$.

So $b = y = 0$. Then $\begin{pmatrix} a & b \\ c & d \end{pmatrix} + \begin{pmatrix} x & y \\ z & w \end{pmatrix} = \begin{pmatrix} a+x & b+y \\ c+z & d+w \end{pmatrix} = \begin{pmatrix} a+x & 0 \\ c+z & d+w \end{pmatrix} \in S$. So S is closed under addition.

36. *Hint*: Exercise 11 will help.

Exercises 4.2

1. (a) $0\cancel{R}0$, so R is not reflexive.
   $2R1$ but $1\cancel{R}2$. Therefore R is not symmetric.
   Let $a, b, c \in \mathbf{Z}$ and suppose aRb and bRc. Then $a > b$ and $b > c$, so $a > c$. Hence aRb. Therefore, R is transitive.
   R is antisymmetric for the same reason that the relation in Example 10 is antisymmetric.

3. (a) R is not reflexive since no set is a proper subset of itself.
   If $X \subset Y$, then X is a proper subset of Y and therefore Y is not a proper subset of X. Thus R is not symmetric.
   R is transitive by a proof similar to the proof of Proposition 2.1.2.
   R is antisymmetric since the statement $X \subset Y$ and $Y \subset X$ is never true.

(d) XRX for all X in $\mathbf{P}(A)$ since $|X| = |X|$. Thus R is reflexive.
   If $|X| = |Y|$, then $|Y| = |X|$. Therefore if XRY, then YRX. Hence R is symmetric.
   If $|X| = |Y|$ and $|Y| = |W|$, then $|X| = |W|$. Thus if XRY and YRW, then XRW. Therefore R is transitive.
   R is not antisymmetric. Let $A = \{1, 2, 3, 4, 5, 6, 7, 8, 9, 10\}$, $X = \{1, 2, 3\}$, $Y = \{2, 3, 4\}$. Then XRY and YRX, but $X \neq Y$.

4. (a) If $f \in \mathbf{F}(\mathbf{R})$, then $f(0) = f(0)$, so fRf. Therefore, R is reflexive.
   Let $f, g \in \mathbf{F}(\mathbf{R})$. Suppose that fRg. Then $f(0) = g(0)$ and so $g(0) = f(0)$. Hence gRf. Therefore, R is symmetric.
   Let $f, g, h \in \mathbf{F}(\mathbf{R})$. Suppose that fRg and gRh. Then $f(0) = g(0)$ and

$g(0) = h(0)$. It follows that $f(0) = h(0)$ and therefore fRh. Hence R is transitive. Therefore, R is an equivalence relation.

If $f \in \mathbf{F}(\mathbf{R})$, then the equivalence class of f is the set of all functions that intersect f at the origin.

5. (a) $[18] = \{x \in \mathbf{Z} \mid x = 4t + 2 \text{ for some } t \in \mathbf{Z}\} = \{2, -2, 6, -6, 10, -10, \ldots\}$.

12. (a) Let $a \in \mathbf{Z}^+$. Then aRa since $a = 1a$. Therefore, R is reflexive.

Let $a, b, c \in \mathbf{Z}^+$. Suppose that aRb and bRc. Then $b = na$ and $c = mb$ for some $n, m \in \mathbf{Z}^+$. Hence $c = mb = m(na) = (mn)a$, so aRc. Thus R is transitive.

Let $a, b \in \mathbf{Z}^+$. Suppose that aRb and bRa. Then $b = na$ and $a = mb$ for some $n, m \in \mathbf{Z}^+$. We get $b = na = n(mb) = (nm)b$. Therefore, $nm = 1$ and since n and m are positive integers, it follows that $n = m = 1$. Therefore, $a = b$, proving that R is antisymmetric.

16. (a) Let $f \in \mathbf{F}(\mathbf{R})$. Clearly, $f(x) \leq f(x)$ for all $x \in \mathbf{R}$, so fRf. Therefore, R is reflexive.

Let $f, g, h \in \mathbf{F}(\mathbf{R})$ and suppose that fRg and gRh. Then $f(x) \leq g(x)$ and $g(x) \leq h(x)$ for all $x \in \mathbf{R}$. Thus $f(x) \leq h(x)$ for all $x \in \mathbf{R}$, fRh. Therefore, R is transitive.

Let $f, g \in \mathbf{F}(\mathbf{R})$ and suppose that fRg and gRf. Then $f(x) \leq g(x)$ and $g(x) \leq f(x)$ for all $x \in \mathbf{R}$. Thus $f(x) = g(x)$ for all $x \in \mathbf{R}$. So $f = g$. Hence R is antisymmetric.

20. (a) $([a, b] + [c, d]) + [e, f]$
$$= [ad + bc, bd] + [e, f,] = [(ad + bc)f + (bd)e, (bd)f]$$
$$= [adf + bcf + bde, bdf] = [a(df) + b(cf + de), b(df)]$$
$$= [a, b] + [cf + de, df] = [a, b] + ([c, d] + [e, f])$$

22. (a) $(0, x)R(0, y)$ since $0 \cdot y = x \cdot 0 = 0$. Therefore, $[0, x] = [0, y]$.

23. (a) $[a, b] + [0, 1] = [a \cdot 1 + b \cdot 0, b \cdot 1] = [a, b]$ for all $[a, b] \in \mathbf{Q}$. Therefore, $[0, 1]$ is the additive identity of $\mathbf{Q}$.

Exercises 5.1

---

1. Let $a, b, c \in \mathbf{Z}$.
   (a) Since $a + (-a) = -a + a = 0$, it follows that a is the additive inverse of $-a$; in other words, $-(-a) = a$.
   (d) $(-1)a + a = (-1)a + 1a$ by Axiom 7
   $$= ((-1) + 1)a \text{ by Axiom 8}$$
   $$= 0a = 0 \text{ by part 1 of Proposition 5.1.1.}$$
   Therefore, $(-1)a$ is the additive inverse of a, implying that $(-1)a = -a$.

5. (a) Suppose that $a < 0$ and $b < 0$. Then $-a > 0$ and $-b > 0$ by part 2 of Proposition 5.1.4. Hence $(-a)(-b) > 0$ by part 3 of Proposition

5.1.4. But, by part 4 of Proposition 5.1.1, $(-a)(-b) = ab$. Therefore, $ab > 0$.

(d) Suppose $a < b$ and $c > 0$. Then $b - a \in \mathbf{Z}^+$ and $c \in \mathbf{Z}^+$. By Axiom 9, $(b - a)c \in \mathbf{Z}^+$. But $(b - a)c = bc - ac$ by Axiom 8. Therefore, $bc - ac \in \mathbf{Z}^+$, so that $ac < bc$.

7. *Hint*: Prove the contrapositive.

11. (a) *Hint*: Assume that $b \notin \mathbf{Z}^+$ and derive a contradiction.

19. (a) *Hint*: Use the fact that if $x \in \mathbf{Z}^-$, then $-x \in \mathbf{Z}^+$ and 1 is the smallest element of $\mathbf{Z}^+$.

Exercises 5.2

1. (a) Let $P(n)$ be the statement "$1 + 3 + 5 + \ldots + (2n - 1) = n^2$." Since $1 = 1^2$, $P(1)$ is true. Suppose now that $P(k)$ is true. Then $1 + 3 + 5 + \ldots + (2k - 1) = k^2$. This is the induction hypothesis. Hence

$$1 + 3 + 5 + \ldots + (2k - 1) + (2(k + 1) - 1)$$
$$= k^2 + 2k + 1 = (k + 1)^2.$$

Thus $P(k + 1)$ is true. Therefore, by induction, $P(n)$ is true for all integers n.

4. (a) Let $P(n)$ be the statement: "$2^n > n^2$." Since $2^5 > 5^2$, $P(5)$ is true.
Suppose that $P(k)$ is true. Then $2^k > k^2$. It follows that $2^{k+1} = 2 \cdot 2^k > 2k^2 > (k + 1)^2$. The last inequality is true since $2k^2 - (k + 1)^2 = k^2 - 2k - 1 > k^2 - 2k - 3 = (k - 3)(k + 1) > 0$ since $k > 4$. This proves that $P(k + 1)$ is true. So by induction, $P(n)$ is true for all positive integers $n \geq 5$.

8. (a) $f_1 = 1, f_2 = 1, f_3 = 2, f_4 = 3, f_5 = 5, f_6 = 8, f_7 = 13, f_8 = 21, f_9 = 34, f_{10} = 55$

(d) Let $P(n)$ be the statement "$f_1^2 + f_2^2 + \ldots + f_n^2 = f_n f_{n+1}$." Since $f_1^2 = f_1 f_2 = 1$, $P(1)$ is true. Now suppose that $P(k)$ is true for a positive integer k. Then $f_1^2 + f_2^2 + \ldots + f_k^2 = f_k f_{k+1}$. Therefore,

$$f_1^2 + f_2^2 + \ldots + f_k^2 + f_{k+1}^2 = f_k f_{k+1} + f_{k+1}^2 = f_{k+1}(f_k + f_{k+1}) = f_{k+1} f_{k+2}.$$

This proves that $P(k + 1)$ is true and so by induction, $P(n)$ is true for all positive integers n.

9. (a) $H_2 = 1, H_3 = 3, H_4 = 6, H_5 = 10, H_6 = 15$

16. (a) The basis for this proof is Proposition 3.3.4, which says that the composition of two surjective functions is surjective. Let $P(n)$ be the statement "If $f_1, f_2, \ldots, f_n$ are surjective functions in $\mathbf{F}(A)$, then the composition $f_1 f_2 \ldots f_n$ is surjective. Clearly $P(1)$ is true. Suppose that $P(k)$ is true for a positive integer k. Let $f_1, f_2, \ldots, f_k,$

$f_{k+1} \in \mathbf{F}(A)$ be surjective functions. By the induction hypothesis, the composition $f_1 f_2 \ldots f_k$ is surjective. By Proposition 3.3.4, the composition $(f_1 f_2 \ldots f_k) f_{k+1}$ is surjective. But $(f_1 f_2 \ldots f_k) f_{k+1} = f_1 f_2 \ldots f_k f_{k+1}$ since composition is an associative binary operation on $\mathbf{F}(A)$. Therefore, $P(k + 1)$ is true. Thus $P(n)$ is true for all positive integers n.

25. *Hint*: Use the previous problem.

27. $B_5 = 0$, $B_6 = 1/42$, $B_7 = 0$, $B_8 = -1/30$

29. $1^5 + 2^5 + 3^5 + \ldots + (n-1)^5 = \dfrac{1}{12} n^2 (n-1)^2 (2n^2 - 2n - 1)$

30. (a) *Hint*: Apply the Binomial Theorem to $(\ell + 1)^{k+1}$.
    (d) $S_4(n) = \frac{1}{5}n^5 - \frac{1}{2}n^4 + \frac{1}{3}n^3 - \frac{1}{30}n$

31. $\zeta(6) = \dfrac{\pi^6}{945}$

36. (a) $f(0) = f(0 + 0) = f(0) + f(0) = 2f(0)$. Thus $f(0) = 0$.

Exercises 5.3

---

1. (a) Let $n \in \mathbf{Z}$. By the Division Algorithm, there exists t and r in $\mathbf{Z}$ such that $n = 3t + r$, where $0 \le r < 3$. So $n = 3t$, $n = 3t + 1$, or $n = 3t + 2$.

4. (a) Let $P(n)$ be the statement: $4 \mid 5^n - 1$. Since $4 \mid 5 - 1$, $P(1)$ is true. Suppose that $P(k)$ is true; that is, $4 \mid 5^k - 1$. Then $5^k - 1 = 4t$ for some t in $\mathbf{Z}$. Hence $5^{k+1} - 1 = 5(5^k - 1) + 4 = 5(4t) + 4 = 4(5t + 1)$. Therefore $4 \mid 5^{k+1} - 1$ and $P(k + 1)$ is true. By induction, $P(n)$ is true for all integers n.

6. (a) If $a \mid b$ and $b \mid a$, then there exist c and d in $\mathbf{Z}$ such that $ac = b$ and $bd = a$. We get $a = bd = (ac)d = a(cd)$, which implies that $cd = 1$. Hence $c \mid 1$ and by part 1 of Lemma 5.3.3, $c = \pm 1$. Therefore, $a = \pm b$.
   (e) Converse of 1: If $a = \pm 1$, then $a \mid 1$. True: Since $(1)(1) = 1$ and $(-1)(-1) = 1$, it follows that both 1 and -1 divide 1.
   Converse of 2: If $a = \pm b$, then $a \mid b$ and $b \mid a$. True: If $a = b$, then $a(1) = b$ and $b(1) = a$, which shows that $a \mid b$ and $b \mid a$. If $a = -b$, then $a(-1) = b$ and $b(-1) = a$, again showing that $a \mid b$ and $b \mid a$.

10. (a) $1840 = (1518)(1) + 322$
    $1518 = (322)(4) + 230$
    $322 = (230)(1) + 92$
    $230 = (92)(2) + 46$
    $92 = (46)(2) + 0$

Therefore, $(1840, 1518) = 46$.

$46 = (1840)(-14) + (1518)(17)$

11. *Hint*: Use the fact that $d = ax + by$ for some $x, y \in \mathbf{Z}$.

14. *Hint*: This is a set theoretic proof. Prove that $n\mathbf{Z} \cap m\mathbf{Z} \subseteq nm\mathbf{Z}$. (Exercise 13 will help.) Then prove that $nm\mathbf{Z} \subseteq n\mathbf{Z} \cap m\mathbf{Z}$.

16. (a) *Hint*: Let $d = (n, n + 2)$. Prove that d divides 2 and then show that d cannot be 2.

20. *Hint*: First show that if x and y are even, then z is even, a contradiction. Then assuming x and y are odd, show that z is even. Writing x as $2t + 1$ and y as $2s + 1$ for some integers t and s, derive the contradiction that 4 divides 2.

21. *Hint*: Assume that x and y are not divisible by 3 and use Exercise 1(a) to derive a contradiction similar to the one for Exercise 20.

Exercises 5.4

2. *Hint*: Prove the contrapositive.

4. *Hint*: Use induction on the integer m.

7. *Hint*: Assume that $\log_3 10$ is rational. Use the definition of logarithm to derive an equation that contradicts Theorem 5.4.7.

8. (a) $594 = 2 \times 3^3 \times 11$
   (d) $191{,}737 = 7^3 \times 13 \times 43$

10. *Hint*: Use Theorem 5.4.7.

12. *Hint*: Write m as $a(2b + 1)$ where a is a power of 2. (Explain why you can do this.) Then show that $2^a + 1$ is a factor of $2^m + 1$ to conclude that $b = 0$.

14. *Hint*: Show that if n is not prime, then $2^n - 1$ can be factored.

18. *Hint*: Prove the contrapositive.

24. *Hint*: Use Theorem 5.4.7.

Exercises 5.5

1. (a) If x is even, then $x = 2k$ for some integer k. Then $x^2 = 4k^2 \equiv 0 \pmod 4$ since 4 divides $4k^2$.

   Alternative proof: if x is even then $x \equiv 0$ or $2 \pmod 4$. Then $x^2 \equiv 0^2 \equiv 0 \pmod 4$ or $x^2 \equiv 2^2 = 4 \equiv 0 \pmod 4$.

3. (a) Suppose that there exist integers x and y such that $x^2 = 8y + 3$. Then $x^2 \equiv 3 \pmod 8$. But by Exercise 2, $x^2 \equiv 0$, 1 or $4 \pmod 8$. Therefore no such integers x and y exist.

8. (a) [12]

12. (a) Let [a], [b], [c] $\in \mathbf{Z}_n$. Then ([a] + [b]) + [c] = [a + b] + [c] = [(a + b) + c] = [a + (b + c)] = [a] + [b + c] = [a] + ([b] + [c]). Therefore, addition is associative.

15. (a) [4]
    (d) [8]

16. (a) 21

17. (a) [15]
    (d) [617]

18. Addition and multiplication tables for $\mathbf{Z}_8$:

+	[0]	[1]	[2]	[3]	[4]	[5]	[6]	[7]
[0]	[0]	[1]	[2]	[3]	[4]	[5]	[6]	[7]
[1]	[1]	[2]	[3]	[4]	[5]	[6]	[7]	[0]
[2]	[2]	[3]	[4]	[5]	[6]	[7]	[0]	[1]
[3]	[3]	[4]	[5]	[6]	[7]	[0]	[1]	[2]
[4]	[4]	[5]	[6]	[7]	[0]	[1]	[2]	[3]
[5]	[5]	[6]	[7]	[0]	[1]	[2]	[3]	[4]
[6]	[6]	[7]	[0]	[1]	[2]	[3]	[4]	[5]
[7]	[7]	[0]	[1]	[2]	[3]	[4]	[5]	[6]

·	[0]	[1]	[2]	[3]	[4]	[5]	[6]	[7]
[0]	[0]	[0]	[0]	[0]	[0]	[0]	[0]	[0]
[1]	[0]	[1]	[2]	[3]	[4]	[5]	[6]	[7]
[2]	[0]	[2]	[4]	[6]	[0]	[2]	[4]	[6]
[3]	[0]	[3]	[6]	[1]	[4]	[7]	[2]	[5]
[4]	[0]	[4]	[0]	[4]	[0]	[4]	[0]	[4]
[5]	[0]	[5]	[2]	[7]	[4]	[1]	[6]	[3]
[6]	[0]	[6]	[4]	[2]	[0]	[6]	[4]	[2]
[7]	[0]	[7]	[6]	[5]	[4]	[3]	[2]	[1]

The elements in $\mathbf{Z}_8$ that have multiplicative inverses are: [1], [3], [5], and [7].

28. (a) 7

Exercises 6.1

6. (a) $f: \mathbf{Z} \to \mathbf{O}, f(n) = 2n + 1$

(d) $f: \mathbf{Z} \to S,\ f(n) = \begin{cases} 2n + 100, & \text{if } n > 0 \\ -2n + 101, & \text{if } n \le 0 \end{cases}$

(g) $f: \mathbf{Z} \to \mathbf{Z} \times \{0\},\ f(n) = (n, 0)$

8. (b) *Hint*: First prove that if $x \notin A$, then $A \cup \{x\}$ is finite. Then, if n is the number of elements of B, use induction on n to prove that $A \cup B$ is finite.

10. $f(10) = 4,\ f(15) = 1/5$

11. (c) *Hint*: First prove that the set $\{e^x \mid x \in \mathbf{Z}\}$ is countably infinite.

Exercises 6.2

1. (a) *Hint*: Define a linear function from $[0, 1]$ to $[a, b]$

5. (a) *Reminder*: To prove that f is surjective, you need first to show that $\operatorname{Im} f \subseteq (0, 1)$ and then show that $(0, 1) \subseteq \operatorname{Im} f$.
   (c) *Hint*: First define a bijection from $\mathbf{R}^+$ to (a, ∞).

7. (a) *Hint*: Show that $[0, 1] \preceq [0, 1] \cup [2, 3] \preceq \mathbf{R}$ and then use Lemma 6.2.4 and the Schroeder–Bernstein Theorem.
   (d) *Hint*: First show that there is an injection $f: (0, 1) \times (0, 1) \to \mathbf{R}$. To do this, recall that any element of $(0, 1)$ can be represented as $.a_1 a_2 a_3 \ldots$, where $a_1, a_2, a_3, \ldots$ are integers between 0 and 9, and that every such expansion is unique if we do not use decimal expansions ending in an infinite string of 9's.
   (g) *Hint*: First, define a function $\vartheta: \mathbf{F}(\mathbf{Z}^+) \to (0, 1)$ as follows. Denote the binary expansion of an integer n by n_b. If $f \in \mathbf{F}(\mathbf{Z}^+)$, let $\vartheta(f) = 0.f(1)_b 2f(2)_b 2f(3)_b 2 \ldots$, interpreted as a decimal expansion. Prove that ϑ is an injection.

   Second, define a function $\lambda: \mathbf{P}(\mathbf{Z}^+) \to \mathbf{F}(\mathbf{Z}^+)$ as follows. If A is a subset of $\mathbf{Z}^+$, then $\lambda(A)$ is the function $f: \mathbf{Z}^+ \to \mathbf{Z}^+$ defined by

   $f(n) = \begin{cases} 1 \text{ if } n \in A \\ 2 \text{ if } n \notin A. \end{cases}$ Prove that λ is an injection.

Exercises 6.3

2. (a) *Hint*: Assume that $(a, b) < (c, d)$ and $(c, d) < (a, b)$ and derive a contradiction.

4. *Hint*: Let $\lambda: \prod_{i \in \mathbf{Z}^+} A \to \mathbf{P}(\mathbf{Z}^+)$ be the function defined as follows: an element f in $\prod_{i \in \mathbf{Z}^+} A$ is a function $f: \mathbf{Z}^+ \to \{0, 1\}$, so let $\lambda(f) = \{n \in \mathbf{Z}^+ \mid f(n) = 0\} \in \mathbf{P}(\mathbf{Z}^+)$. Prove that λ is a bijection.

Exercises 7.1

3. (a) *Hint*: Use induction on n.

6. (a) *Hint*: With m fixed, use induction on n.

10. (a) *Hint*: Assume that $x^{-1} \notin P$ and derive a contradiction.

15. *Hint*: First prove that $\mathbf{Z}^+ \subseteq P$.

17. (a) *Hint*: Assume that $-x \notin N$ and derive a contradiction.

Exercises 7.2

1. (a) lub S does not exist; glb $S = -1$
   (b) lub $S = 1$; glb $S = 0$

2. *Hint*: Consider the set $T = \{-x \mid x \in S\}$.

4. (a) *Hint*: Use the fact that if $a < x < b$, then there is a real number y such that $x < y < b$.

15. *Hint*: Use Exercise 12 of Section 7.1.

Exercises 7.3

2. (a) $22 - 21i$
   (d) $-234 - 415i$
   (g) $-\dfrac{34}{61} - \dfrac{8}{61}i$

7. (a) $3\sqrt{2}(\cos(3\pi/4) + i \sin(3\pi/4))$
   (d) $2(\cos(3\pi/2) + i \sin(3\pi/2))$
   (g) $\sqrt{26}(\cos \theta + i \sin \theta)$ where $\theta = \sin^{-1}(1/\sqrt{26})$

9. (a) $648\sqrt{3}$
   (d) $2^{28}(1 + i)$
   (g) $-2^{30}3^{18}$

10. (a) The fourth roots of z are:

$$z_1 = \sqrt[8]{108}\left(\frac{1}{2} + \frac{\sqrt{3}}{2}i\right), \qquad z_2 = \sqrt[8]{108}\left(-\frac{\sqrt{3}}{2} + \frac{1}{2}i\right),$$

$$z_3 = \sqrt[8]{108}\left(\frac{\sqrt{3}}{2} - \frac{1}{2}i\right), \qquad z_4 = \sqrt[8]{108}\left(-\frac{1}{2} - \frac{\sqrt{3}}{2}i\right).$$

12. The fifth roots of unity are:

$$\omega_0 = 1, \omega_1 = \frac{\sqrt{5} - 1}{4} + \frac{\sqrt{10 + 2\sqrt{5}}}{4} i,$$

$$\omega_2 = \frac{-(\sqrt{5} + 1)}{4} + \frac{(\sqrt{5} - 1)\sqrt{10 + 2\sqrt{5}}}{8} i,$$

$$\omega_3 = \frac{-(\sqrt{5} + 1)}{4} - \frac{(\sqrt{5} - 1)\sqrt{10 + 2\sqrt{5}}}{8} i,$$

$$\omega_4 = \frac{\sqrt{5} - 1}{4} - \frac{\sqrt{10 + 2\sqrt{5}}}{4} i.$$

Exercises 8.1

4. (a) *Hint*: Prove the contrapositive.

6. (a) $f(x) + g(x) = x^4 + 9x^3 - 6x^2 - 10x + 7$

7. (a) (i) $f(x) + g(x) = [4]x^3 + [3]x^2 + [6]x + [5]$
   (ii) $f(x)g(x) = [1]x^5 + [4]x^4 + [2]x^3 + [2]x^2 + [2]x + [6]$

9. (a) $q(x) = x^3 + x - 1, r(x) = 9x + 6$

11. (a) -1 is the only real zero.

12. (a) $-1, -\frac{1}{2} + \frac{\sqrt{3}}{2} i$, and $-\frac{1}{2} - \frac{\sqrt{3}}{2} i$

14. (a) Any integer x such that $x \equiv 4$ or $7 \pmod{11}$ is a solution.

15. (a) $x^2 \equiv a \pmod 7$ is solvable for $a = 1, 2, 4$ and not solvable for $a = 3, 5, 6$.

Exercises 8.2

2. (a) $x^2 - 5$ is irreducible over $\mathbf{Q}$.
   $x^2 - 5 = (x - \sqrt{5})(x + \sqrt{5})$ over $\mathbf{R}$ and $\mathbf{C}$.

14. (a) The g.c.d. of $f(x)$ and $g(x)$ is the constant function 1.

20. The only irreducible polynomial of degree 2 in $\mathbf{Z}_2[x]$ is $[1]x^2 + [1]x + [1]$. The irreducible polynomials of degree 3 are $[1]x^3 + [1]x^2 + [1]$ and $[1]x^3 + [1]x + [1]$.

Exercises 8.3

4. (a) There are three real zeros: $x_1 \approx -3.423$, $x_2 \approx .378$, and $x_3 \approx 1.545$.

12. (a) 2 is the only rational zero of $f(x)$.
    $f(x) = (x - 2)(x^2 + x - 1)$ in $\mathbf{Q}[x]$.
    $f(x) = (x - 2)(x + 1/2 - \sqrt{5}/2)(x + 1/2 + \sqrt{5}/2)$ in $\mathbf{R}[x]$ and $\mathbf{C}[x]$.

BIBLIOGRAPHY

[1] Boyer, C. B. *A history of mathematics.* 2d ed. John Wiley & Sons, 1991.

[2] Cipra, B. "Number theorists uncover a slew of prime imposters," *What's happening in the mathematical sciences 1993.* Providence, RI: American Mathematical Society.

[3] Cipra, B. "Fermat's theorem—At last!" *What's happening in the mathematical sciences 1995–1996.* Providence, RI: American Mathematical Society.

[4] Cox, D. "Introduction to Fermat's last theorem," *American Mathematical Monthly,* 101 (January 1994): 3–14.

[5] Eves, H. *Foundations and fundamental concepts of mathematics.* 3d ed. PWS-Kent, 1990.

[6] Faber, R. *Foundations of Euclidean and non-Euclidean geometry.* Marcel Dekker, 1983.

[7] Fine, B., and G. Rosenberger. *The fundamental theorem of algebra.* Springer-Verlag, 1997.

[8] Gardner, M. *aha! Insight.* Scientific American Inc./W.H. Freeman, 1977.

[9] Grabiner, J. "Who gave you the epsilon? Cauchy and the origins of rigorous calculus." *American Mathematical Monthly,* 90 (March 1983): 185–194.

[10] Ireland, K., and M. Rosen. *A classical introduction to modern number theory.* 2d ed. Springer-Verlag, 1990.

[11] Kleiner, I. "Evolution of the function concept: A brief survey." *The College Mathematics Journal,* vol. 20, no. 4 (1989).

[12] Kleiner, I. "Rigor and proof in mathematics." *Mathematics Magazine* 64 (December 1991): 291–314.

[13] Kleiner, I., and N. Movshovitz-Hadar. "The role of paradoxes in the evolution of mathematics." *American Mathematical Monthly,* 101 (December 1994): 963–974.

[14] Kline, M. *Mathematical thought from ancient to modern times.* Oxford University Press, 1972.

[15] Laugwitz, D. "On the historical development of infinitesimal mathematics." *American Mathematical Monthly,* 104 (May 1997): 447–455.

[16] Niven, I. *Irrational numbers.* Mathematical Association of America, 1963.

[17] Vilenkin, N. Y. *Stories about sets.* Academic Press, 1970.

INDEX